4th Consortium of Asia-Pacific Education Universities (CAPEU 2017)

IOP Conference Series: Materials Science and Engineering
Volume 296

Surabaya, Indonesia
22 - 23 May 2017

Editors:

Abadi
Ali Mustofa
Setya Chendra Wibawa

ISBN: 978-1-5108-5835-0
ISSN: 1757-8981

Printed from e-media with permission by:

Curran Associates, Inc.
57 Morehouse Lane
Red Hook, NY 12571

Some format issues inherent in the e-media version may also appear in this print version.

Printed by Curran Associates, Inc. (2018)

Institute of Physics
Dirac House, Temple Back
Bristol BS1 6BE UK

Phone: 44 1 17 929 7481
Fax: 44 1 17 920 0979

techtracking@iop.org

Additional copies of this publication are available from:

Curran Associates, Inc.
57 Morehouse Lane
Red Hook, NY 12571 USA
Phone: 845-758-0400
Fax: 845-758-2633
Email: curran@proceedings.com
Web: www.proceedings.com

Table of contents

Volume 296

The Consortium of Asia-Pacific Education Universities (CAPEU)

22–23 May 2017, Universitas Negeri Surabaya, Indonesia

Accepted papers received: 8 January 2018
Published online: 21 February 2018

Preface

Papers

IOP Conf. Series: Materials Science and Engineering **296** (2017) 011001 doi:10.1088/1757-899X/296/1/011001

Preface of
The 4th CAPEU Conference 2017

The 4th Consortium Asia-Pacific Education Universities (CAPEU) 2017 is one of a series of international conferences held by Universitas Negeri Surabaya (Unesa), Indonesia, in collaboration with some universities in Asia, South East Asia, and Pacific as well as with other institutions. Unesa held this event as its commitment to research activities and collaborations to strengthen the academic climate in Asia and beyond. This event is a perfect platform for senior and junior researchers in the fields of sciences, technology, engineering, and mathematics to exchange experiences and ideas and to build a network. In addition, this conference will certainly help improve scientific publication of researchers and lecturers.

The theme of this year event is "Innovations for Human Talents". This two-day conference aims to bring together educators, graduate students, researchers, and faculty members of the CAPEU and of other countries to contribute to the concept of the STEM in education and to formulate goals on how this can be applied and implemented in their respective countries. The researchers can do their bit with their research findings in promoting better achievement of research innovation for the betterment of human life.

This conference has attracted much attention from scientific communities from diverse background of research interests. There are 130 research-based papers submitted and presented at the conference. Out of the submitted papers, the selected ones, around 56 papers, will be published by IOP Publishing. The papers have a wide range of topics, namely education, teaching and learning, assessment, community engagement, job opportunities related to STEM, and parents' involvement in enhancing STEM. The topics are not limited to, but mostly of the above-mentioned areas of interest.

Editors:

1. Abadi (Universitas Negeri Surabaya)
   abadi@unesa.ac.id
2. Ali Mustofa (Universitas Negeri Surabaya)
   alimustofa@unesa.ac.id
3. Setya Chendra Wibawa (Universitas Negeri Surabaya)
   setyachendra@unesa.ac.id

Organize Commitee:

Abadi (Universitas Negeri Surabaya)
Ali Mustofa (Universitas Negeri Surabaya)
Setya Chendra Wibawa (Universitas Negeri Surabaya)
Raya Sulistyowati (Universitas Negeri Surabaya)
Noviyanti (Universitas Negeri Surabaya)
Yuni Lestari (Universitas Negeri Surabaya)
Dimas Avian Maulana (Universitas Negeri Surabaya)
Muhamad Ro'is Abidin (Universitas Negeri Surabaya)

Peer review statement

All papers published in this volume of *IOP Conference Series: Materials Science and Engineering* have been peer reviewed through processes administered by the proceedings Editors. Reviews were conducted by expert referees to the professional and scientific standards expected of a proceedings journal published by IOP Publishing.

IOP Conf. Series: Materials Science and Engineering **296** (2018) 012001 doi:10.1088/1757-899X/296/1/012001

Development of short Indonesian lesson plan to improve teacher performance

B Yulianto , Kamidjan, A Ahmadi and P V Asteria

Department of Indonesian Language and Literature, Universitas Negeri Surabaya

bambangyulianto@unesa.ac.id

Abstract. The developmental research was motivated by the results of preliminary study through interviews, which revealed almost all of the teachers did not create lesson plan themselves. As a result of this load, the performance of the real learning in the classroom becomes inadequate. Moreover, when lesson plan was not made by teachers themselves, the learning process becomes ineffective. Therefore, this study designed to develop a prototype of the short lesson plan, in particular, Indonesian language teaching, and to investigate its effectiveness. The participants in the study were teachers who were trained through lesson study group to design short model's lesson plan. Questionnaires and open-ended questions were used, and the quantitative and qualitative data obtained were analyzed accordingly. The analysis of the quantitative data, aided with SPSS, were frequency, percentage, and means, whereas the qualitative data were analyzed descriptively. The results showed that the teachers liked the model, and they were willing to design their own lesson plan. The observation data revealed that the classroom learning process became more interactive, and classroom atmosphere was more engaging and natural because the teachers did not stick to the lesson plan made by other teachers.

1. Introduction

The curriculum has a vital role as a blueprint for teachers and students in achieving learning objectives [1]. So it is regarded as a dogmatic powerful grip so that teachers do not dare to do re-creations [2]. Lesson plan (LP) as a component of the operational curriculum implementation is also considered as a rigid document by many teachers. It means the existing components in the LP is a fixed. On the contrary, in the teaching process, teachers have academic autonomy in determining teaching materials, methods, assessments, etc [3]. Even so, the academic autonomy is not intended that the teacher can be as free as possible without proper preparation.

The National Education System Law (NES Law) has ordered that teachers are obliged to design LP. It at least consists of learning objectives, what is taught (material), how to teach (method), and learning assessment. Based on that, it is necessary to rethink the direction of LP that the law requires. Only four components are at least required in the LP. Meanwhile, the LP developed by the government currently is very complex, contains many components, and each component is described in detail so it is like a drama scenario that the implementation will be rigid and less contextual.

It is very burdensome to the teachers. It is revealed from the preliminary study results through interviews in February-March 2016 from twenty Indonesian Secondary School teachers in Surabaya, who among them is also a student of Master Program at State University of Surabaya. The results show that indeed all teachers (100%) have LP, but none (0%) of teachers make it themselves. They

IOP Conf. Series: Materials Science and Engineering **296** (2018) 012001 doi:10.1088/1757-899X/296/1/012001

copy and paste the existing LP from their friends. Among these respondents, 60% feel overwhelmed in designing the LP, while the 40% feel not. Those who claim unencumbered because of its preparation through copy paste. Because the preparation of this LP is an obligation as The NES Law, according to them (100%) the burden is reasonable but they want a simpler and easier LP. Most respondents (85%) promised to make their own LP if the format is simpler. According to them, LP made by the teacher will be implemented more effectively because the teachers are aware of the state of the students and the class.

It is in line with the Regulation of the Ministry of Education and Culture No. 65 of 2013 [4] on Process Standards which reveals that every teacher is obliged to develop a complete and systematic LP so that learning takes place interactively, inspiration, fun, challenging, motivates learners to participate, and provide sufficient space for initiative, creativity, and independence in accordance with the talents, interests, and physical and psychological development of learners. Thus, it is very important that there is an effective Indonesian LP of secondary school to improve teacher performance.

Based on these facts, this study aims to develop a more concise LP prototype but does not violate the applicable law of the NES Law so that it is expected that teacher performance will increase. In addition, it is also studied the effectiveness of model LP after being applied by teachers in the classroom.

2. Research Methods

2.1 Research design
This research is a development research using Fenrich cycle model [5], covering 5 major phases, namely analysis, planning, design, development, and implementation. Each phase is always followed by evaluation and revision phase. This phase is very important because it is a continuous activity carried out at each major phase throughout the cycle. At each of the main phases of the examination activities in the form of evaluation and revision to produce a product that is steady. Thus, a subsequent major phase will not be carried out before the prior main phase produces a steady product after the follow-up phase above.

2.2 Subject
The subjects of this study are 40 Indonesian language teachers in secondary school at Surabaya formed in 8 groups of lesson study (LS). Each group consisted of 5 teachers with 1 teacher who acted as a model teacher, while 4 other teachers became partners doing the discussion (reflection) after doing an observation on learning performance of a model teacher.

2.3 Procedure
The analysis, planning, and design phase were carried out by the research team (RT). Nevertheless, in these phases the RT involved the officials of the Education Office at Surabaya and Secondary School Indonesian Teacher Discussion Group (ITDG), especially to present the model LP as well as the teacher engagement plan in the development phase. The result is support from both institutions for the implementation of this research.

The development phase begins to involve the teacher. This phase was carried out in several workshops: workshop I (introduction of prototype model and guided LP), workshop II (preparation of independent LP), workshop III (LS preparation), workshop IV (LS act), and V workshop (program evaluation).

Training activities with this LS model have also been done by Janjai [6], but the respondents are college teacher students. This LS model uses the basic theory of constructivism and metacognition. LS is basically chosen to solve problems and improve the quality of learning in the classroom by way of coaching by lecturers or peers of teachers in a collaborative and sustainable activity based on the principles of a colleague and the formation of professional learning community [7–9]. Cerbin & Kopp

[10] and Haithcock [7] added that LS activities are also beneficial for lecturers or colleagues in order to obtain an analysis of learning objectives, teaching practices, assessment of learning outcomes, and observation of students' thought processes in the classroom. About the advantages of this LS ever reported Murata & Takahashi [11] that teachers in Japan feel the benefits in the long term in developing learning. A survey of 125 LS actors found that 98% of them said that LS had helped them develop the learning and teaching process and 91% of respondents believed that LS was very effective in developing teacher professionalism.

2.4 Data Analyzing

The data collected in the form of qualitative and quantitative data. Qualitative data (in an overview of the model LP development process through several workshops and supporting and inhibiting factors obtained through observation and field notes) were analyzed using content analysis techniques, which considered its context and content. Meanwhile, the quantitative data related to the effectiveness of the model (from questionnaires and observations) were analyzed using descriptive statistics.

3. Results and Discussion

3.1 Profile of The Short LP

In the Regulation of the Ministry of Education and Culture No. 65 of 2013 [4] mentioned that teachers are obliged to develop a complete and systematic LP. What does it mean here? Is it written in detail the contents of the LP component? Does complete reference to the adequacy of LP components? Or is it understood that the LP should be accompanied by the teaching materials, assessment sheets, etc.?

In the Regulation of the Ministry of Education and Culture No. 81A of 2013 [12] on the implementation of the curriculum mentioned that the LP developed in detail of a subject matter or a specific theme that refers to the syllabus. LP includes: (1) school data, subject, and class/semester; (2) subject matter; (3) time allocation; (4) learning objectives, basic competencies and indicators of achievement of competence; (5) learning materials; learning methods; (6) media, tools and learning resources; (7) steps of learning activities; and (8) assessment.

Meanwhile, the previous document states that as a follow up of the NES Law, the Government Regulation No. 19 of 2005 [13] was made. In it was revealed that the LP contains learning objectives, learning materials, learning methods, learning resources, and assessment of learning outcomes.

Based on a preliminary study that revealed teachers' complaints due to the complexity of the LP, both the editorial and the content, and the consideration of non-violation of The NES Law, this study developed a short LP with only the main components. In detail, the components in question are as revealed in the following table. To give a complete picture of the short LP profile in the table presented the difference with the current LP.

A short LP contains components a) identity, b) learning objectives, c) learning materials, d) learning resources/materials, d) methods/learning models, e) assessment, and f) learning activities steps. The learning objective is formulated with just one line with all indicators (without repetition as usual). Component A (*audience*) and C (*condition*), as ideally formulate a learning objective, in this LP are not used. The material is also only written the subject matter. Sources/materials contain student books with reference to how many pages, such worksheets (if any), and other materials (eg *audio recording*). In the method written what the model (eg *scientific*). In the assessment is written the type of assessment and the level (eg written *assessment of knowledge*). For learning activities steps are filled in accordance with the model syntax. However, each step describes its activity and its learning resources (eg *reading individually to find the object and purpose of the description text in SB: 3-5*).

IOP Conf. Series: Materials Science and Engineering **296** (2018) 012001 doi:10.1088/1757-899X/296/1/012001

Table 1. The differences Current LP with Short LP

Description	Current LP	Short LP
Components and sequences	a) identity, b) core competencies, c) basic competencies and indicators of achievement of competencies, d) learning objectives, e) learning materials, f) learning approaches, g) learning resources, h) learning media, i), and j) assessment.	a) identity, b) learning objectives, c) learning materials, d) learning resources/materials, d) methods/learning models, e) assessment, and f) learning activities steps.
Identity	Yes	Yes
Core competence	Yes	No
Basic competencies and indicators	Yes	No
Learning objectives	• Formulated with four elements (A, B, C, and D) • Each indicator is a goal statement	• Formulated with two elements (B and D) • Several indicators in one objective formula
Learning materials	Presented the topic title and details	Presented the topic (subject matter)
Approach/learning method	Mentioned the name of method/ learning model	Mentioned the name of method/ learning model
Learning Resources	• Mentioned titles of books used • Expressed separately from instructional media (independently)	• Mentioned titles of books (abbreviated) and page numbers • Served a series of teaching materials/media used
Instructional Media	• Mentioned media used • Expressed separately from learning resources (independently)	• Mentioned teaching materials used • Served a series of teaching materials/media used
Learning activities steps	• Divided into three parts of activities: preliminary activities, core activities, and closing activities • The syntax component of the learning model used is explicitly written • Presented in complete sentences, including mentioning the actors, such as teachers or learners • Time allocation for each part of the activity is written down • *Learning activities steps* is written **before** the *assessment* component	• Only presents steps according to the syntax of the learning model without mentioning the introduction, core, and closing • The syntax component of the learning model used is not explicitly written • Presented in concise but complete sentences (containing the types and forms of activities), without mentioning the actors, such as teachers or learners • Time allocation for each part of the activity is not written down • *Learning activities steps* is written **after** the *assessment* component
Assessment	• Written the types, techniques, and references to assessment instruments in attachments in a separate form • *Assessment* is written **after** the *learning activities steps* component • Not explicitly indicated in the *learning activities steps*	• Written type, technique, and reference of assessment instruments in attachments in an integrated form • *Assessment* is written **before** the *learning activities steps* component • Explicitly demonstrated in the *learning activities steps*
Number of pages	2-3 pages	1 page

3.2 Effectiveness of Application of Model LP Through LS

The results of LS group reflection concluded that a) model LP is easy to apply, b) teachers rarely see LP at the time of teaching, c) the learning direction is systematic, d) the learning process in the class seems smooth, e) the students are passionate about questioning, f) teachers are more animated what is taught, g) tasks assigned to students seem natural, and h) textbooks take role in learning.

Questionnaire data showed that this LP is easy to use in learning. All teachers (100%) stated that the test LP is easier to design than the LP currently in use. All teachers (100%) said that it was because there was no need to write core competencies, 70% of teachers stated the formula was more concise because some parts were similarly put together, and 85% of teachers said sentences were not repeated.

According to eight model teachers, all teachers (100%) stated that LP of the trial was easy to use in the classroom because the textbook page presentation was clear (according to 87.5% of teachers), all prepared by teachers (100% teachers), and the short LP was easy to understand and do not need to see often when using it (87.5% of teachers). In this case, 25% of the model teachers still saw the LP at the time of teaching because they find obstacles and the rest (75% of teachers) said that they do not see the LP at the time of teaching. It showed that the teacher was very animating what will be taught.

The questionnaire data from the partner teacher who acted as an observer (32 teachers) showed that according to 85% observer, the learning process went smoothly, 85% observer thought the class was alive and natural, and 100% observer said that the weld is not rigid. With respect to student activity in the classroom, all observers (100%) agreed that teacher-student communication was smooth, 93% observers stated that students actively used textbooks, and 85% observers saw that students readily understood teacher commands.

Table 2. The Required Time Duration for Preparing an LP (In Minutes)

Group	LP 1	LP 2	LP 3	LP 4	LP 5	Total	Average
1	70	54	40	35	33	232	46,4
2	62	50	46	37	35	230	46
3	67	60	52	45	37	261	52,2
4	63	51	43	37	31	225	45
5	60	57	46	38	36	237	47,4
6	73	60	57	42	36	268	53,6
7	57	50	42	37	30	216	43,2
8	66	61	52	45	38	262	52,4
Total	518	443	378	325	276	1940	
Average	64,75	55,38	47,25	40,62	34,5		48,5

The data and chart above show the average teacher only takes 48.5 minutes to produce an LP. If at the initial time the teacher took 64.75 minutes to produce an LP, when making the 5th LP, they took only 34.5 minutes.

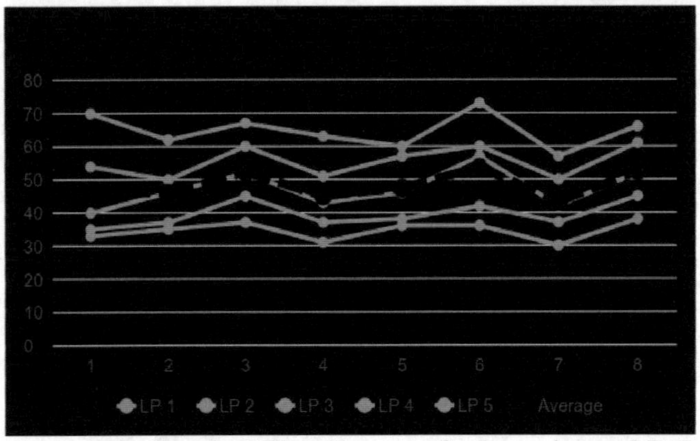

Figure 1. Chart of The Required Time Duration for Preparing an LP (In Minutes)

4. Conclusion

Based on the exposure in front it can be concluded that teachers, in general, will not require a long time when preparing their own LP when textbooks are available. Therefore, it is recommended that the LP format that has been used should be reviewed and further simplified as in the results of the development of this research so that the teachers are eager to make themselves. This fact shows that teachers generally will not take long to preparing their own LP if the textbook has been available. Thus, a contextual class climate, planned, passionate, and conducive can be realized.

References

[1] Richard J C 2010 Curriculum Approaches in Language Teaching: Forward, Central, and Backward Design *ELC J.* **44** pp 5—33

[2] Musyaddad K 2013 Problematika Pendidikan di Indonesia *Edu-Bio* **4** pp 51–7

[3] Kuscu S and Unlu S 2015 Teaching Translation: A suggested lesson plan on translation of advertising through the use of authentic materials *Procedia – Soc. Behav. Sci.* **199** pp 407–14

[4] Peraturan Menteri Pendidikan dan Kebudayaan Nomor 65. 2013 *Tentang Standar Proses Pendidikan Dasar dan Menengah*

[5] Fenrich P 1997 *Practical Guidelines for Creating Instructional Multimedia Application* (Fort Wort: The Dryden Press)

[6] Janjai S 2012 Improvement of The Ability of The Students in an Education Program to Design The Lesson Plans by Using an Instruction Model based on The Theories of Constructivism and Metacognition *Procedia Eng.* **32** pp 1163—8

[7] Haithcock F 2010 *Lesson Study for District and School Leadership Teams in Differentiated Accountability Schools*

[8] Hiebert J, Gallirmore R and Stigler J 2002 A knowledge base for the teaching profession: What would it look like and how can we get one? *Educ. Res.* **31** pp 3–15

[9] Indprasidh and Al M et 2004 The mathematics subject, using the lesson study approach *Proceedings of a workshop on instructional development project under the 1000 national education reform act* (Faculty of Education, Khon Kaen University)

[10] Cerbin W and Kopp B 2006 Lesson Study as a Model for Building Pedagogical Knowledge and Improving Teaching *Int. J. Teach. Learn. High. Educ.* **18** pp 250–7

[11] Murata A and Takahashi A 2002 District-level lesson study. How Japanese Teacher Improve Their Teaching of Elementary Mathematics *Paper presented at a Research Pre-Session of The Annual Meeting of The National Council of Teachers of Mathematics* (Las Vegas)

[12] Peraturan Menteri Pendidikan dan Kebudayaan Nomor 81A 2013 *Tentang Implementasi Kurikulum*

[13] Peraturan Menteri Pendidikan dan Kebudayaan Nomor 19 2005 *Tentang Standar Nasional Pendidikan*

IOP Conf. Series: Materials Science and Engineering **296** (2018) 012002 doi:10.1088/1757-899X/296/1/012002

The development of guided inquiry-based learning devices on photosynthesis and respiration matter to train science literacy skills

Choirunnisak[1], M Ibrahim [2] and Yuliani[2]

[1]Student of Science Education Program, Graduate Program of Universitas Negeri Surabaya
[2]Lecturer of Department of Biology Universitas Negeri Surabaya

choirunnisakchoirunnisak@mhs.unesa.ac.id

Abstract. The purpose of this research was to develop a guided inquiry-based learning devices on photosynthesis and respiration matter that are feasible (valid, practical, and effective) to train students' science literacy. This research used 4D development model and tested on 15 students of biology education 2016 the State University of Surabaya with using one group pretest-posttest design. Learning devices developed include (a) Semester Lesson Plan (b) Lecture Schedule, (c) Student Activity Sheet, (d) Student Textbook, and (e) testability of science literacy. Research data obtained through validation method, observation, test, and questionnaire. The results were analyzed descriptively quantitative and qualitative. The ability of science literacy was analyzed by n-gain. The results of this research showed that (a) learning devices that developed was categorically very valid, (b) learning activities performed very well, (c) student's science literacy skills improved that was a category as moderate, and (d) students responses were very positively to the learning that already held. Based on the results of the analysis and discussion, it is concluded that the development of guided inquiry-based learning devices on photosynthesis and respiration matter was feasible to train students literacy science skills.

1. Introduction

Science literacy becomes a must for everyone. Science literacy becomes very important for a person because the reverse of a nation one of them is determined by the quality of human that has the science literacy and technology skills [1].

The results of the 1988-2008 survey show that the increase in science literacy among college students in America is insignificant, only 10% -15% [2] and the science literacy of prospective teachers in Turkey is also low [3]. The average scientific skills of Indonesian learners has only come to the ability to recognize a number of basic facts, but they have not been able to communicate and link these abilities to various topics of science, let alone up to the application of concepts [4]. The main problems in the lectures in Indonesia in general due to the existence of some lecturers who are less understanding or less concerned about learning achievements, strategies and methods of learning, and how to appraise the right. The lessons dominated by lectures make it difficult for students to understand the essence of learning materials, so that their activities are limited to making notes and tend to be less able to listen due to the dependence on the photocopy of the presentation material from the lecturers, as well as the opportunity to uncover the learning material obtained in the real world/society is very limited [5].

With the release of Permendikbud No. 49 of 2014 on National Standards of Higher Education which emphasizes the characteristics of the learning process in universities must be interactive, scientific, contextual, collaborative, and student-centred, the Study Program is required to produce graduates in accordance with the curriculum of Indonesia National Qualification Framework (KKNI). Therefore, the development of science literacy is needed to help prospective teachers understand the content of science literacy and the elements of science literacy, and be able to use appropriate teaching methods as the key mechanism that will lead students to develop their science literacy through the process of learning in the classroom [6]. Indicators of prospective teachers who understand science literacy are aware of and understand the impact of science and technology in everyday life, making personal decisions about things including science, health, the use of energy resources, reading and understanding important matters of media reports about science materials, critiquing information automatically , participating in discussions with full confidence about science issues [2].

Universitas Negeri Surabaya (Unesa) has committed itself to a national literacy centre, with plans and pilot programs in and out. Unesa acts as a centre for literacy studies to emerge as an institution that deepens and develops literacy from conceptual and theoretical realms, through various research, discussions, seminars, and so on. Unesa also acts as a centre of literacy movement. In this context, Unesa will cultivate literacy internally in advance, including cultivating reading-writing for lecturers, students, and (if necessary) employees. Programs with measurable performance targets are being matured, so that achievement is satisfactory. Proverbially, Unesa residents must educate themselves before launching the literacy culture movement for the wider community. The principle of exemplary is expected to accelerate the realization of the dream of literacy [7].

The development of literacy from the conceptual and theoretical realms is still lacking. This can be seen from the analysis of tasks and interviews that have been done with lecturers of general biology course which is one of the prestigious programs of plant physiology course in the biology department of Unesa. The results of study shown by students on the midterm exam which includes photosynthesis and respiration are still low. The values range from 44 to 74 with an average value of 53.4. The low learning outcomes of students indicate that students have difficulties in photosynthesis and respiration study, besides that based on the result of a practical report on plant physiology course shows that students do not have the ability of science literacy. Students are still less able to assemble the problem of the two variables they specify (variable manipulation and response variables), identify variables, design experimental design, create table titles and tables of experimental results, analyze data and discuss problems that arise and relate them to theory which is relevant, and concludes the results of the experiment even though the skill has actually been trained since the basic biology course.

Seen from the base of the problem it turns out to be prepared prospective teachers who have the ability of science literacy so that when students become educators can teach students how to use the ability of science literacy to face the demands of the 21st century and global competition. The ability of science literacy can be taught through guided inquiry method. According to Piaget's cognitive development theory, children of 18 years old to adulthood have achieved formal operational development. This is in line with the stage of cognitive development of students majoring in biology is the formal operational stage. The main characteristic of development at this stage is the child is able to think abstractly and logically by using the pattern of "possible" thinking. Scientific thinking models with hypothetico-deductive and inductive types have begun to possess children, with the ability to draw conclusions, interpret and develop hypotheses [8]. The learning device for scientific literacy on photosynthesis and respiration material is designed using guided inquiry model and is a practicum worksheet that train the science literacy skills. Based on the above description, one of the efforts that need to be done is by arranging a device that can guide the lecturers and students in the learning process, so it is necessary to do a research development about learning devices in the form of "The Development of Guided Inquiry-Based Learning Devices on Photosynthesis and Respiration Matter to Train Science Literacy Skills".

2. Methods

This type of this research was research development, namely the development of guided inquiry-based learning device to facilitate student's science literacy. The development that will be carried out refers to the 4-D development model (four-D models). Testing of learning devices on photosynthesis and respiration matter was used one-group pretest-posttest design study. Learning devices are arranged in the form of Semester Lesson Plan, Lecture Schedule, Student Activity Sheet, Student Textbook and science literacy skills test. Learning devices are then validated by experts before implemented in the classroom. Subjects in this study were 15 Biology Education students in 2016 Department of Biology, Universitas Negeri Surabaya.

The Instruments that used in data collection consist of learning device validation sheet, lesson plan implementation observation sheet, student activity observation sheet, student response questionnaire, and science literacy skill test. Data from learning device validation, observation of lesson plan implementation, student activity, student response, and literacy test of student's science literacy were analyzed quantitatively

3. Results and Discussion

3.1 Learning Device Validity

Validation of learning devices as a whole is categorized as excellent. Thus the learning device developed feasible to use in learning.

3.2 SAP Implementation

The observation of lesson implementation in learning is done by two observers. Observations were made at three meetings, meeting I on photosynthesis, meeting II differences in plants C3, C4 and CAM and III respiration meetings. Average assessment of attendees at meetings I, II, and III amounted to 4.00 with the percentage of compliance between observers of 100%. Implementation of all activities is also accompanied by the management of the time and atmosphere of a class well conditioned.

3.3 Student Activity

Recording of student activities at the time of learning will provide an overview of student activities at the time of learning with guided inquiry. Students activities during learning will affect students opportunities in learning. Broophy and Good (in Muijs & David, 2008) [9] argue that the important factor that affects the opportunity to learn is time on tasks, that is, the number of time students spend on learning and not with other activities such as walking in the classroom.

Student activity that most students do during the learning with guided inquiry-based is doing the experiment At least students do activities that are not relevant to the percentage 1.8%.). The average student activity at each meeting as shown in Figure 1 below.

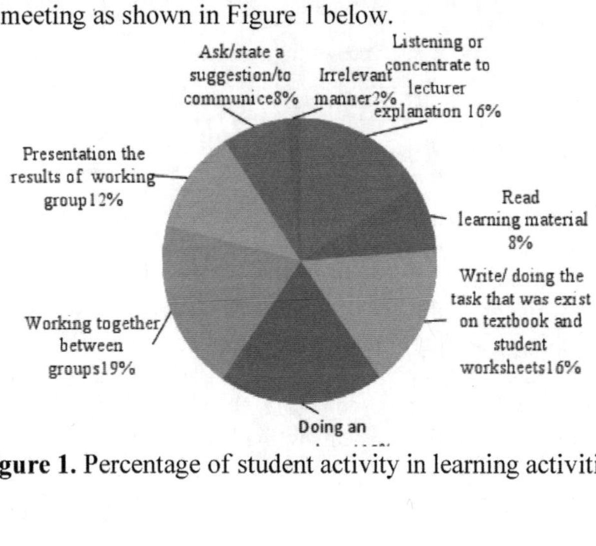

Figure 1. Percentage of student activity in learning activities

High student activity during learning shows that learning is student-centred. Students are able to find their own concept as in Ausubel's meaningful learning theory which emphasizes the importance of students associating experiences, phenomena, and new facts into the student's sense system. Student-centered teaching enables important concepts in learning to be retained in the minds of students. This is in accordance with the information processing theory that student involvement actively in learning leads to the storage of information into long-term memory [10].

3.4 Student responses

Student responses include interest in the use of the device, the newness of the device, the ease of the device, the continuation of the use of the device, the role of the lecturer in teaching as well as the ease of test questions. The students responded positively to the components.

The student's interest response using guided inquiry-based learning devices on the next material scored 80%. This means that students support, feel happy, and are interested in learning by using guided inquiry-based development devices on photosynthesis and respiration matter to train science literacy skills. Students are very enthusiastic when doing student activities worksheets, this is evidenced by 100% student activity related to guided inquiry-based learning. According to Maher and Zuhro (2009) [11], some students have learning goal orientation, these students are internally motivated such as satisfaction to learn something new [12]. Students feel interested in guided inquiry learning because inquiry learning still feels new.

3.5 Students Science literacy skills

The student's science literacy test refers to the student science literacy skills score indicator. Indicator of science literacy skills used as a reference in preparing the test, which is to identify scientific opinion; Conducting an effective scientific search; Designing the elements of the research design and how it impacts the findings/conclusions; Make a precise graph of the data; Solving problems using quantitative skills including basic statistics; Interpret basic statistical data; Conducting inferences and predictions based on quantitative data; Making conclusions based on quantitative data [13]. The result of science literacy skills test analysis seen from scores obtained for each indicator of science literacy ability can be seen in Figure 2.

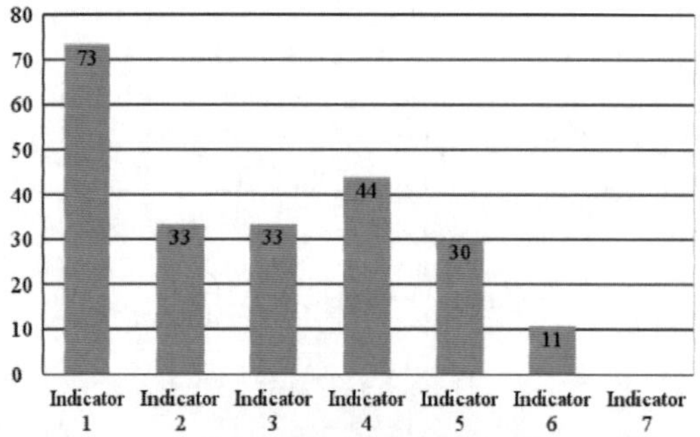

Figure 2. percentage of increasing in scores of each science literacy indicator

The highest increase indicator is the problem-solving indicator using quantitative skills, including the basic statistic of 73% while the indicators make the exact graph of the data still does not increase. This is because the teaching-learning process with guided inquiry will involve students in an investigation, helping them identify conceptual or inquiry issues, helping them identify conceptual or methodological problems in research and engaging them by designing ways to manganate the problems all of which can be done well if students trained critical thinking [14].

4. Conclusions and suggestions

Based on the results of the analysis, and discussion, it can be concluded that the devices based on guided inquiry learning on photosynthesis and respiration mater that has been developed were valid, practical, and effective to train students science literacy so it feasible to use in learning.

References

[1] UNESCO. (2008). *Science Education Policy-Making* Eleven *Emerging issues.* s.l. : UNESCO
[2] Impey, C. (2013). *Science literacy of undergraduates in the United States*. Organizations People and Strategies in Astronomy 2 (OPSA 2). Department of Astronomy, University of Arizona.
[3] Akenngi, H & Sirin, A. (2013). A Comparative study upon determination of scientific literacy level of teacher candidates. *Academic journals, Vol. 8(19), 1882-1886.*
[4] Toharuddin, Uus., Hendrawati, Sri dan Rustaman, Andrian. (2011). *Membangun Literasi Peserta didik.* Bandung: Humaniora.
[5] Sunarti, T.,(2015). Pemahaman Literasi Sains Mahasiswa Calon Guru Fisika Universitas Negeri Surabaya. Seminar Nasional Fisika Dan Pembelajarannya 2015 Universitas Negeri Malang
[6] Udompong, L., Traiwicitkhun, D. And Wongwanich, S. (2014). *Causal model of research competency via scientific literacy of teacher and student*. Procedia-Social and Behavioral Sciences, Vol. 116, 1581-1586.
[7] Khoiri. (2015). Unesa Sebagai Pusat Literasi oleh Much. Khoiri. Kompasiana.com.htm
[8] Budiningsih, Asri. (2004). *Belajar dan Pembelajaran*. Yogyakarta: Rineka Cipta.
[9] Muijs, Daniel and David Reynold. (2008). *Effective Teaching Evidence and Practice.* London: Sage Publication Ltd London
[10] Slavin, E. R. (2011). *Educational psychology.theory and practice.* USA: Pearson.
[11] Maher,A. Dan Zuhro.(2009). Learning Outcomes in Higher Education: Implementation of Curriculum Design and Student Learning. Journal of
[12] Hospital,Leisure,Sport and Tourism Education. 3(2): 47 Arrends, Richard I. (2012). *Learning to Teach Ninth Edition.* New York: McGraw-Hill.
[13] Gorrmally, Cara., Bricman, Peggy., and Lutz, Marry. (2012). " Developing a Test of Scientific Literacy Skill (TSOSL): Measuring Undergraduates' Evaluation of Scientific Information and Arguments". *CBE- Life Sciences Education.* Vol. 11, pp. 364-377
[14] Kardi, S. (2013). *Pengantar Pengembangan Kurikulum & Rencana Pelaksanaan Pembelajaran.* Surabaya: Universitas Negeri Surabaya

Initial draft of CSE-UCLA evaluation model based on weighted product in order to optimize digital library services in computer college in Bali

D G H Divayana[1], A Adiarta[2] and I B G S Abadi[3]

[1]Department of Informatics Education, Universitas Pendidikan Ganesha,
[2]Department of Electronics Education, Universitas Pendidikan Ganesha,
[3]Department of Elementary School Teacher Education, Universitas Pendidikan Ganesha

hendra.divayana@undiksha.ac.id

Abstract. The aim of this research was to create initial design of *CSE-UCLA* evaluation model modified with *Weighted Product* in evaluating digital library service at Computer College in Bali. The method used in this research was developmental research method and developed by Borg and Gall model design. The results obtained from the research that conducted earlier this month was a rough sketch of *Weighted Product* based *CSE-UCLA* evaluation model that the design had been able to provide a general overview of the stages of weighted product based CSE-UCLA evaluation model used in order to optimize the digital library services at the Computer Colleges in Bali.

1. Introduction

Good quality of education would not be conducted if there were no adequate supporting facilities and infrastructures. Facilities and infrastructures were used in supporting the implementation of education, especially in college, such as laboratories, libraries, study rooms, seminar rooms, tables, chairs, air conditioner, internet facilities, and others. From those several facilities and infrastructures, one of the most important facilities that must be provided in a college was the library as a place to find or obtain all the information related to books, literature, and collections that support the learning process.

Along with the development of information technology, the physical of the library has begun to change from initially requiring large rooms/buildings to accommodate a large number of collections, into digital form which does not require a large room/building. However, it only requires relatively small physical size with a large digital collection capacity storage. With the digital library, the number of documents or collections can be stored well and also can be accessed easily by visitors without any space and time limit.

In general, for the computer college, the existence of digital library becomes the most important things which must be provided because it becomes the characteristic of a Computer College engaged in the field of information technology, in which all the educational processes from the learning system and supporting learning infrastructure are expected to be developed based on technology information. Thus, it will produce a good quality of education.

In fact, it is not fully following that general view. However, in particular case, the existence of digital libraries in some college computers in Bali could not be implemented optimally. This was due to several obstacles, such as: there were the lack of strong legality about the implementation of digital library services in the colleges, human resources who were not skilled in the field of digital library services, the lack of full support and awareness by the parties who concerned on the importance of optimizing digital library services for the advancement and education quality, incomplete supporting facilities, and so on.

From these obstacles, it was necessary to conduct an evaluation of digital library services at Computer Colleges in Bali so that there would be a recommendation for service improvements. In principle, the evaluation is an activity conducted by the evaluator to collect, analyze, and present complete and accurate information about a particular object/program/service/policy being studied. Thus, the results could be used as a recommendation in making a decision [1]. Evaluation is an activity for collecting, analyzing, and presenting information about a particular object to be used for a consideration in making an appropriate and accurate decision [2]. Evaluation is an activity to collect, analyze, and present information about an object to be evaluated, where the results of these evaluations are used for consideration in making a decision that is precise, accurate, and reliable [3]. Evaluation is an activity conducted by an evaluator in collecting, analyzing, and presenting information related to the program/object/policy that the results can be used to take a decision [4]. Evaluation is an activity for collecting, analyzing, and explaining comprehensively information about a particular object/program/policy being studied and the results of an evaluation can be used for the consideration in making a decision to continue or to stop the object/program/policy [5]. Evaluation is an activity of data collection, data processing, data analysis, presentation of data into information that used as a recommendation in taking a right decision [6]. Evaluation is an activity that collects, analyzes, and presents data into useful information in making decisions based on recommendations obtained from these activities [7]. From these opinions, in general, evaluation can be interpreted as an activity to collect, analyze, and present related information about a particular object/program/policy/ service based on measurement using appropriate evaluation model and appropriate and valid evaluation instrument so that the results could be used as a recommendation in making a decision.

The appropriate evaluation model that could be used to evaluate digital library services in computer colleges in Bali is *CSE-UCLA*. This is supported by Divayana's opinion [8], which stated that "*CSE-UCLA* model is an evaluation model that has five dimensions of evaluation (*system assessment, program planning, program implementation, program improvement, and program certification*) and it is suitable for evaluating service programs that help people's life, Such as library programs, banks, economic enterprise, e-government, e-learning, and others".

Although the *CSE-UCLA* model was stated to be a suitable model for evaluating digital library services, it still had some weaknesses in terms of determining the level of optimization/effectiveness in sequence from the highest to the lowest category of the programs/the services evaluated based on quantitative calculations on each evaluation component. To overcome these obstacles, it was necessary to design a new innovation name a *weighted product* based on *CSE-UCLA* evaluation model. By using this model, the evaluation aspects used were still based on the components of *CSE-UCLA* evaluation model, while the quantitative calculation results would use the *weighted product* method.

Based on several problems that had been explained, there was one question that should be sought in this research, that was: How was the initial design of *Weighted Product* based *CSE-UCLA* evaluation model used in order to optimize the digital library service at the Computer College in Bali? The main purpose of this research was to find out the initial design of *CSE-UCLA* evaluation model based on the weighted product used in order to optimize digital library service at Computer College in Bali.

There were several previous studies that underlie to this research, such as, a research that had been implemented by Divayana, *et al* [9] in 2015 which was about expert-based digital library system at the Indonesia University of Technology. The advantages obtained in this research was the formation of a digital library application program that applied the concept of an expert system which contained the knowledge-base for data storage collection/digital library and inference motor to search/browse the collections/libraries online. The weakness found in the research was not able to measure the level of quality/optimization/effectiveness of digital library services. Research conducted by Divayana [10] in

2016 about the evaluation of digital library program based expert system at Indonesia University of Technology, aimed to know the extent of service quality in digital library program based on expert system applied at University of Technology Indonesia through evaluation activity using CSE-UCLA model. The results obtained from the research as an advantage that was the quality of digital library service program viewed from the component *assessment system* was 79.80% so that became good category, *program planning* component was 69.69%, belong to good category, *program implementation* component was 57.40% belong to good enough category, *improvement program* component equal to 74.90% belong to good category, and *certification program* component equal to 66.80% so that belong to good category. Whereas, the weakness found in the research was the CSE-UCLA model could not show the accurate calculation result of each evaluation component in sequence from highest to lowest category in determining optimization/effectiveness level of digital library service. According to several studies that had been conducted in 2015 and 2016, then the research that began to be implemented in 2017 to 2019, it is required the findings of new models that can overcome the problems of the previous research. The new finding model was *Weighted Product* based CSE-UCLA evaluation model, which was a CSE-UCLA evaluation model which is modified by using the weighted product method.

Based on the existing problems and purpose of this research, the researcher was interested to conduct a research on an initial design of *Weighted Product based* CSE-UCLA evaluation model in order to optimize digital library service at Computer College in Bali.

2. Literature Review

There were several important concepts/theories which need to be explained as the sources/references in this research, including the concepts of *Digital Library*, *CSE-UCLA*, and *Weighted Product Method*.

2.1. Digital Library

Digital library is all about the provision of digital collections, services, and infrastructure to support continuous learning, research, scholarly communication as well as preservation and conservation of recorded knowledge accessible anywhere and anytime [11]. Digital libraries are set of library activities and services which facilitate electronic means the processing, transmission, and display of information [12]. Based on those several definitions, it could be concluded that a digital library is a library integrated with a global computer network or the internet which is used as a facility to store large number collection of digital books and other documents and also can be accessed easily whenever and wherever the library users are located.

2.2. CSE-UCLA (Center for the Study of Evaluation-University of California in Los Angeles) Evaluation Model

CSE-UCLA model is an evaluation model that has five evaluation dimensions, which include system assessment, program planning, program implementation, program improvement, and program certification that is suitable to be used to evaluate service programs that help human life [13]. *CSE-UCLA* model evaluation was accomplished in several phases, namely: system assessment, program planning, program implementation, program improvement, and program certification [14]. Based on definition above, it could be taken a general conclusion that CSE-UCLA model is an evaluation model developed by Alkin consists of five evaluation components, namely: *system assessment*, *program planning*, *program implementation*, *program improvement*, and *program certification*, then this evaluation model is appropriate and suitable to be used in order to evaluate programs/policy in educational and other areas related to the service programs which help human life.

2.3. Weighted Product Method

Weighted Product method is one method of decision supporting system that uses multiplication to relate the attribute rating, where the rating of each attribute must be raised first with the attribute weight. This process is similar to the normalization process. Preferences for alternative A_i are given as follows [15,16]:

$$S_i = \prod_{j=1}^{n} x_{ij}^{w_j} \tag{1}$$

with i=1,2,...,m; and $\sum w_j = 1$.
w_j is a positive-valued rank for the attribute of gain and is negative-valued for the cost attribute.

3. Research Method

A research method in this research was development method design developed by Borg and Gall. The Borg and Gall model have 10 stages. Particularly in this research that was conducted in 2017, the development stage using the Borg and Gall model was limited to few stages only, including *research and collecting information*, research *planning,* design development, *preliminary field testing*, and preliminary *product revisión*. The reason for limiting the stages of development, because the main purpose of research in this 2017 was only limited to know the design of weighted product based CSE-UCLA evaluation model.

The object of this research was the design of Weighted Product based CSE-UCLA evaluation model. Research subjects who were involved in the preliminary test on this evaluation model design were 2 experts of Informatics, especially those who master the decision support system and 2 experts of education, especially those who master the field of educational evaluation. Research location which was conducted in this research was in STIKOM Bali because STIKOM Bali was one of Computer College in Bali that already implemented digital library and it had complete facilities to conduct digital library research.

The main instrument that used in collecting data in this research was a questionnaire while complementing instrument that used in collecting the data were interview guide and photo/documenting picture.

The data analysis technique which was used in the accurate calculation of each evaluation component of Weighted Product based CSE-UCLA model from the highest category to the lowest category in this research was quantitative descriptive analysis technique. While data analysis about the constraints found in the design of Weighted Product based CSE-UCLA evaluation model used qualitative descriptive analysis technique.

4. Results and Discussion

The results obtained from the research and field data collecting stage during one month the beginning of this study was the rough sketch of Weighted Product based CSE-UCLA evaluation model design. These aspects were measured using (core/main/key) instruments and complementary instruments. The main instrument was a questionnaire used by respondents to evaluate each aspect of the evaluation through filling the questionnaires toward digital library services, while complementary instrument were interview guides and photographs/documentation pictures that used by researchers in conducting direct observations in the field to obtain evidence in order to reinforce the research results and derive the constraints found in digital library services.

The assessment results that were carried out by the respondents for each evaluation aspect using that main instrument, was given input weight value preferences from decision makers so that it could be calculated using the weighted product to determine the conception about optimization digital library services based on each evaluation component sequentially from highest to the lowest category. In addition, there was also an illustration about constraints found in digital library services that lead to the results sequenced calculation on each evaluation component. From the constraints illustration, it could be recommended the appropriate solutions in order to improve/repair digital libraries service, so that, digital library service at Computer College in Bali could run optimally.

The rough sketch of *Weighted Product* based *CSE-UCLA* model design could be seen in Figure 1 below.

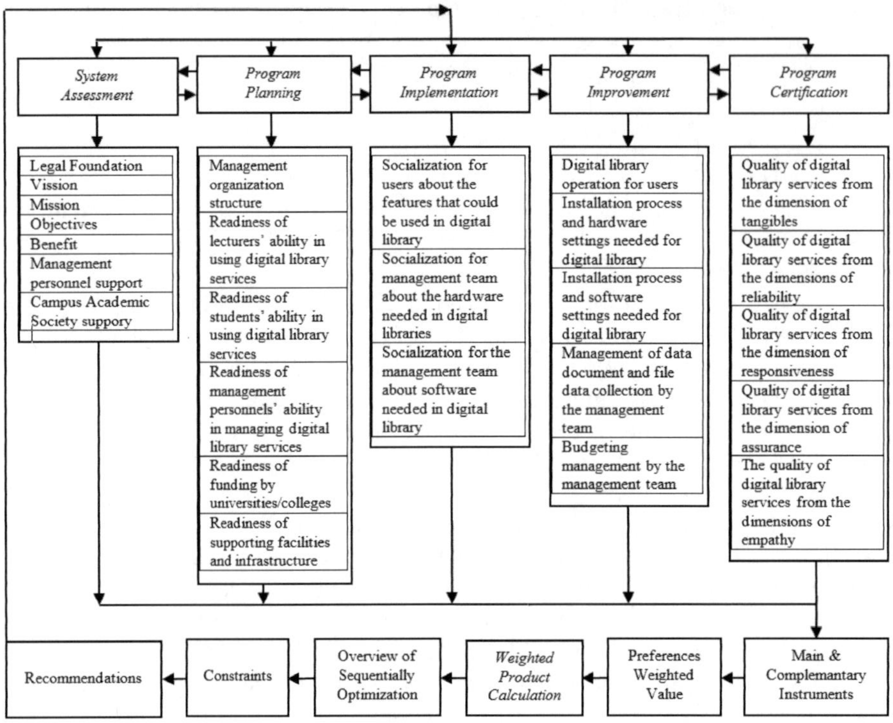

Figure 1. The rough sketch of *Weighted Product* based *CSE-UCLA* model design

Related to the rough sceptical design of the Weighted Product-based CSE-UCLA model shown in Figure 1 showed that digital library services were evaluated using five components of CSE-UCLA model: *system assessment*, *program planning*, *program implementation*, *program improvement*, and *program certification*. Each evaluation component had evaluation aspects used to determine the optimum level of digital library services.

The obstacles found in this research was not able yet to find the evaluation model design which was able to input new aspects automatically if in the future it would be found new aspects that influence the optimization of digital library services.

5. Conclusions

Based on the problems stated, the results of research and discussion, it could be concluded several things, such as 1) the research which was conducted at the beginning of this month had produced a rough sketch of Weighted Product based CSE-UCLA evaluation model design shown in Figure 1. Although it was still a sketch, the design had been able to provide a general overview with the stages of weighted product based CSE-UCLA evaluation model used in order to optimize the digital library services at the Computer Colleges in Bali; 2) To overcome the obstacles found in this research, then in the future was needed to develop evaluation model design which was able to input new aspects automatically.

References

[1] Divayana D G H, Marhaeni A A I N, Dantes N, Arnyana I B P and Rahayu W 2017 Evaluation of blended learning process of expert system course program by using a cse-ucla model based on mobile technology *J. Theor. Appl. Inf. Technol.* **95** pp 3075–86

[2] Divayana D G H and Sanjaya D B 2017 Mobile phone-based cipp evaluation model in evaluating the use of blended learning at school in bali International *J. Interact. Mob. Technol.* **11** pp 149–59

[3] Divayana D G H, Sanjaya D B, Marhaeni A A I N and Sudirta I G 2017 CIPP evaluation model used based on mobile phone evaluating the use of blended learning platforms at vocational schools in bali *J. Theor. Appl. Inf. Technol* **95** pp 1983–95

[4] Divayana D G H, Adiarta A and Abadi I B G S 2017 Conceptual and physical design of evaluation program for optimizing digital library services at a computer college in bali based on cse-ucla model modification with weighted product *J. Theor. Appl. Inf. Technol* **95** pp 3767–82

[5] Divayana D G H, Ardana I M and Ariawan I P W 2017 Measurement of the effectiveness of a lecturer in transferring algebra knowledge through of multimedia facilities by using certainty factor-formative-summative model *J. Theor. Appl. Inf. Technol.* **95** pp 1962–73

[6] Suandi I N, Putrayasa I B and Divayana D G H 2017 Compiling a dictionary of loan words in balinese: the evaluation result of effectiveness testing in the field aided by mobile technology *J. Theor. Appl. Inf. Technol.* **95** pp 3186–95

[7] Divayana D G H 2017 Utilization of cse-UCLA model in evaluating of digital library program based on expert system at universitas teknologi indonesia: a model for evaluating of information technology-based education services *J. Theor. Appl. Inf. Technol.* **95** pp 3585–96

[8] Divayana D G H 2015 Penggunaan model cse-ucla dalam mengevaluasi kualitas program aplikasi sistem pakar *Seminar Nasional Teknologi Informasi & Aplikasinya (SNATIA)* (Denpasar Bali) pp 165–8

[9] Divayana D G H, Ariawan I P W, Sugiarta I M and Artanayasa W 2015 Digital library of expert system based at indonesia technology university *Int. J. Adv. Res. Artif. Intell.* **4** pp 1–8

[10] Divayana D G H 2016 *Evaluasi Program Perpustakaan Digital Berbasis Sistem Pakar Pada Universitas Teknologi Indonesia* (Jakarta: Program Pascasarjana Universitas Negeri Jakarta)

[11] Eze U F and Ekpe E S 2014 A pragmatic virtual library system for universities academic enhancement in the developing country *Int. J. Inf. Technol. Bus. Manag* **24** pp 70–82

[12] Towolawi and Oluwakerni K 2013 School library media specialist's awareness and perception of digital library services: a survey *Ozean J. Soc. Sci* **6** pp 77–89

[13] Ardana I M, Ariawan I P W and Divayana D G H 2017 Measuring the effectiveness of blcs model (bruner, local culture, scaffolding) in mathematics teaching by using expert system-based cse-ucla *Int. J. Educ. Manag. Eng* **7** pp 1–12

[14] Kurniawan D 2013 Evaluation on foreign language development program *J. Eval. Pendidik* **4** pp 1–11

[15] Divayana D G H and Sugiharni G A D 2016 *Sistem Pakar dan Sistem Pendukung Keputusan-Bahan Ajar Pendidikan Teknik Informatika* (Singaraja: Undiksha Press)

[16] Divayana D G H 2014 Development of duck diseases expert system with applying alliance method at bali provincial livestock office *Int. J. Adv. Comput. Sci. Appl.* **5** pp 48–54

IOP Conf. Series: Materials Science and Engineering **296** (2018) 012004 doi:10.1088/1757-899X/296/1/012004

Development of the living thing transportation systems worksheet on learning cycle model to increase student understanding

E Rachmawati, S Nurohman and A Widowati

Faculty of Mathematics and Science Universitas Negeri Yogyakarta

ekarachma95@gmail.com

Abstract. This study aims to know: 1) the feasibility LKPD review of aspects of the didactic requirements, construction requirements, technical requirements and compliance with the Learning Cycle. 2) Increase understanding of learners with Learning Model Learning Cycle in SMP N 1 Wates in the form LKPD. 3) The response of learners and educators SMP N 1 Wates to quality LKPD Transportation Systems Beings. This study is an R & D with the 4D model (Define, Design, Develop and Disseminate). Data were analyzed using qualitative analysis and quantitative analysis. Qualitative analysis in the form of advice description and assessment scores from all validates that was converted to a scale of 4. While the analysis of quantitative data by calculating the percentage of materializing learning and achievement using the standard gain an increased understanding and calculation of the KKM completeness evaluation value as an indicator of the achievement of students understanding. the results of this study yield LKPD IPA model learning Cycle theme Transportation Systems Beings obtain 108.5 total scores of a maximum score of 128 including the excellent category (A). LKPD IPA developed able to demonstrate an improved understanding of learners and the response of learners was very good to this quality LKPD IPA.

1. Introduction

In 2013 curriculum, there is a Learning Cycle model where the learning process of learners can master the competencies that must be achieved by participating actively participate. The success of the learning cycle is examined by experts from the beginning of its development. The results show that the learning cycle can improve the learners' learning outcomes, improve the reasoning abilities and skills of the science process and learners have better conceptual retention [1].

Understanding concepts are considered important in the learning process because concepts are the building blocks of thinking and the basis for higher mental processes to formulate principles and generalizations [2]. If in the process of learning to teach learners, we have problems in understanding the concept, this will result in low learning achievement of science. Although, not all students experience it because each learner has different abilities. To overcome these problems required an innovation in learning so that the learning process learners run effectively and learning objectives can be achieved completely. Benny A Pribadi [3] suggests that the learning process can be called successful if learners interact with learning resources seriously and continuously, doing exercises for competency mastering, getting feedback immediately after learning, applying skills in real context, and interacting in acquiring knowledge And skills. To realize a successful learning process and in

IOP Conf. Series: Materials Science and Engineering **296** (2018) 012004 doi:10.1088/1757-899X/296/1/012004

accordance with the objectives of learning, appropriate approaches, models and learning methods are needed.

The process of learning in the class, of course, cannot be separated from the role of teachers as educators. To find a concept, educators need to guide learners toward the discovery of concepts. One method that can be used to find a concept that is experimental activities, learners positioned as if a scientist who performs a process to connect with the symptoms he encounters. Trial activities require activity guidance, that is LKPD (Student Activity Sheet) which can be arranged as a guide for learners in doing activities. According to Trianto [4], the worksheet of learners activities is the learners' guides used to carry out investigative or problem-solving activities. LKPD has a big role in the learning process because it can help educators to lead learners to discover concepts through their own activities. This material science is very complex, such as the material in the Transportation System of Living Things. Make teachers overwhelmed to teach it in the classroom. Usually, the teacher will explain the material with the theory without any experiments conducted by learners.

2. Research method

2.1. *Type of research*
This research type is *Research and Development* (R & D).

2.2. *Time and place*
The study was conducted at SMPN 1 Wates, 15-23 February 2016.

2.3. *Subject and objects research*
The subjects of this study were 28 children in class VIII C SMP N 1 Wates as respondents. The object of this research is LKPD IPA theme "Transportation of Life" in Learning Cycle model to improve students' understanding.

2.4. *Procedure*
The method used in this research is development. The model used is the development of the 4-D model. The 4-D development model was developed by S. Thiagarajan, Dorothy S. Semmel and Melvyn I. Semmel [5] includes four stages: *define*, *design*, *develop* and *disseminate*. The defining stage did by problem analysis, student analysis, task analysis, concept analysis, and formulate learning objectives. In the design stage, the preparation of instruments, selection of teaching materials, format selection, and initial product design. The development stage includes the review phase by the supervisor, expert judgment (validation by the expert lecturer and science teacher), and product trial. In the disseminated stage (dissemination) is only done in a limited, considering the realm of R & D research is very broad.

2.5. *Technique of data analyze*
Validation/feasibility analysis, students' response, and observation sheet of learning cycle implementation are done by calculating the average score, the average score is then converted to scale 4 presented in Table 1.

IOP Conf. Series: Materials Science and Engineering **296** (2018) 012004 doi:10.1088/1757-899X/296/1/012004

Table 1. Conversion Score to Value on a Four Scale [6]

No.	Score Range	Value	Category
1.	$X > Xi\,Xi + 1 \cdot Sbi$	A	Very Good
2.	$\underline{Xi}\,\underline{Xi} + 1 \cdot Sbi > X \geq \underline{Xi}\,\underline{Xi}$	B	Good
3.	$\underline{Xi}\,\underline{Xi} > X \geq \underline{Xi}\,\underline{Xi} - 1 \cdot Sbi$	C	Bad
4.	$X \leq \underline{Xi}\,\underline{Xi} - 1 \cdot Sbi$	D	Very Bad

While the increase in understanding of learners is calculated by using gain score

$$\text{Gain score} = \frac{posttest\ score - pretest\ score}{maximum\ score - pretest\ score} \tag{1}$$

Table 2. Completeness Criteria Based on KKM [9]

Range Score	Category
$g > 0.7$	High
$0.3 \leq g \leq 0.7$	Medium
$g < 0.3$	Low

Based on the completeness of the final test result (posttest) of the whole students, the percentage of learners' complete mastery can be calculated as follows.

$$\text{Percentage of Mastery: } \frac{the\ number\ of\ succeeded\ students}{total\ number\ of\ students} \tag{2}$$

3. Results and discussion

3.1. *Feasibility of LKPD IPA development results*

The researcher compiled the operational definition of the theory that has been studied then compiled the LKPD grid. The grid is used as a development guide and measurement instrument on LKPD.

LKPD was developed as an initial product which then carried out a series of tests to get constructive criticism, suggestions, and inputs, so as to produce products that meet the criteria that can be used in the learning process. Revisions are made after the validation stages to make the product ready for use for development testing.

After providing inputs and responses to the developed products, expert lecturers and teachers as validators also provide an assessment of LKPD products in order to determine the feasibility of product development. The assessment contains two aspects: material aspect and media aspect. The results of the scores by the validator are shown in Figure 1.

The assessment which is done by the teacher is a validation of LKPD product by referring to LKPD validation questionnaire. The teacher's assessment of LKPD products includes the components of didactic conditions, terms of construction, technical requirements and conformance with Learning Cycle.

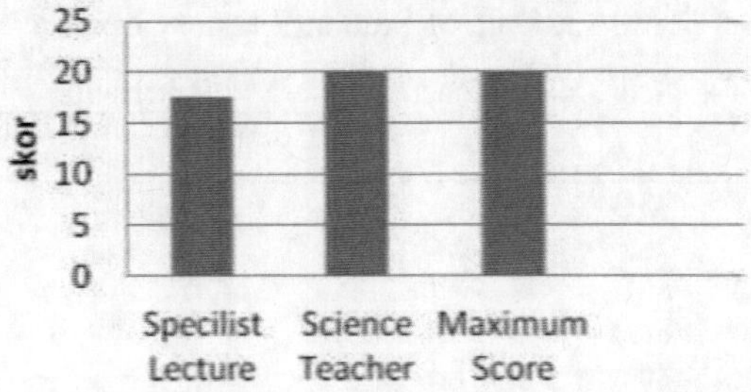

Figure 1. Lecturer and Teacher Rating Score Diagrams

3.2. *Student's Response to LKPD*

Response data of students to LKPD IPA Learning Cycle model using questionnaire given at the end of learning activities. Questionnaire response of learners consists of four aspects namely didactic aspects, construction, and technical and conformity aspects of the Learning Cycle model. Right statements are tailored to the level of understanding of students. Questionnaire response of learners distributed to 28 students class VIII C SMP N 1 Wates. The calculation of student's questionnaire responses uses the same way as media validation score calculation. The result of the average score of didactic condition requirement equals to 15,9, the aspect of construction condition equal to 35,4, the aspect of technical requirement equals to 9,07 and the aspect of conformity with LC equals to 3,2. Based on the scores, the response of student to LKPD included in the category of "Very Good". It shows that LKPD is feasible to be used and can attract students' interest in their learning activities. The diagram of students response of LKPD presented in Figure 2. It shows that LKPD is feasible to be used and can attract learners' interest in their learning activities.

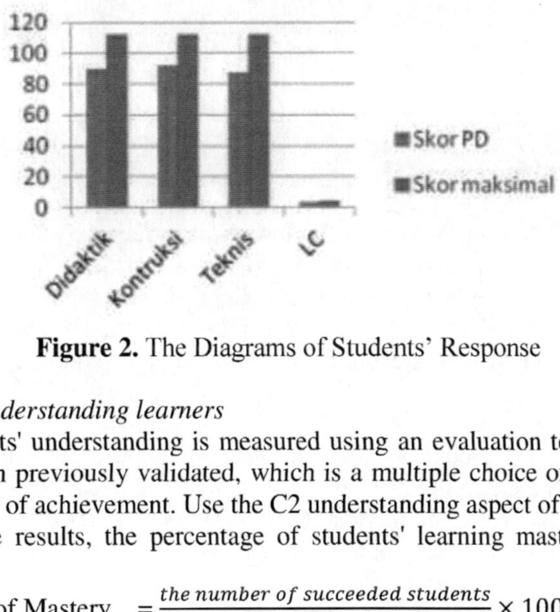

Figure 2. The Diagrams of Students' Response

3.3. *Enhancement of understanding learners*

Enhancement of students' understanding is measured using an evaluation test (pretest-posttest). The test instrument has been previously validated, which is a multiple choice of 30 items that match the predetermined indicator of achievement. Use the C2 understanding aspect of Bloom Taxonomy.

Based on these results, the percentage of students' learning mastery can be calculated as follows.

$$\text{the Percentage of Mastery} = \frac{\textit{the number of succeeded students}}{\textit{total number of students}} \times 100\%$$

$$= \frac{24}{28} \times 100\% = 85.7\%$$

IOP Conf. Series: Materials Science and Engineering **296** (2018) 012004 doi:10.1088/1757-899X/296/1/012004

Table 3. Data of Student Mastery Lesson

No	Score	Frequency	Description	
			Complete	Incomplete
1	73,3	-	-	-
2	76,6	4	-	√
3	80	7	√	-
4	83,3	11	√	-
5	86,6	3	√	-
6	90	-	-	-
7	93,3	1	√	-
8	96,6	1	√	-
9	100	1	√	-
Total		28	24	4

Based on the calculation of the evaluation test has increased from the previous material which achieved the mastery below 75% to 85.7%. Based on the calculation of gain score obtained a value of 0.6 that belongs to the category "medium".

3.4. *Observation of learning cycle in learning activities*

The implementation of learning with Learning Cycle model is done by observation technique. Observations were made by two observers during three meetings. This observation sheet fits the lattice of the Learning cycle learning model. The data of learning outcomes using Learning Cycle can be seen in Table 4.

Table 4. Percentage of Learning Implementation

Activity	Percentage of Implementation (%)			Average (%)
	P 1	P 2	P 3	
Teacher	100	100	100	100
Student	72,2	100	100	90,7

4. Conclusion and suggestion

4.1. *Conclusion*

The feasibility of LKPD IPA with the theme "Living Transportation System" in the Learning Cycle learning model that has been produced according to the assessment of expert lecturers and IPA Teachers is included in the excellent category (A). Enhancement of learners' understanding with Learning Cycle Learning Model at SMP N 1 Wates in 3 meetings gets percentage of complete evaluation value based on KKM that is equal to 85,71% and gain score category including medium. The response of learners and educators SMP N 1 Wates to the quality of LKPD Living Transportation System included in the category very well.

LKPD with Learning Cycle model is said to be feasible and included in the category very well because in accordance with the purpose of learning cycle model that can help teachers in developing a conceptual understanding that can accommodate learning opportunities of teachers and learners. The learning cycle can be implemented flexibly and meet the real needs of teachers and learners. Judging from the dimensions of the teacher application of this model broaden the horizons and improve the creativity of teachers in designing learning activities. In terms of students 'dimensions, the application of this model provides benefits as (1) improves learning motivation because learners are actively involved in the learning process, (2) helps develop the students' scientific attitudes, (3) more meaningful learning. (Fajaroh dan Dasna dalam Buku Guru IPA 2013: 15).

4.2. *Suggestion*
With the arrangement of LKPD IPA "Living Transportation System" in Learning Cycle learning model is expected teachers can use it as a medium of learning while doing learning on the material. And it is expected that the use of this LKPD in addition to the learning process, it is expected that learners can use it outside of class hours as supporting independent activities.

References
[1] Zubaidah 2017 Pembelajaran Kontekstual Berbasis Pemecahan Masalah Untuk Mengembangkan Kemampuan Berpikir Kritis
[2] Ratna Wilis Dahar 1988 Teori- Teori Belajar (Jakarta: Erlangga)
[3] Pribadi B A 2009 *Model Desain Sistem Pembelajaran* (Jakarta: Dian Rakyat)
[4] Trianto 2014 Model Pembelajaran Terpadu
[5] Thiagarajan, Sivasailam, Semmel and Melvyn I 1974 Instructional Development for Training Teachers of Exceptional Children (Bloomington, Indiana: Indiana University)
[6] Mardapi, Djemari 2008 *Teknik Penyusunan Instrumen Tes dan Nontes* (Yogyakarta: Mitra Cendekia Press)
[7] Anderson, Lorin W and Krathwohl David R 1956 *A Taxonomy for Learning, Teaching and Assessing: a Revision of Bloom's Taxonomy* (New York: David McKay Company, Inc)
[8] Benny A Pribadi 2009 *Model Desain Sistem Pembelajaran* (Jakarta: Dian Rakyat)
[9] Hake Richard R 1999 *Analyzing Change/Gain Scores* (Retrieved from http://www.physics.indiana.edu/~sdi/AnalyzingChange-Gain.pdf. at 2015-12-09, 18.50 WIB)
[10] Hendro Darmodjo and Jenny R E Kaligis 1992 *Pendidikan IPA 2* (Jakarta: Depdikbud)
[11] Ratna Wilis Dahar 1988 *Teori- Teori Belajar* (Jakarta: Erlangga)
[12] Thiagarajan, Sivasailam, Semmel, Dorothy S, Semmel and Melvyn I 1974 *Instructional Development for Training Teachers of Exceptional Children* (Bloomington, Indiana: Indiana University)
[13] Trianto 2014 *Model Pembelajaran Terpadu* (Jakarta: Bumi Aksara)
[14] Zubaidah S 2017 *Pembelajaran Kontekstual Berbasis Pemecahan Masalah Untuk Mengembangkan Kemampuan Berpikir Kritis* (Retrieved from https://www.google.com/url?sa=t&rct=j&q=&esrc=s&source=web&cd=5&ved=0ahUKEwiF9_yMi7bXAhWIU7wKHUhVBtkQFghBMAQ&url=https%3A%2F%2Fwww.researchgate.net%2Fprofile%2FSiti_Zubaidah5%2Fpublication%2F318013668_PEMBELAJARAN_KONTEKSTUAL_BERBASIS_PEMECAHAN_MASALAH_UNTUK_MENGEMBANGKAN_KEMAMPUAN_BERPIKIR_ KRITIS.pdf at 2017-11-11, 15.20 WIB)

Engaging pre-service teachers to teach science contextually with scientific approach instructional video

E Susantini [1], I Kurniasari[2], A N M Fauziah[3], T Prastowo[4], A Kholiq[4] and L Rosdiana[3]

[1]Department of Biology, Universitas Negeri Surabaya, Surabaya, Indonesia
[2]Department of Mathematics, Universitas Negeri Surabaya, Surabaya, Indonesia
[3]Department of Natural Sciences, Universitas Negeri Surabaya, Surabaya, Indonesia
[4]Department of Physics, Universitas Negeri Surabaya, Surabaya, Indonesia
Kampus Unesa Ketintang, Surabaya 60231, Indonesia

endangsusantini@unesa.ac.id

Abstract. Contextual teaching and learning (CTL) present new concepts in real experiences and situations, where students can find out the meaningful relationship between abstract ideas and practical applications. Implementation of CTL using scientific approach fosters teachers to find constructive ways of delivering and organizing science contents in science classroom settings. An instructional video for modelling by using a scientific approach in CTL was then developed. Questionnaires with open-ended questions were used to, asking whether modelling through instructional video could help them to teach science contextually with a scientific approach or not. Data for pre-service teachers' views were analyzed descriptively. The aims of this research are to engage pre-service teachers in learning how to teach CTL and to show how their responses to learning and how to teach CTL using the video. The study showed that ten pre-service teachers in science department were involved, all observed through videos that demonstrated a combined material of CTL and scientific approach and completed worksheets to analyze the video contents. The results show that pre-service teachers could learn to teach contextual teaching and make use of scientific approach in science classroom settings with the help of model in the video.

1. Introduction

Contextual teaching presents new concepts in real experiences and situations, where students can discover the meaningful relationship between abstract ideas and practical applications in the context of their usage. Scientific approach makes the way of the science taught was found through objective data rather than conventional lecture [1,2]. Hence, implementing contextual teaching together with scientific approach will foster teachers to find constructive ways of delivering and organizing science content. In this paper, we developed an instructional video to become the model which is an adequate example of contextual teaching integrated with a scientific approach. We proposed that the use of this instructional video is capable to engage pre-service teachers to learn how to teach science contextually as well as corroborate scientific approach along with learning material taught.

Contextual teaching is supported by cognitive psychology theory, stating that "cognition is contextually dependent and must be described in that context before the material is understood at all." There are two complementary usages of word context [3]. First, context used to denote domain specificity which relates to the disciplinary knowledge that the learners wish to acquire, while the later used to denote signal task that contains true-to-life problems for the learners in the process of acquiring or applying the knowledge [4]. Contextual learning occurs when students process new information or knowledge in a way that it makes sense to them in their own frames or reference (their own memory, experience, and response) [5]. There are seven components of contextual teaching and learning: a) constructivism; b) modelling; c) questioning; d) learning community; e) inquiry; f) authentic assessment; and g) reflection. In constructivism, students build their own knowledge by testing ideas based on prior knowledge or experience, applying these ideas to a new situation, and integrating the new knowledge gained with pre-existing knowledge [6]. As a result, students are actively engaged in hands-on activities and encouraged to gain knowledge through exploration [7]. The inquiry has its origins in the practices of scientific inquiry, emphasizing students to pose questions, to gather and analyze data, and to construct problem-solving or conclusion based on evidence [6]. Modeling in contextual teaching can be performed by the teacher or even involving students to give example to other students relating to the material taught. Questioning aims to stimulate and foster students' thinking skills; this can be conducted by encouraging students to ask questions and the questions must be answered by other students or simply asking the students about what they have not understood or they wanted to know. Learning community refers to increase the use of group work and to help low-achieving students to learn with the assistance of high-achieving students. Authentic assessment utilized tasks that are the real examples of extended criteria performances of actual learning goals [8]. This kind of assessment also used rubrics and other criteria checklists as standards to improve learning and teaching, providing multiple opportunities for students to learn and practice the desired outcomes as well as receiving feedback and reflection. The last component, reflection engages students and teachers to review, think, and evaluate their learning process and what they have just learned.

In science, contextual teaching can be conducted by illuminating theoretical practices and providing an opportunity for hands-on investigation. The contextual hands-on investigation should increase the degree of openness and confront students using raw phenomena with opened problems, answers, and methods [9]. Contextualized science should be taught like scientist' science or open-ended research in which the task is simplified and students' motivation and engagement are enhanced by the perception that their practical work is authentic. Essentially, contextual teaching in science should also present concepts in familiar contexts and tangible examples or experiences compared to abstract conceptual models [10].

Bandura's Social Learning Theories used as the theoretical framework for using instructional video in pre-service teacher education. Instructional video integrates modelling and video as visual cues in the expectation that the teachers will engage themselves in specific behaviour which is planned to teach [11]. Video with classroom instruction also builds bridges between learning theory and teaching practice in the actual classroom. It helps the students to raise insightful ideas relating to good teaching practice [12]. The use of video in teacher education is likely to increase the ability to plan a lesson and observe authentic classroom sequences [13].

In investigating how the instructional video affects learning process of pre-service teachers in applying contextual teaching along with scientific approach, this study addressed these research questions:

- Could the instructional video engage pre-service teachers in learning how to teach CTL with the scientific approach?
- What were pre-service teachers' responses on how to teach CTL using instructional video?

2. Research methods

2.1. *Participants*
Participants of this study were ten (10) pre-service teachers in Department of Natural Sciences, Surabaya State University, Indonesia. These pre-service teachers had been taught about the general overview of contextual teaching and learning as well as scientific approach and had no experience to teach in the actual classroom. All participants' names used in this study were pseudonyms.

2.2. *Classroom procedure*
All participants observed the instructional video which demonstrated contextual teaching and learning combined with the scientific approach as they completed a worksheet to analyze the video content. During this activity, they were supervised by a lecturer (one of the authors). The worksheet asked pre-service teachers replayed the video on their own to discuss: (1) which scene showed characteristics of contextual teaching and learning, (2) which scientific approach steps appeared at each phase of it, (3) what social skills could be gained from learning activities and (4) what strengths and weaknesses of learning activity could be noticed.

2.3. *Instructional video*
Instructional video used in this study was developed to provide adequate modelling of contextual teaching and learning in the science classroom with a scientific approach. This instructional video was validated by three independent educational experts. A lesson in the video was taught by one of the authors on the subject of 'The Motion of Human Body' to a class of 13-14 years old students. The purpose of this lesson was to provide an opportunity to students to construct their own knowledge how simple machines work through contextual teaching and learning activities combined with a scientific approach. The instructional video was also featured with captions, showing which characteristic of contextual teaching was demonstrated at specific scenes and what element of scientific approach inserted along.

2.4. *Data collection and analysis*
We monitored how pre-service teachers planned their learning activities through their designed lesson plan and observed their teaching skills through microteaching using criteria and rubrics performances. Observers assessed pre-service teachers' teaching skills using a Likert scale: 1 = 'poor', scale 2 = 'fair', 3 = 'good', and 4 = 'advanced'). These teaching skills involve two criteria: (1) performing contextual teaching and learning and (2) integrating scientific approach into science topics. Score 3.0 or above indicates adequate teaching skills. Data for participants' responses were collected using questionnaires with open-ended questions, asking whether modelling through instructional video could help them to teach science contextually with a scientific approach. Data for pre-service teachers' views were analyzed descriptively.

3. Results and discussions

3.1. Could pre-service teachers teach contextually and scientifically with instructional video?
According to their average score (Table 1), pre-service teachers showed adequate skills in teaching science contextually and incorporated the learning activities to the scientific approach. We examined

their teaching performance based on essential characteristics of contextual teaching learning and scientific approach elements.

Table 1. Pre-service teachers' performance in conducting contextual teaching with scientific approach

Performance aspect	Average score
Teaching science contextually	
a. Learning activities were relevant with seven components of contextual teaching and learning:	3.4
b. Characteristics of contextual teaching and learning were conducted systematically	3.7
c. The material was relevant to be taught with contextual teaching and learning	3.5
Learning activities with scientific approach	
a. Learning activities were focused on teaching scientific approach	3.8
b. Developing scientific approach	3.7
c. The applied scientific approach was relevant to be used with contextual teaching and learning	3.6

In their consistency of well-contextualized teaching, pre-service science teacher number 5, 8, 7, 9 and number 6 correctly identified all of model teacher's behaviours representing characteristics of contextual teaching and scientific approach elements. As pre-service science teacher number 7 asserted:

> "Model teacher started *constructivism* approach as she played a show of badminton match. Later, as part of *modelling*, she asked several students to demonstrate how a badminton player served. In this situation, students *observed* both of badminton show and live-show of serving and make them interested and *questioned* how badminton techniques related to the human locomotion, model teacher constructed the knowledge further by asking them to *make groups* (learning community) and *investigated* the problem (inquiry) through an open-ended worksheet and portfolios. After that, students *collected data* and *made the association* to their prior knowledge about the motion. The students then being asked *to communicate* their investigation in portfolios that became part of the assessment. Lastly, model teacher *reflected* on learning process with her students, reviewing how motion system could work in the human body and how this principle can be useful to score a serving point."

Interestingly, although some of the pre-service teachers, such as pre-service science teacher number 4, 10, 3 and number 2 misidentified inquiry and authentic assessment during video-analysis, our preceding description about their microteaching indicated that they showed adequate skills within those criteria. For instance, this misidentification could be seen as pre-service science teacher number 2 asserted, "Inquiry can be seen as students work on their worksheet. Authentic assessment occurs as students generate research questions, invent hypotheses, and perform investigations." In her microteaching, although pre-service science teacher number 2 was still poor at contextualizing the material, she conducted an inquiry and authentic assessment well. Pre-service science teacher number

4 even had a good idea of authentic assessment by involving her students to do peer-assessment toward their friends' presentation. The evaluation and reflection conducted by the lecturer upon the video analysis provided great help in noticing correct scenes.

3.2. What did they perceive?

All pre-service teachers perceived that the instructional video helped them to identify and how to teach science contextually with the use of the scientific approach (Table 2), including each characteristics representing contextual teaching and stages of the scientific approach.

Table 2. Pre-service teachers' views about learning with instructional video

Response	Response was given (%)
The video helps me to:	
Identify seven essential characteristics of contextual teaching and learning	100
Learn how to teach contextual teaching and learning	100
Learn how to teach scientific approach	100
The video was easy to understand	90

In line with Table 2, pre-service science teacher number 4 also gave positive comments, "I love the idea of learning to teach contextually with the [instructional] video. It encourages me to apply this approach in my future classroom." pre-service science teacher number 7 added, "[This] video is very useful reference for teaching practice." pre-service science teacher number 1 suggested,

> "It is very convenient to learn with the video because the caption helps me to identify specific scenes which are related to characteristics of contextual teaching. However, it would be nice if the video can show a clearer stage of authentic assessment."

In general, this study showed that pre-service teachers could learn how to teach science contextually along with a scientific approach to the instructional video that we had developed. They noticed the scenes in the instructional video as visual cues, that attaching mental images in their thinking system; this could help pre-service teachers to construct more concrete knowledge and build greater interests in learning than verbal description does [11]. Taking notes (on the worksheet) and discussing model teachers' teaching behaviour, as performed by pre-service teachers in this present study, also helps them to increase classroom awareness. Previous work reported that video lesson analysis improves pre-service teachers' ability to pay attention to noteworthy events of classroom interactions [14]. Higher scores on the video analysis task are also associated with better instructional quality [15]. Furthermore, pre-service teachers who write their own observations at video clips show better classroom awareness [16]. A recent study also indicated that peer dialogue during the use of video potentially heightens teacher's awareness about their own teaching practice [17].

The tendency of learning to teach contextually using instructional video was also affected by pre-service teachers' reflective behaviours. The video made them evaluate that model teacher should be more open to the students about authentic assessment and interact with proper classroom language. Teachers often evaluate model teacher's pedagogical approaches as viewed in the video, offering advice on what the teacher should have done differently [12]. For instance, pre-service teachers decided to apply peer-assessment during group presentation to strengthen authentic assessment and spoke grammatically corrected in Bahasa Indonesia.

We also find that these reflective behaviours may depend on the interaction between the facilitator and the pre-service teachers during the learning process. At the early microteaching session, when pre-service science teacher number 1 and pre-service science teacher number 2 taught with the poor contextual framework, the lecturer evaluated their techniques and strategies for teaching and asked the pre-service teachers to focus on how their material would be learned within real-world situations which were familiar to students. In the preceding description, we noticed that they used recent news on newspaper to contextualize their material and left the problem, method, and answers open within practical work to emphasize constructivism as well as scientific approach. Previous research also revealed that facilitator supports video-mediated teacher education by focusing issues to the learning goal, managing the noteworthy scene to be viewed and rewind, and explored understanding and interpretation upon viewing the video [12].

4. Conclusion

We concluded that pre-service teachers could learn to teach contextual teaching and to use the scientific approach in a science classroom with the help of model in the instructional video which has been developed. Pre-service teachers' skills in teaching showed that they are reflective about their own teaching practice and believed upon viewing the video and receiving an evaluation from the facilitator. However, the pre-service teachers may be weak at some criteria, such as utilizing teaching aids incorrectly, selecting irrelevant teaching aids, and associating one teaching ideas which mismatched with the real world situation. It would be unrealistic to expect that pre-service teachers could teach contextually with no failure or weakness after a session of learning with the video. Long periods of training are needed to sustain the way they learn to teach and improve their pedagogical content knowledge. We consider that facilitator takes important parts to help pre-service teachers focus on essential characteristics of targeted-performance and manage peer-dialogue during video analyses.

References

[1] Wieman C and Gilbert S 2015 Taking a Scientific Approach to Science Education, Part I-Research *Microbe* **10** pp 152–6
[2] Wieman C 2007 *Why not try a scientific approach to science education?* (Change)
[3] Giere R N 1992 *Cognitive Models of Science* vol XV (Minneapolis, MN: University of Minnesota Press)
[4] Ebenezer J V and Gaskell P J 1995 Relational conceptual change in solution chemistry *Sci. Educ.* **79** pp 1–17
[5] Crawford M L 2001 *Teaching Contextually Research, Rationale, and Techniques for Improving Student Motivation and Achievement in Mathematics and Science* (Waco: CCI Publishing)
[6] HMELO-SILVER C E, DUNCAN R G and CHINN C A 2007 Scaffolding and Achievement in Problem-Based and Inquiry Learning: A Response to Kirschner, Sweller, and Clark (2006) *Educ. Psychol.* **42** pp 99–107
[7] Berns R G and Erickson P M 2001 *Contextual Teaching and Learning: Preparing Students for the New Economy* (The Highlight Zone: Research @ Work)
[8] Sears S 2003 *Introduction to Contextual Teaching and Learning* (Bloomington: Phi Delta Kappa Educational Foundation)
[9] Roth W and Roychoudhury A 1993 The development of science process skills in authentic contexts *J. Res. Sci. Teach.* **30** pp 127–52
[10] Klassen S 2006 A Theoretical Framework for Contextual Science Teaching Interchange **37** pp 12–62
[11] Bellini S and Akullian J 2007 A meta-analysis of video modeling and video self-modeling

intervention for children and adolescents with autism spectrum disorders *Except. Child.* **73** pp 264–87
[12] Van Es E A and Sherin M G 2008 Mathematics teachers' "'learning to notice'"in the context of a video club *Teach. Teach. Educ.* **24** pp 244–76
[13] Seidel T, Blomberg G and Renkl A 2013 Instructional strategies for using video in teacher education *Teach. Teach. Educ.* **34** pp 56–65
[14] Susantini E, Kurniasari I, Fauziah A N M and Prastowo T 2016 *Laporan penelitian unggulan perguruan tinggi Pengembangan Video Pembelajaran Pendekatan Saintifik untuk Mengembangkan Keterampilan Mengajar Calon Guru MIPA* (Surabaya)
[15] Kersting N B, Givvin K B, Sotelo F L and Stingler J W 2010 Teachers' Analyses of Classroom Video Predict Student Learning of Mathematics: Further Explorations of a Novel Measure of Teacher Knowledge *J. Teach. Educ.* **61** pp 172–81
[16] Fadde P and Sullivan 2010 Using interactive video to develop teachers' classroom awareness *Contemp. Issues Technol. Teach. Educ.* **13** pp 156–74
[17] Kourieos S 2016 Video-Mediated Microteaching – A Stimulus for Reflection and Teacher Growth *Aust. J. Teach. Educ.* **41** pp 65–80

Acknowledgement
The study was fully funded by Ministry of Research and Technology, Directorate General of Higher Education of Indonesia and Islamic Development Bank (IDB) as Excellent Research of Higher Education "The Development and Upgrading of Seven Universities in Improving the Quality and Relevance of Higher Education in Indonesia" given to Universitas Negeri Surabaya in 2016.

Developing workshop module of realistic mathematics education: *Follow-up workshop*

E L W Palupi, S Khabibah

Department of Mathematics, Faculty of Mathematics and Natural Science, Universitas Negeri Surabaya, Indonesia.

evangelistapalupi@unesa.ac.id

Abstract. *Realistic Mathematics Education* (RME) is a learning approach which fits the aim of the curriculum. The success of RME in teaching mathematics concepts, triggering students' interest in mathematics and teaching high order thinking skills to the students will make teachers start to learn RME. Hence, RME workshop is often offered and done. This study applied development model proposed by Plomp. Based on the study by RME team, there are three kinds of RME workshop: start-up workshop, follow-up workshop, and quality boost. However, there is no standardized or validated module which is used in that workshops. This study aims to develop a module of RME follow-up workshop which is valid and can be used. Plopm's developmental model includes materials analysis, design, realization, implementation, and evaluation. Based on the validation, the developed module is valid. While field test shows that the module can be used effectively.

1. Introduction

Since 1971 the Freudenthal Institute has developed Realistic Mathematics Education (RME). The RME group in the Netherlands reviews about math, how students learn mathematics, and how math can be taught [1]. The principle underlying RME is influenced by Freudenthal's idea that mathematics is a human activity.

In the view of RME, the subject matter that is to be mathematized should, in all cases, be experientially real for the students [1]. Thus the teaching materials which are to be mathematized and they can be the real examples for students. That is why this approach is called RME. This does not mean that RME always uses real-life problems [2], but it can also use things that students have experienced or understood or students can imagine [3]. There are three RME principles, namely: 1) rediscover, 2) didactic phenomena, and 3) self-developed models. Based on these three principles, the RME approach has five characteristics: 1) real-world use, 2) modelling, 3) use of production and construction, 4) use of interaction and 5) intertwining [4–6].

RME in Indonesia is an instructional approach adapted from RME. It holds RME principles and characteristics. This makes RME is considered to be one of the suitable approaches to teach mathematics based on the aims of the curriculum which is focused on teaching high order thinking skills to the students, includes constructing their own knowledge, problem-solving, and critical and creative thinking.

The success of PMR in teaching mathematics concepts, triggering students' interest in mathematics and teaching high order thinking skills to the students will make teachers start to learn PMR. Hence, PMR workshop is often offered and done, one of them was offered by RME team Unesa.

Based on the study of the RME team of Unesa, RME workshop can be done in three types, they are startup workshop, follow-up workshop, and quality boost. Startup workshop is intended for schools or teachers who are interested and want to implement mathematics learning with RME approach. At the startup workshop, participants are introduced to RME and invited to analyze mathematical learning using RME approach and those who do not use it. While the quality boost is intended for schools or teachers who have experienced in implementing mathematics learning with RME approach. The purpose of quality boost is to broaden teachers' insight into learning by RME approach by training them to conduct Classroom Action Research.

The follow-up workshop which is the focus of this research is intended as a training to strengthen the implementation of RME which has been done in pilot schools which accompanied by the model teacher [7]. Before the workshop is held, there will be a visit to the partner school with the model teacher who will see directly (through video) the RME implementation directly by the RME assistant and the RME expert team. Therefore the successful experience that emerges during the implementation of RME can be shared with other teachers so it can motivate other teachers to apply RME and indirectly encourage the dissemination of RME in all schools in Indonesia.

In order to hold a good workshop, the module should meet the standards of RME workshop. Hadi [8]formulates that the RME workshop standard is: (1) focus on process and product, (2) facilitate participants to experience RME characteristics directly, (3) synergy with applicable curriculum, (4) gives opportunity to reflect RME theory and practice, and (5) to strengthen sustainability of RME implementation In schools. In addition, Hadi et al [8]states that the standard of a workshop must meet: (1) the existence of a strategy that can be applied practically, (2) offer cognitive experience through pedagogic content, (3) contain learning experience from the learner's point of view as a reflection material , (4) facilitating the network as a supporter of RME implementation, and (5) able to prepare teachers as leaders of change in their schools. However, particularly in Unesa, there is no standardized module of RME workshop. This study aims to develop a follow-up workshop module of RME which meet the standard, valid and can be used.

2. Methodology

The development model used in this research is the development model proposed by Plomp [9]. The steps of the general model of development are 1) problem analysis, 2) design, 3) realization, 4) implementation and 5) evaluation. The five development steps are described as follows:

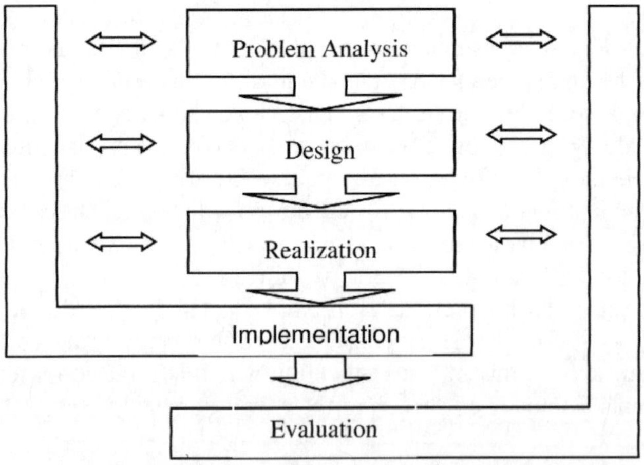

Figure 1. Developmental model of Plomp [9]

In detail, this study is conducted based on the proposed developmental model by Amin [10] which includes the developmental steps proposed by Plomp [9] (Figure 2).

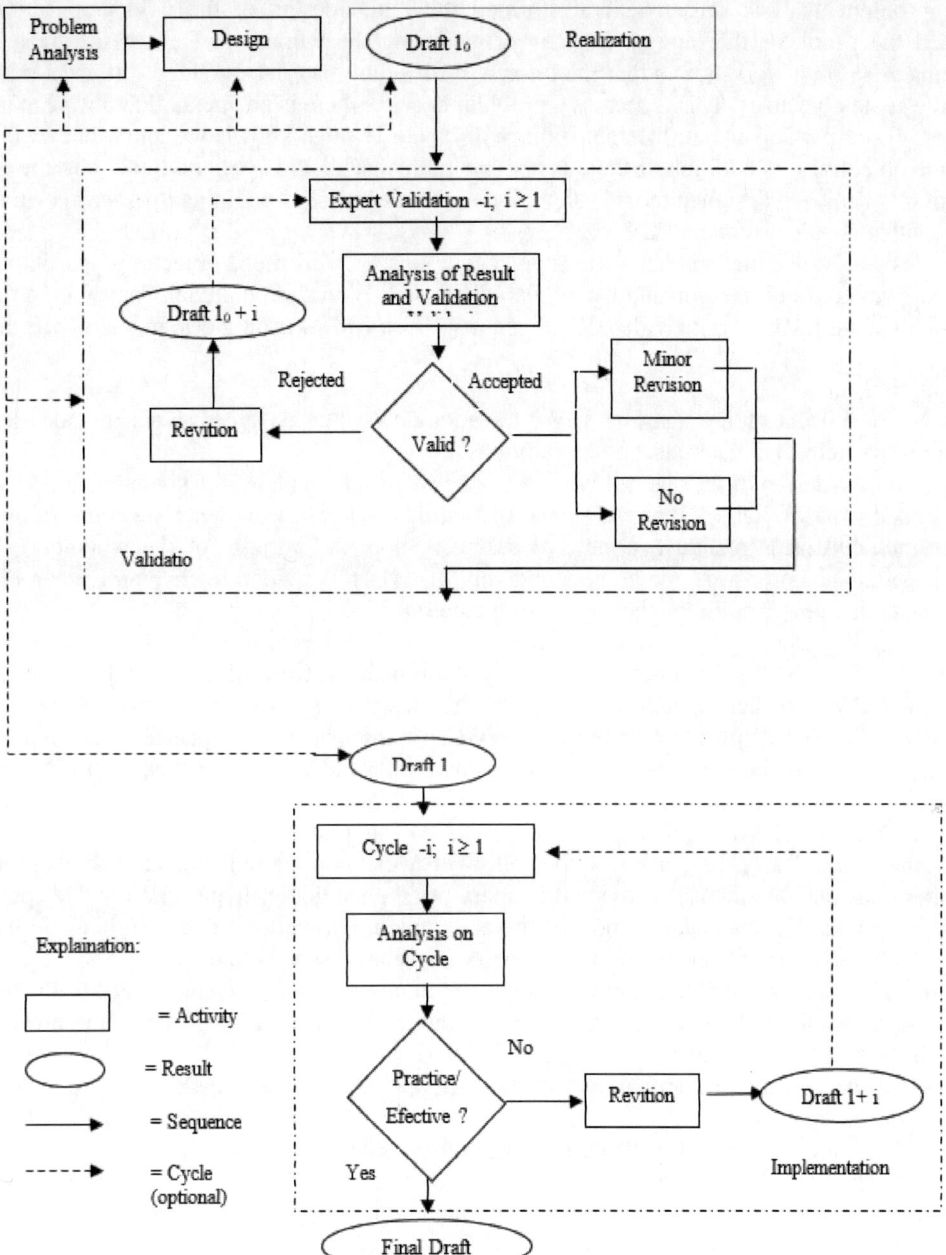

Figure 2. Flowchart [10]

The validity of the workshop material is known through the validation results by the validator that refers to the standard workshop RME [8]. While the practical criteria refer to whether or not the developed workshop materials can be used in the training class.

IOP Conf. Series: Materials Science and Engineering **296** (2018) 012006 doi:10.1088/1757-899X/296/1/012006

3. Results and discussion

3.1. Problem analysis
In the problem analysis stage, we brainstormed the initial design of the workshop activities and analyzed the problems that emerge or may arise during the workshop. One of the most common problems arise on a workshop is that the tutor will different materials, conduct different activities to teach the same objectives. This is seen as a problem because there is no standard and we can not assess whether all the participants at different room will have the same knowledge and whether they reach the same objectives. In addition, if there is another tutor that should continue the lesson then it would be difficult. Thus, the designed module should give the clear direction of the workshop, materials, and activities that should be conducted.

Besides analyzing the problems that may occur during workshop, objectives, participants, and technical provisions of the workshop were also discussed. Workshop materials that will be developed and used include RME learning video, RME activity sheet, RME learning design, and RME media.

3.2. Design
Based on the results of the analysis stage, the module is then designed. This includes its layout, objectives, participants, materials, and its activities.

Based on the design of the planned workshop, the workshop module is designed in accordance with the designed workshop and follow the format of writing modules by Higher Education. In the module will be included on a general overview of the workshop, the purpose of the workshop, technical guidelines, allocation of time, move the workshop, materials like video RME Plan Lesson Plan RME (RPP), and slide power point to other supporting material.

Follow-up workshop is focused on the sharing of successful experiences in implementing RME in school. This successful experience can be derived from the successful experience of the workshop participants, RME teachers, and/or experience of the facilitator and experience that can be seen in the video RME learning. With these activities, workshop participants are expected to be motivated and inspired to design realistic mathematics learning activities and implement them in the classroom or school. In addition, the realistic mathematics learning activities are expected by seeing directly or through a video how RME applied in the classroom, workshop participants can identify and how RME used in the class, clearly. In general, in the follow-up workshop, participants actively examined. The facilitator does not continuously provide the theory but the facilitator invites and involves participants in practicing what has learned. So, most of the activities in the workshop should focus on works and practices. The activities planned on follow up workshop can be seen below:

- Introduction: The facilitator delivers the background of the purpose and outline of the workshop
- The principles and characteristics RME: the facilitator recalls what RME is and how, what are the principles and characteristics of the learning
- Share success story 1: facilitator or teacher who has implemented RME in class tells the success of the learning done
- Reflection on success story 1: participants and facilitators reflect on the success of constraints and solutions when implementing RME
- Success story 2: participant videos and facilitators display RME learning videos and ask participants to observe the learning on the video
- Video analysis: workshop participants analyzed the learning-related videos and the appropriateness of learning with RME principles and characters
- Exposure to the results of the analysis: the participants present the results of the analysis that has been done, the discussion focused on the principle and character RME
- Hypothetical learning trajectory (HLT): the facilitator explains the learning path on RME and ICEBERG as well as question and answer
- RME simulation: facilitator simulate RME learning so that participants experience firsthand how RME

- Discussion: the focus of the discussion is how the learning loop on learning is simulated by the facilitator. any major mathematical ideas that arise, the context used, the goals and the activities
- Designing HLT: with facilitator's guidance, participants in groups are asked to design a learning path with a specific topic
- Gallery: participants display the design results on poster paper. facilitators and participants give each other opinions and suggestions
- Reflection workshop: facilitator and participants reflect on their workshop. participants fill out a response questionnaire

3.3. *Realization and validation*

The designed module then is written in the Realization phase. The first draft is produced and has been validated. Draft 1 of follow-up workshop module was validated by a lecturer who is competent in implementing RME. The aim of the validation is to get feedbacks to revise the module and to know whether Draft 1 is already suitable for the standard of follow-up workshop RME. Validation results of Draft 1 reveal that the designed module meets the standard of RME workshop, yet the needed materials are not yet attached.

Draft 1 then is revised based on the validation results. Materials needed for powerpoint slide of introduction activity, the lesson plan is added. The revision of Draft 1 then is called as Draft 2 which is then to be validated again. Form the validation results of Draft 2, it is known that there are still many things that need to be revised such as (1) background should be focused on follow-up workshop, do not write it as a general workshop. (2) Language used should consider the characters of the participants, (3) Participants character should be clear, (3) Workshop is for 2 days, 180 hours is too hectic, and (4) it would be better to add slide about character and principles of RME to be used as a guideline in evaluation whether the implementation of the RME is a success or not. These feedbacks are used to revise Draft 2 to Draft 3. Draft 3 which is assessed and validated by the validator then is implemented in a real workshop to collect more feedback and to see whether the module can be used in a real workshop or not.

3.4. *Implementation and evaluation*

As we revised Draft 2 become Draft 3 then we try to use the module in a real workshop. Test field involves 40 mathematics teachers of elementary school who know about RME. During the implementation, it happens that the module can be used by the facilitator, success stories of implementing RME in a lesson by the facilitators successfully motivate workshop participants to implement and disseminate RME. Furthermore, we collect some data on Hypothetical learning trajectories (HLT) on mathematics topics created by the participants.

HLT which is created by the participants shows mathematics is the big idea of the topic, mathematics contexts, learning the purpose, and activities. However, most of the mathematics context is not yet a meaningful real problem (Figure 3). This suggests that explanation about mathematics problem, the real problem, and the meaningful problem should be added to the workshop activity and materials. Thus, participants can design meaningful activity for their students.

IOP Conf. Series: Materials Science and Engineering **296** (2018) 012006 doi:10.1088/1757-899X/296/1/012006

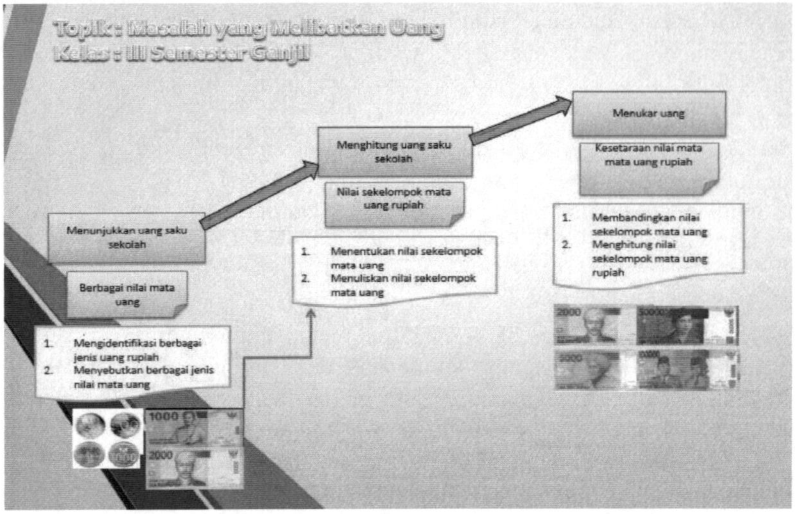

Figure 3. HLT produced by workshop participants

The positive response from the participants is given for the conducted workshop. Ninety percent of participants stated that they enjoyed the workshop and learned more about RME and its implementation from the workshop. Only 10% percent who said that they were sad because of external factors like missing their families. Even, there is one participant that argue that he is sad because that he realized that there is so much thing about RME and teaching mathematics that he did not know. All participant write that they wish that the similar workshop could be conducted again next time.

4. Conclusion and discussion

The developed module is valid and can be used. Workshop conducted based on the developed module got positive responses from the participants. However, the reliability of the module was not be tested. We still do not know whether the developed module can be used by other facilitators without briefing and whether the workshop can be conducted independently. It means the participants are teachers that do not know about RME and/or do not attend the start-up workshop.

References

[1] Fred D and Dolk M 1995 *Freudenthal Institute* (Ultrecht: Universiteit Ultrecht)
[2] Lange D and Jzn J 1978 *Mathematics Insight and Meaning* (Utrecht: OW & OC)
[3] Slettenhaar 2000 Adapting Realistic Mathematics in the Indonesian Context *Maj. Ilm. Himpun. Mat. Indones.*
[4] Treffers 1991 Didactical Background of a Mathematics Program for Primary Education *Realistic Mathematics Education in Primary School* (Ultrecht: Freudenthal Institute)
[5] Streefland L 1991 *Realistic Mathematics Education in Primary School* (Ultrecht: CD Press)
[6] Heuvel-panhuizen marja van den 1998 Freudenthal Institute *Work Prog.*
[7] S H, M D and E Z 2010 The Role of Key Teachers in RME Dissemination *A decade of RME in Indonesia* ed R Sembiring, K Hoogland and M Dolk (Bandung, Utrecht: APS International)
[8] S H, Zulkardi and Hoogland K 2010 Quality Assurance in RME - Design of Standards for RME *A decade of RME in Indonesia* ed R Sembiring, K Hoogland and M Dolk (Bandung, Utrecht: APS International)
[9] Plomp T 2001 Development Reseach in/on Education and Training *National Conference of Matematika Realistik Indonesia* (Yogyakarta: Universitas Sanata Darma)
[10] Amin S M 2006 *Pengembangan Buku Panduan Guru untuk Pembelajaran Matematika yang Melibatkan Kecerdasan Intrapribadi dan Interpribadi* (Universitas Negeri Surabaya)

IOP Conf. Series: Materials Science and Engineering **296** (2018) 012007 doi:10.1088/1757-899X/296/1/012007

Media and Information Literacy (MIL) in journalistic learning: *strategies for accurately engaging with information and reporting news*

F Inayatillah

Faculty of Language and Arts,Universitas Negeri Surabaya, Indonesia

fafiinayatillah@unesa.ac.id

Abstract. In the era of digital technology, there is abundant information from various sources. This ease of access needs to be accompanied by the ability to engage with the information wisely. Thus, information and media literacy is required. From the results of preliminary observations, it was found that the students of Universitas Negeri Surabaya, whose major is Indonesian Literature, and they take journalistic course lack of the skill of media and information literacy (MIL). Therefore, they need to be equipped with MIL. The method used is descriptive qualitative, which includes data collection, data analysis, and presentation of data analysis. Observation and documentation techniques were used to obtain data of MIL's impact on journalistic learning for students.This study aims at describing the important role of MIL for students of journalistic and its impact on journalistic learning for students of Indonesian literature batch 2014. The results of this research indicate that journalistic is a science that is essential for students because it affects how a person perceives news report. Through the reinforcement of the course, students can avoid a hoax. MIL-based journalistic learning makes students will be more skillful at absorbing, processing, and presenting information accurately. The subject influences students in engaging with information so that they can report news credibly.

1. Introduction

Journalistic is an important science for students [1]. There are two reasons why journalism is important to learn. First, journalism influences a person in seeing the news. It can construct or shape the true reality. Second, technological advances give impact on the growth of new media. The growth of new media needs to be accompanied by journalistic reinforcement so that public lies can be avoided. It can be said that journalistic has an important role in shaping the ideal social order.

In the era of information, one's survival is determined by how wisely a person gives respond to any information. Information will flow through the various channels. Under these conditions, journalistic is needed as a filter of any information available. Journalism has a variety of functions for everyone. For media users, it serves as a filter to sort out the good and the bad news so that users do not go wrong in consuming news. For journalists, this serves as a signpost in producing news so that when the information is going to be presented to the public, it turns out to be really feasible to be consumed and to be accounted for. If a journalist does not cling to the ethics and the rules of journalism [2,3], the news which is presented becomes lies, slander, and provocation; the chaotic will happen next in the community.

IOP Conf. Series: Materials Science and Engineering **296** (2018) 012007 doi:10.1088/1757-899X/296/1/012007

In Indonesian Language and Literature department of Universitas Negeri Surabaya, Journalism is the course of creative skill. After graduating from the subject, students are expected to be able to understand some concepts and skills related to various works in media and information. To support students' journalism mastery, MIL research is conducted.

2. Theoretical Basis

Etymologically, journalistic means correspondence or matters related to it. This is the art and skill of finding, collecting, processing, composition, and presenting news about events that occur accurately. In this case, it takes the ability to absorb and present information in the media. Thus, MIL is a required skill in the work process of journalists. MIL is one of UNESCO's main strategies to be implemented in the learning process. Literacy skills are widely projected by experts as 21st-century skills that can be used as a password to survive in the information society. MIL is a basis for enhancing access to information and knowledge. It describes the skills and attitudes of the media and other information providers in societies. In addition, it evaluates and produces information and media content [4].

MIL is a concept that includes knowledge, skills and attitudes that enable people to: (1) understand the role and the function of media and other information providers in democratic societies, (2) understand the conditions in which they can be met, (3) recognize and articulate the need for information, (4) critically evaluate information and content of media and other information providers, (5) extract and organize media information and content, (7) synthesize or employ ideas that are abstracted from the content, (8) ethically and responsibly communicate a person's understanding of the knowledge created to readers by appropriate forms and media, (8) implement ICT skills to process information and produce user-generated content, and (9) engage with media and other information providers [4].

Furthermore, in International Federation of Library Associations and Institutions (IFLA) [5] explained that MIL is closely related to lifelong learning. Lifelong learning allows individuals, communities, and nations to achieve their goals and take advantage of emerging opportunities in a growing global environment for the benefit of all individuals, not least. It helps them and their institutions and organizations to meet their technological, economic and social challenges, to remedy flaws, and to improve the well-being of individuals.

Information literacy is more related to the ability to recognize the information that suits people's needs and the source. Media literacy is the ability to access, analyze, evaluate, and communicate information in various formats, both printed and non-printed. The purpose of media literacy education is to protect students from the bad perceptions of the media as well as to educate them to appreciate the positive effects of the media [6]. Media literacy does not merely teach students about the technical aspects of the production of news in media, but also the consequences that arise from the power of the media. Media literacy teaches students about the wise use of media and critical judgments of media content.

3. Methods

MIL-based journalistic learning on Indonesian Literature students is carried out by the learning group model. Group learning is believed to provide more effective results. The steps used are (1) need [identify information needs], (2) access [search for required information], (3) locate [track information source], (4) synthesize [criticize information received], and (5) create [formulate information obtained]. This research was conducted in class of Indonesian Literature major batch 2014 in Journalistic course which consisted of 20 students. The data source of this research is the result of observation, the result of the questionnaire, and the journalistic writing document that has been made by the students. Data collection techniques used in this study were observation techniques, questionnaire techniques, and documentation techniques. Observation techniques and questionnaires were used to extract data of the role of MIL's importance for students.

4. Discussion

From the results of observations and questionnaires, it was obtained information that students needed MIL to support the ability to write journalistic work. After mastering the MIL, they were more competent in absorbing the correct information so that the news they wrote was more accurate. Before studying the MIL, students absorbed information based on what was seen and heard, without checking the truth. This could be seen in lecture events with the topic of feature writing. At that time the students were asked to find a partner to learn. They were asked to interview their partner and write the results of the interview in a feature news report. When a student named RA interviewed IM (the name of his partner), IM did not give a real answer (IM lied).

RA interviewed IM about the reasons he went to college as well. IM confessed that money from his parents was enough. He worked to earn money for his wedding plans. In fact, IM worked to enjoy her leisure time, the income was used for fun, such as buying clothes, bags, shoes, food, and for hangouts. RA immediately wrote the feature based on interview results. RA news reports were inaccurate because they did not fit the actual facts.

Conversely, when IM interviewed RA, RA told the truth. He worked because there were job opportunities. He performed entrepreneurship with his friends. They opened a bookstore at home. In addition, they also sold books online and in important events. He used his salary for extra pocket money. From the interview results, IM could write a feature with accurate data. From the results of the reflection of learning written in the questionnaire, the average students looked at the facts from actual events. If they were interviewing people, whatever the person's response was considered facts. Generally, they were not thinking critically, not analyzing and assessing information received.

In the following class session, they learned about MIL, strategies to engage with information, and make news reports accurately. Then they were given the task of making a wall magazine project. They were asked to work in groups. Each consisted of five students. In the class session to follow, they presented their assignments. There were five wall magazines with different themes: (1) "Tourism", (2) "Environment", (3) "Art in Surabaya", (4) Culinary, and (5) "Fashion". Afterwards, they were also asked to interview the team of wall magazine makers from another group. Then, they wrote the result of the interview in the form of the news report.

They had taken the advantage of MIL in writing news reports. They began the work step by identifying the required information, searching for needed information, tracking information sources, criticizing information received, formulating information obtained, assessing information (for refinement). After mastering the MIL, they were more skillful in absorbing the correct information so that their news reports were written more accurately.

5. Conclusion

Journalistic is a science that is essential for students because it affects how a person perceives news report. Through the reinforcement of MIL, students can avoid a hoax. MIL-based journalistic learning makes students more skillful at absorbing, processing, and presenting information accurately. MIL influences students in engaging with information so that they can report news credibly.

References

[1] Meyer P 2002 *precision Journalism: A Reporter's Introduction to Social Science* (New York: Rowman & Littlefielfd Publishers)
[2] Ward S J A 2005 Philosophical foundations for globals journalism ethics *J. Mass Media Ethics Explor. Qurstion Media Moral.* **20** pp 3–21
[3] Sanders L 2003 *Ethics & Journalism* (London: SAGE Publications)
[4] Grizzle, Alton and et al 2013 *Media and Information Literacy: Policy and Strategy Guidelines* (Paris: UNESCO)
[5] IFLA 2014 Media and Information Literacy Recommendations
[6] Jolls T and Thoman E 2004 Media Literacy—A National Priority for a Changing World *Am. Behav. Sci.* **48** pp 18–29

IOP Conf. Series: Materials Science and Engineering **296** (2018) 012008 doi:10.1088/1757-899X/296/1/012008

Differentiated strategy, business performance, and intellectual capital: *Evidence small medium enterprise from Indonesia*

Hariyati, L Venusita and Z F Dyani
Department of Accounting, Faculty of Economics, Universitas Negeri Surabaya, Indonesia.

hariyati@unesa.ac.id

Abstract. Small and Medium Enterprises (SMEs) have a very important position in Indonesian economics. Implementation of the differentiated strategy has been impacted on improving the business performance of SMEs where the role of intellectual capital strongly supports the success of the implementation of the differentiated strategy. This study applied quantitative research which used survey method. This research examines the relationship between differentiated strategy to the performance of SMEs with mediated by intellectual capital. The results of this study show that intellectual capital mediates the relationship between differentiation strategies and business performance of SMEs. This study theoretically proves the importance of contextual variables in contingency theory. The practical results of this study contribute to raising awareness of business unit managers or other equivalent positions, especially managers in East Java to understand the importance of the role of intellectual capital, this is because intellectual capital meets the criteria as a unique source of the company that is able to create competitive advantage and increase the firm's value.

1. Introduction

Based on data from Badan Pusat Statistik, the production of large and medium manufacturing industries in East Java in the second quarter of 2016 (y on y) grew by 1.71 percent compared with production in the same quarter of 2015, while the production of large and medium National manufacturing industries in the second quarter of 2016 (y on y) also grew by 5.54 percent. This production growth shows that East Java's production growth (y on y) is 3.83 percent lower than national production growth. In the development of the manufacturing industry, there is a very high fluctuation.

Manufacturing business performance in East Java fluctuated in several sectors. Such fluctuations move upward which impacts on improved business performance, but fluctuations that move down will certainly have an impact on business performance that is not profitable. Therefore, a fit strategy for manufacturing companies in East Java is needed. The strategy is a driver that directs the company's activities in positioning companies and products or services in the competition in an effort to improve company performance. Companies in maintaining their presence in the midst of intense business competition will choose and apply strategies that fit with the company's character and environmental conditions. The right company in choosing a strategy will be able to outperform its competitors in growth and profitability and be able to survive in the business life cycle in the long run. One of the efforts made is the selection of strategies applied adapted to the core competencies that are owned and the external conditions of the company. Core competence and environmental external conditions are used to achieve competitive advantage. There is a theoretical gap in achieving competitive advantage in business performance. Theory of Industrial Organization (I/O) emphasizes that to achieve organizational or corporate performance must pay attention and study the factors and the external environment [1]. According to Resource-Based Theory (RBT), stressed to achieve organizational performance is largely determined by the characteristics and internal factors of the company [2]. To cover the gap theory of Industrial organization and resource-based theory in this research used Contingency Theory.

The differentiated strategy is a strategy that prioritizes innovation process both product innovation and process innovation. Innovation plays an important role as one of the key success factors for a company to achieve a sustainable competitive advantage. Increasing competition, unceasing turbulence, the pace of globalization, the pace of technological development, change and uncertainty have led to the introduction of innovations in various fields. The role of innovation is an important factor in gaining competitive advantage.

SMEs have an important role in the formation of strategies for economic development and recovery in many countries. The development of SMEs provides an opportunity to reduce unemployment rates. In Indonesia, especially in East Java, the development of SMEs in the reform era has gained a great deal of attention from the government, the concern is quite reasonable because of the increasing role of small and medium entrepreneurs in developing the economy. The role of small and medium entrepreneurs in economic development can be seen from the characteristics of the entrepreneurs, the labor-intensive production process is able to absorb labor as well as to increase income and able to survive in times of economic crisis. Increased revenue from SMEs is dominated by SMEs that export their products.

2. Theoretical review

2.1 The industry-organization (I/O) paradigm, in relationship between differentiated strategy and business performance

Several studies empirical pertaining to the model I/O have been done by researchers. Environmental factors external had an important role in business conditions, because of an environmental factor is determined a strategy to be used [3]. The relationship between environmental changes by planning strategies are really powerful [4], large amounts to anticipate change and disarray. [5] suggested that complexity and changes in the neighbourhood of an industry might affect the intensity of strategic planning, but in research conducted by Hopkins & Amp; [6] concluded that strategic planning does not affect financial performance, but financial performance-enhancing strategy planning. I/O approach done by [7], about the environmental impact of industry on tourism in Malaysian business. The industry becomes a very important point in reaching business performance.

2.2 Resource-based theory in relation to strategy and business performance

Empirical studies related to Resources-Based Theory, have been done by [2], the result is a sustainable competitive advantage only if competitors' attempts fail to imitate those advantages. RBT explains that competitive advantage can only arise in situations of corporate heterogeneity and resource immobility. [2] also stated that in RBT, companies can not expect to buy or take on the sustainable competitive advantage that other organizations have because they are a scarce, immeasurable, and irreplaceable resource. The RBT states that internal resources are more important for companies than external factors in achieving and maintaining competitive advantage [8].

According to [9] organizational performance is determined by internal resources, namely physical resources, human resources and organizational resources, which includes corporate structure, planning process, and corporate strategy. The RBT draws attention to managerial strategies and practical strategies for the development of new competitive advantage and wealth creation [10-12]. This opinion is supported by [10] who argues that if it can manage resources and capabilities strategically and structurally, the competitive advantage will increase. With resource-based theory and attention focused on the process of creating competitive advantage, this view is a fundamental principle which determines the difference in the creation of wealth. RBT is the starting point used to identify and explain differences in well-being. The competitive advantage of the company flows by taking into account its resources. Further, [10] argues that more valuable, scarce, imperfectly imitable and non-substitutable resources than competitors are an important source of increasing competitive advantage. Therefore, some of the previous researchers suggested that the competitive advantage built on internal resources.

2.3 Contingency theory in relation to strategy and business performance

Based on the theory of contingencies there are other situational factors that interact with each other in a certain condition. Beginning with this contingency approach, it is likely that decentralization will also lead to differences in management accounting information needs. Management accounting design, strategic approach uncertainly interesting enough to be examined in testing the reliability of management accounting system influential not at any condition based on other determinant variables interact with the conditions encountered. [13] says that the development of an organization is influenced by differences in environmental features. Furthermore, the hypothesis states that the success of an organization depends on uncertainty, internal factors, feedback with other organizations, and the external interaction of the organization.

From theoretical and methodological reviews, basically, a contingency approach is a response to a universalistic approach, which argues that the relationship between organizational design and outcomes variables is consistent with all conditions. While the contingency approach assumes that the relationship is conditional. Contingency theory intuitively helps reconcile the existence of empirical discovery conflicts based on universalistic theoretical models. Management Accounting Studies based on contingency approach have focused on 2 (two) dimensions, they are (1) information characteristic and management accounting system and (2) use of management accounting technique.

Contingency approach in organizational design research can explain the influence of unexpected empirical evidence in the development of theory using a universalistic approach. [7] criticizes the simple linear model that has been the basis of most previous research. With a simple linear model or a universalistic model, it appears that the related research approach focuses on the selection of contextual variables (contingencies) that may directly affect organizational design. Organizational design variables are tested for their influence on the type of information system and then the type of information system in the test of its effect on the output variables.

In a simple linear model or a universalistic model, it is still directed in the effort of selecting contextual variables to test its effect on organizational variables but has not attempted to test the interaction effects between contextual and organizational variables on output variables such as organizational performance, individual performance, attitude motivation, and managerial satisfaction.

3. Development of hypothesis

Implementing a differentiated strategy requires the role of competent intellectual capital as it affects the company's performance. The existence of innovation in the implementation of the differentiated strategy will increase the productivity of the entity. The hypothesis in this research are:

- Differentiation strategies affect business performance directly
- Differentiation strategies have an effect on intellectual capital directly
- Intellectual capital affects business performance directly
- Differentiation strategies affect the performance mediated by intellectual capital.

4. Methodology

4.1 Research approach and design

This study is a quantitative research. This study was designed to examine the effect of differentiated strategy implementation on performance mediated by the intellectual capital requirement. Therefore this research includes verification research (hypothesis testing study). Independent variable in this research is differentiated strategy and the dependent variable is company performance and mediation variable in this research is intellectual capital.

4.2 Data collection
This research uses survey method. To collect the data, the questionnaire was sent to the respondent selected as a sample determined by a particular procedure. Furthermore, after the data obtained then performed the analysis and test the hypothesis and the result is empirical findings.

Data analysis was performed using Structural Equation Modeling (SEM) based on the variant, using PLS tool. The use of SEM-based variants because this research is exploratory or extension of existing theories. This study expands the contextual variables in contingency theory.The population is an individual or unit or element that has certain characteristics set by the researcher. The population in this study is the business unit of medium and large manufacturing companies in East Java. The business unit is the part of the company that is treated as a profit center. The unit of analysis of this study is a business unit, if assumed from one company taken one business unit, then the population of this study amounted to 389 business units.

4.3 Operational variables
The differentiated strategy is developed by Porter that focuses on innovation. *Intellectual Capital.* Intellectual Capital is defined as intellectual material (knowledge, information, intellectual property, experience) that can be used to create wealth. *Business Performance* is defined as a consequence of an economic decision taken from an economic action.

4.4 Hypothesis test

- Differential strategy affect the business performance, seen from the value of the coefficient of 0.229 with a p-value less than 5%.
- Differential strategy influence on intellectual capital, seen from the coefficient value of 0.649 with a p-value less than 5%.
- Intellectual capital effect on business performance, seen from the value of the coefficient of 0.366 with a p-value less than 5%.
- Differential strategy affects the business performance mediated by intellectual capital.

The coefficient value generated equal to 23.8% which means the magnitude of indirect influence Differential strategy on business performance through intellectual capital of 23.8%.

4.5 Testing of moderation variables
Before added IC variable, this research proves that SD has significant effect to KB, seen from a coefficient value equal to 0,497 and p-value less than 5%. After added IC variable, this research proves that:

- SD has significant effect to KB, seen from a coefficient value equal to 0,229 and p-value less than 5%.
- SD significant effect on IC, seen from a coefficient value equal to 0,649 and p-value less than 5%.
- IC significant effect to KB, seen from a coefficient value equal to 0,366 and p-value less than 5%.

Based on the description above, it can be concluded that the IC variable proved as partial mediation variable.

IOP Conf. Series: Materials Science and Engineering **296** (2018) 012008 doi:10.1088/1757-899X/296/1/012008

5. Discussion

As it was already explained in the discussion, intellectual capital is the driver of corporate value in increasing the company's competitive advantage. Manufacturing companies in East Java are facing the development of technology-driven business environments. If it was in the past economics, the economy is heavily dependent on resources such as land, natural resources, equipment, and capital to create value, then in the millennium economy, the economy is heavily dependent on the knowledge that has more value than physical assets for the organization. Human resources (HR) involved in every activity, the company will do less physical jobs and more work by using the capital of the brain, which we call the intellectual capital.

Changing the strategic orientation in knowledge asset requires an understanding that the creation of a company's competitive advantage depends largely on a company's ability to create, use and transfer, and utilize intangible assets that are rare, non-tradable and extremely difficult to replicate. In today's dynamic business environment changing conditions, resource-based valuation of assets emerges in response to the management of forms of intellectual property. Through the assessment of intellectual capital, companies can manage and develop their owned assets so as to benefit the efforts of achieving sustainable competitive advantage.

Related to the explanation above, the role of intellectual capital in the company of manufacturing companies in East Java in the implementation of the differentiated strategy is very important. The condition of manufacturing companies in East Java is currently experiencing a significant increase that requires the role of intellectual capital is qualified. The relationship of differentiated strategy with intellectual capital shows the statistically significant relationship. This is because the implementation strategy requires the role of intellectual capital that has competence. Human capital, structural capital, and customer capital are needed in the implementation of an innovation strategy that is implemented continuously. In accordance with the opinion of [14] who argued that the success of the company is strongly influenced by the company's routine efforts to maximize the values of the company's intellectual capital. Intellectual capital provides a diversity of different organizational values such as increased profits, acquisitions of innovations from other companies, consumer loyalty, cost reduction, and productivity improvements in innovation and implementation of information technology.

Innovation is a process within an organization to utilize the skills and resources to develop new products and/or services or to build new production and operational systems so that the company is able to answer customer needs. The discussion of the positive effect of innovation on performance strategy based on Resources-Based Theory is innovation is a critical factor for companies to compete effectively in domestic and global markets and is considered one of the most important components of organizational strategy [15,16]. This is also in line with the results of research [17] who discovered the importance of human resource management when building innovation strategies for product and process innovation. Knowledge management supports innovative performance if simultaneous approaches from soft human resource management practices and hard information technology practices are implemented together so that they can synergize well.

The strategy that the company implements will determine the need for intellectual capital in order to achieve good performance. In globalization era, product innovation and information technology and processes, as well as intense business competition in this century, forced companies to change the way they do business. To win the competition, companies must quickly change their strategy from a labor-based business to knowledge-based business, so that the main characteristics of the company become science-based companies. A resource can be said to have a competitive advantage if it meets the following criteria (a) The resource enables the company to capture business opportunities and overcome challenges, (b) The resource is unique and difficult to obtain in the market and is owned by only a few players Business only, (c) Such resources can be utilized by the company to provide benefits to the company. Resource-Based Theory concluded that the resources owned by the company affect the performance of the company that will ultimately increase the value of the company. One of the company's resources of intangible assets expressed is intellectual capital.

The relationship between intellectual capital and customer performance showed statistically insignificant results. This indicates no intellectual capital relationship with customer performance. Customer performance that includes product attributes, customer relationships, and brand image has no impact on intellectual capital needs.

6. Conclusion

Based on research findings and study limitations, the following suggestions can be given for future research improvements as well as for the results of this study to inform the manufacturing companies in East Java about the importance of continuous innovation and information technology in the implementation of strategy for the company has competitive advantage and it is competitive in facing globalization.

References

[1] Porter M E 1980 *Competitive strategy Techniques for analyzing industries and competitors* (New York)

[2] Bamey J B 1991 Firm Resources and Sustained Competitive Advantage *J. Manage.* **17** 99–120

[3] Covin J G and Covin T 1990 Competitive Aggressiveness, Environmental Context, and Small Firm Performance *Entrep. Theory Pract.* **14** 35–50

[4] Ansoft H I 1991 Critique Henry Mintzberg, The Design School : Reconsidering the Basic Premise of Strategies Management *Strateg. Manag. J.* **12** 449–61

[5] Bird A 1990 On Strategic Planning *Bankers Mag.* 66–9

[6] Hopkins and Hopkins 1997 Strategic Planning Financial Performance Relationship in Bank: A Causal Examination *Strategy. Manag. J.* **18** 635–52

[7] Otley D T 1980 The Contingency Theory of Management Accounting: Achievement and Prognosis *Accounting, Organ. Soc.* **5** 413–28

[8] Brahmana S S 2007 Resources-Based View: The Effect of Product Innovation on Market Orientation and Performance Relationship *DeReMa J. Manaj.* **2** 94–110

[9] David F R 2005 *Strategic Management: Concepts and Cases* (Prentice Hall)

[10] Ireland R D, A Hitt M and G Sirmon D 2003 A Model of Strategic Entrepreneurship: The Construct and Its Dimensions *J. Manage.* **29** 963–89

[11] Phiem R and Butler J 2001 Tautology in The Resource Based View and The Implications of Externally Determined Resource Value: Further Comments *Acad. Manag. Rev.* **26** 57–66

[12] Teece D J, Pisano G and Shuen A 1997 Dynamic Capabilities and Strategic Management. *Strateg. Manag. J.* **18** 509–33

[13] Hirst M K 1987 The Effects of Setting Budget Goals and Task Uncertainty on Performance: A Theoretical Analysis *Accounting Review* pp 774–84

[14] Harrison S and P H Sullivan Sr 2000 Profiting from Intellectual Capital: Learning from Leading Companies *Ind. Commer. Train.* **32** 139–48

[15] Davila T 2000 An empirical study on the drivers of management control systems' design in new product development *Accounting, Organ. Soc.* **25** 383–409

[16] Hitt M A, D Ireland R, M Camp S and L Sexton D 2001 Strategic Entrepreneurship: Entrepreneurial Strategies for Wealth Creation *Strategy. Manag. J.* **22** 479–91

[17] Gloet M and Terziovski M 2004 Exploring The Relationship Between Knowledge Management Practices and Innovation Performance *J. Manuf. Technol. Manag.* **15** 402–9

IOP Conf. Series: Materials Science and Engineering **296** (2018) 012009 doi:10.1088/1757-899X/296/1/012009

Worm Disease Profile of Primary School Children

Hartati, Destriani and A R Victoria

University of Sriwijaya

hartati@fkip.unsri.ac.id

Abstract. The result of an elementary study of relations between disease symptoms and signs with the wormy occurrence child at Elementary School in Ilir Barat II Subdistrict of Palembang in concerned in this research amount to 200 people from the population of 4200 people of a child in elementary school. The result indicates that the child roommates infested worm marked with a few symptoms like passion eat to Decrease the puking pain in the home of stomach after eating, diarrhea, defecate with the mucus and bleed the purities at the anus before controlled from some worm type of there no difference having a meaning between child roommates worm infested with the child roommates do not infest worm (P> 0.05). Symptom vomit got more amount by having a meaning a child by infested is Trichuris (RO> 2.669, P <0/005). The result of infested Tricuris generate the more amount of symptoms lust to eat to Decrease by having a meaning of Compared to the which do not infest Trichuris (RO = 3.772; CI 95% = 1.214 to 11.726; P = 0.016) symptoms lust to eat to Decrease and more amount diarrhea got at Infest oxyuris with the special sign of purities at nighttime anus (RO = 0.557; 85% CI = 0.166 to 2.168). The risk of the happening of unfavorable nutrition According to BB / U and Also TB / U growing niche to more amount (having a meaning) at child roommates worm infested by a child Compared to the which do not infest worm.

1. Introduction

Worm disease prevalence is very high, especially in the tropical and sub-tropical. This disease is the cause of much pain in the world. Three and a half billion people globally are infected with intestinal parasites, including roundworm *(Ascaris lumbricoides, Thricuris trichuira, Ancylostoma duodenale, Necator americanus).* Four hundred and fifty million of them on children [1]. Around the world an estimated 1.47 billion *lumbriocides Ascaris-infected* children, *trichuira Thricuris* interference 1.3 billion *and* 1.05 billion infected with hookworm *Ancylostoma duodenale Necator* [2]. The prevalence of worm diseases in Indonesia is still quite high at 30.4% for *Awaris lumbriocides,* 21.25% and 6.5% *Thricuris trichuira Necator americanus* and *A-flycostoma duodenale* [3,4]. In Indonesia, the worm disease is most of the health problems after malnutrition. The prevalence and the highest intensity found among children of primary school age [5]. The prevalence among primary school children the province (Sum-Cells, Jakarta, and Yogyakarta) in Indonesia is 12.9% percent to Ascaris lumbricoides to thricuris tricuira 19.8% and 7.8% for hookworm [3]. According to the World Bank development report, in developing countries is estimated between 5-14 years old girls worm disease is 12% of the total morbidity burden (boys 11%); therefore the worm is the largest single contributor to the burden of illness in this age group, [3]. Worms including macroparasites, which means that parasite from multiplying in the body of its host [6]. Therefore, the infection is not the same as the disease. Pain directly related to the intensity of infection, so most of its host, especially with worm infestation depends on the parasite (virulence and number) and the host (immune factors) so that the child is healthy with good nutrition can suffer infestation of light without clinical manifestations altogether worm infestation can heavily cause symptoms in healthy children while clinical symptoms in mild clan moderate infestation will appear in children with malnutrition or malnutrition [2].

The learning achievement of primary school children, one of which is influenced by physical factors, especially a disease - a disease that causes disruption consent. According to Watkins and Pollit [6], the disruption caused by the symptoms of the disease will engage worm learning concentration. Keller, Casparis and Leather (1931) as quoted Watkins (1997), conduct cross-sectional study in children whites and Hitman in Tennesse that have economic and sanitary conditions are similar. White child infested caring to give the experience more discomfort in the abdomen (70% vs 7%), sleep disorders (60% vs. 15%), dental Gigis (32% vs. 11%) and tiredness (27% vs 7 %) compared with the control group. This situation will engage learning achievement of children in primary school [5].

Given the primary school children provide human resources in the future, and learning achievement is also greatly influenced by physical factors, then it is necessary to identify the symptoms associated with worm infestation can disrupt the concentration of children's learning.

2. Methods

The study ran between April 2013 - June 2013 in the District of Ilir Barat II Palembang, South Sumatra. This study is part of a research relationship between infestation and academic achievement of elementary school children in the same place. Based on the estimated prevalence of worm infestation among Naak-school base in Indonesia is 40% (P) and accuracy relative to expect 7% (d), a = 0.005, Z = 1.960, then the required number of samples of at least 188 primary school children (Pollitt E, 2006; 123). In this study, the number of subjects included 200 children who were randomly selected from the 4200 primary school children in the District of Ilir Barat II Palembang, South Sumatera. The method of randomization using a random table, by means of simple randomization [7].

Stool elementary school children have collected feces when, because of the presence of worm eggs in the stool is not affected time [1]. Stool elementary school children were collected stool while. Measurement with fecal worm egg examination by the direct method using eosin staining (2% in distilled water) and viewed under a microscope with 10 times magnification objective lens. Test results indicated by the presence or absence of worm eggs (qualitative). Examination of the worm eggs done by a senior laboratory technician at the General Hospital Muhammad Hoesin, who had attended a refresher course one year ago at Airlangga University, Surabaya.

Signs measured worm disease is nutritional status. The nutritional status was determined by anthropometric, i.e. by measurement of weight and height. Body weight was measured using a spring balance stand *(spring scale)* scale 0-100 kg that has been calibrated using the known weight of goods. The accuracy of measurement up to 100 g. measure height by using microtome, the vertical board with a headwind standard gauge (0-200 cm) and a horizontal board for limiting head. Measurement accuracy up to 0.1 cm. so measurement results or height compared with the standard WHO-NCHS based on the age of the child. Nutritional status was measured by weight loss of weight/age (W/A) and height/age (H/A) and expressed as a percent to avoid *a recall/anamnestic* bias, noting complaints symptoms last for 1 month only. Interviews were conducted with mother's pupil with a guided questionnaire. Symptoms are asked is decreased appetite, abdominal pain after eating, diarrhea, vomiting, cough with pull worms, worms secrete big bowel, bowel movement with blood and itching in the anus.

Statistical analysis of the relationship between symptoms and signs of disease with an incidence wormy worm using chi-square test and risk analysis research results [8].

3. Research result

Of the 200 subjects studied, two subjects were not analyzed because the data is incomplete. Hookworm infestation was not analyzed because the number of subjects was only two.

Worm infestation is characterized by the presence of some symptoms and signs such as lack of appetite, vomiting, abdominal pain after eating, diarrhea, irritable bowel out worms, cough out worms, bowel movements with mucus and blood and itching in the anus. Given some of the symptoms can be owned by more than one type of worm, then the control of the respective types of worms. Both before and after controlling for each type of worm, gekada, and signs of disease in children infested worm was not significantly different compared with that not trifectas ($p > 0.05$). Vomiting obtained more significantly in children with *Trichuris* infestation (RO = 2.669, $p < 0.05$). After *Ascaris* infestation is not included in the analysis, showed that *Trichuris* infestation nagsu cause more symptoms of eating was significantly reduced compared to non-infested *Trchuris* (RO = 3.772; IK95% = 1.214 to 11.726; $p = 0.016$).

In Sisko analysis, although inconclusive, it was found that some of the symptoms and signs of disease are more prone to worms found in children infected with worms (odds ratio> 1) (Table 1, 2, 4 and 4). Symptoms that can be associated with *Ascaris* infestation is diarrhea and bowel movements issued a worm. *Trichuris* centering provide more symptoms of lack of appetite vomiting, abdominal pain after eating, diarrhea and bowel movements and diarrhea more obtained at *Oxyuris* infestation. *Oxyuris* Khai symptoms in the form of itching in the anus at night time is not a distinctive mark in this study (RO = 0.553, IK95% = 0, from 166 to 2.168). Symptoms of bowel movements with blood which can occur at *Trichuris* infestation by removing the worm clan cough symptomatic ticlak Ascaris infestation was found in the study sample population. The risk of malnutrition, according to both W/A and H/A, were more likely to be obtained on worm infestation. In fact, at the risk of Ascaris infestation, undernourishment based TB/U more significantly than those who are not infested. In *Oxyutiasis,* the risk of malnutrition based TB/U cannot be rated as one of the cell number zero.

4. Discussion research

Some of the symptoms and signs associated with infestation showed an increased risk even if the results are inconclusive (RO> 1, IK95% passing 1). The results are inconclusive can associate with less specific symptoms and signs of worm disease, because it can be caused by other diseases are also often suffered by children of primary school age

Decreased appetite tends to be more common in children with *Trichuris* infestation and *Oxyuris* on *trichuriasis* are significant differences, Trichuriasis is the most many give symptoms than those lain. Indication of worm infestation is a lack of appetite, vomiting, diarrhea, abdominal pain after eating and defecating issue a worm. These results are more or less according to a case report in Yogyakarta (1998) and in Jakarta (2000). In a series of 81 reports of cases without complications *trichuiura Trichuris* infestation in Yogyakarta (1997), obtained the most frequent symptoms are an epigastric pain, lumbar and abdominal muscles, vomiting, constipation, abdominal distension, bloating and systemic intoxication [7].

Decreased appetite *Oxyuris* infestations associated with children less rest at night [2]. But often also associated with behavioral disorders, such as low self-esteem, anxiety, cannot concentrate, hyperactivity and others. In this study, the symptoms of itching in the anus at night can be connected with *Oxyuriasis* with IK 95% to RO = 0.166 to 2.168.

Ascaris infestation more at risk to cause diarrhea and wasting large are issued worms, while the symptoms of abdominal pain after eating, which, according to Faust et al [9] is the most frequent complaint in Ascaris infestation, were inconclusive (Table 3). Diarrhea occurred in *Ascariasis* can be caused by reduced absorption in the intestine as a result of the surface covered by a mass *of Ascaris* in the bowel lumen.

In this study, although inconclusive, children with worm infestation have more risk for malnutrition (RO for undernourished <1, IK95% including 1). These results are consistent with research Hadju et al. (2007) who get the relationship between the incidence of infestation with a decrease in good nutritional status based on BB/U and TB/U. Reviewed obtained Ascaris infestation is improving nutritional status by TB / U after deworming, the cross-sectional study that did Hadju et al [10].

According as quoted by Watkins and Pollitt [7,11] , the incidence of undernourishment on worm infestation can be explained in several ways: (a) the direct consumption of nutrient, (b) digesting the blood or intestinal bleeding, which causes loss of iron and nutrient other, (c) malabsorption usu against the carbohydrate, protein and fat, (d) *protease inhibitors,* generated worm to protect yourself from digestive enzymes would reduce the uptake of amino acids, and (e) respoons immune to infection, which causes loss of nutrients (proteins of immunoglobulins, lymphocytes, and other inflammatory cells) were anorexia.

5. Conclusion

- Although inconclusive, children with worm infestations tend to be more experienced and tana worm disease symptoms than children who are not infested,
- Children with Trichuirs trichiura infestation significantly showed a higher risk of experiencing symptoms of vomiting and lack of appetite,
- Children with Ascari asis has a greater risk of experiencing chronic undernourishment.

6. Suggestion

Eradication program in order to be considered for inclusion in the school health program in primary schools. An educational program of parents, especially mothers and dissemination of information about worm diseases incorporated into the worm eradication program in elementary school.

References

[1] WHO 2008 *Physical status: the use interpretation of anthropometry* (Genewa)
[2] WHO 2010 *Epidemiological data of intestinal parasites, WHO. World development report: Investing in health* (New York: OXFORD University Press)
[3] Miraza EM G 2005 Parasitic infestation in children at the three kindergartens in Medan, North Sumatera, pediatric Indonesia 312
[4] Indrawati V 2005 *Suplementasi zat besi dan asam folat serta riboflavin terhadap pengkatan kadar Hb yang mempengaruhi konservasi belajar anak sekolah dasar* (Surabaya: Pasca Sarjana Universitas Airlangga)
[5] Adriani M and Wijatmadi B 2012 *Pengantar Gizi Masyarakat* (Jakarta: Kencana Prenada Media Group)
[6] Smith G, Basanes M G and Dietz K 1995 *Macroparasite group report: problem in modelling the dinamic of macropasitic system* (Cambridge, UK: Cambridge University Press)
[7] Pollitt E and Marthew R 2006 Breakfast and cognition: an integrative summary *Am. J. Clin. Nutr.* 123
[8] Sastroasmoro S and Ismael S 2005 *Dasar-dasar penelitian klinis* (Jakarta: Bagian ilmu kesehatan anak FKUI)
[9] Faust K, Sathirapongsasuti J F, Izard J, Segata N, Gevers D and Raes J 2012 Microbial Co-occurrence Relationships in the Human Micrfobiome
[10] Hadju V, Stephenson L, Abadi K, Mohammed H, Bowman D and Parker R 2007 Improvement in appetite and growth in helminth-infected schoolboys three and seven weeks after a single dose of pyrantel pamoate, parasitology *pubmed.gov*
[11] Watkins W E and Pollit E 1997 Stupidity of Worm: do intestinal worm impair mental performance? *Psychol Bull* **121** 171–91

IOP Conf. Series: Materials Science and Engineering **296** (2018) 012010 doi:10.1088/1757-899X/296/1/012010

Error analysis of mathematical problems on TIMSS: *A case of Indonesian secondary students*

H A Priyani and R Ekawati

Department of Mathematics, Universitas Negeri Surabaya

rooselynaekawati@unesa.ac.id

Abstract. Indonesian students' competence in solving mathematical problems is still considered as weak. It was pointed out by the results of international assessment such as TIMSS. This might be caused by various types of errors made. Hence, this study aimed at identifying students' errors in solving mathematical problems in TIMSS in the topic of numbers that considered as the fundamental concept in Mathematics. This study applied descriptive qualitative analysis. The subject was three students with most errors in the test indicators who were taken from 34 students of 8th graders. Data was obtained through paper and pencil test and student's' interview. The error analysis indicated that in solving *Applying level problem*, the type of error that students made was operational errors. In addition, for *reasoning level problem,* there are three types of errors made such as conceptual errors, operational errors and principal errors. Meanwhile, analysis of the causes of students' errors showed that students did not comprehend the mathematical problems given.

1. Introduction

There is an extensive literature on error analysis in mathematics problem-solving. The main concern is to improve students' learning and understanding of Mathematics. Several researches on error analysis on mathematics problem solving were done such as in the topic of the fraction. Abdullah et.al [1] shared several errors in solving HOTS problems such as comprehension error, transformation error, process skill error, and encoding error.

Problem-solving is an activity that involves various actions in the mind of thought including accessing and using knowledge and experience [2]. Thus Polya in 1973 [3], teaching strategies that involve the use of non-routine problems in the classroom give students the opportunity to develop higher order thinking skills in the process of understanding, exploration, and application of mathematical concepts. Since problem-solving involves the activities of processing, the knowledge used in the process of solving the problems is different. As stated by Mayer [4], the knowledge covers such as knowledge of language and facts, knowledge of schemes, knowledge of algorithms, and strategic knowledge. In short, students need to equip themselves with various knowledge and high skills in problem-solving. Quoting Widiharto in 2008 [5], mathematical concepts and skills that are not fully mastered by students have led to difficulties and errors in solving mathematical problems. A study conducted by Susanti et al. [6] found out that students are difficult to solve problems that involve the use of HOTS and among the difficulties faced by them are reading and interpreting data, determining and delegating data, and making conclusions and arguments.

Herholdt and Sapire in 2014 [7] stated that error analysis, also referred to error pattern analysis, is the study of errors in learners' work with a view to finding explanations for these reasoning errors. Not all errors can be attributed to reasoning faults; some are simply careless errors [8], identified as "slips" from Olivier [9], which can easily be corrected if the faulty process is pointed out to the learner. Slips are random errors in declarative or procedural knowledge, which do not indicate systematic misconceptions or conceptual problems [10]. Error analysis is concerned with the pervasive errors (or 'bugs') which learners make, based on their lack of conceptual or procedural understanding (Ketterlin-Geller & Yovanoff 2009).

As a result, the performance of Indonesian students in the international assessments that test students' thinking skills, namely Trends in International Mathematics and Science Studies (TIMSS) and Programme for International Student Assessment (PISA) was not at the satisfactory level. Based on the data of TIMSS 2011, Indonesian students' mathematics achievement was in the ranked 38 out of 42 countries which followed the TIMSS with a score of 386 of the benchmark score of 500 [11]. Generally, one of the causes was that they were lack of knowledge in resolving TIMSS problems [12].

There are two domains that are being tested in TIMSS assessments, namely content and cognitive domains. For the tested cognitive domain, it covers applying, analyzing, and reasoning, which the components of higher-order thinking are in Bloom's Taxonomy. Meanwhile, the tested content domain covers four areas of mathematics learning namely Numbers, Geometry, Algebra, Data, and Probability. The number is one of the fundamental subjects that is tested in TIMSS which percentages is 30% becomes the greatest percentages in the content domain on TIMSS (Mullis et al., 2013). The concept of number is connected and as the basis and prerequisite for the understanding of the next concept [13]. Therefore, it is important to know the errors of number problems on TIMSS.

This study was carried out to identify students' errors in solving mathematics problem in TIMSS in the topic of numbers which considered as the fundamental concept in Mathematics.

2. Methodology

This study is a descriptive study that used a qualitative approach. The subject was three students with most errors in the test indicators who were taken from 34 students of 8th graders in a secondary school of the districts Sidoarjo, East Java. Data was obtained through paper and pencil test students' interview. The instrument used for this study was a set of test questions to identify the types of students' errors.

The items contained in the instrument have been adopted with the questions of international assessments of Trends in International Mathematics and Science Study (TIMSS). Four items were built in a subjective form, and the items contained the elements of Applying and Reasoning level problems as shown in Table 1.

The subjects of the study were required to answer the questions contained in the instruments that have been prepared under the supervision of individual mathematics teacher who taught the respective class. The time allocated to answer questions was 60 minutes. The errors were analyzed using three types of errors that are conceptual errors, operational errors, and principal errors.

IOP Conf. Series: Materials Science and Engineering **296** (2018) 012010 doi:10.1088/1757-899X/296/1/012010

Table 1. Items developed based on Applying and Reasoning level problem on TIMSS

Item	Question	Level
1	Penny had a bag of marbles. She gave one-third of them to Rebecca, and then one-fourth of the remaining marbles to John. Penny then had 24 marbles left in the bag. How many marbles were in the bag to start with?	Applying
2	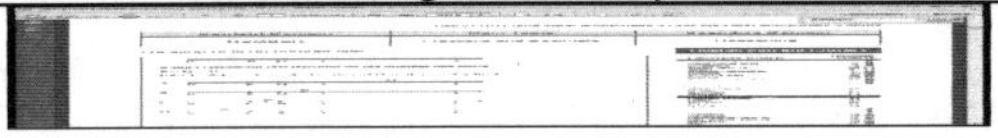 P and Q represent two fractions on the number line above. P × Q = N. Shows the location of N on the number line?	Reasoning
3	Place the four digits 3, 5, 7, and 9 into the boxes below in the positions that would give the greatest result when the two numbers are multiplied. 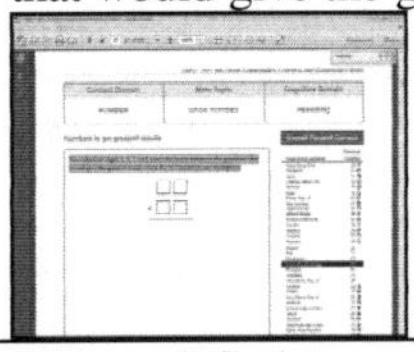	Reasoning
4	John and Cathy were told to divide a number by 100. By mistake, John multiplied the number by 100 and obtained an answer of 450. Cathy correctly divided the number by 100. What was her answer?	Applying

3. Data analysis

3.1 Analysis of the types of errors

This section discusses the errors made by the students. There are three types of students' errors; they are conceptual error, operational error, and principal error.

3.1.1. Conceptual error. Conceptual error occurs when the students are not able to apply the concept of number. This error recorded on *Reasoning level problems* only.

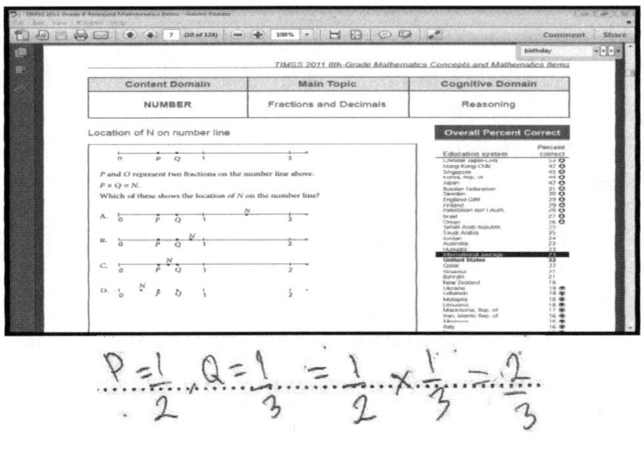

Figure 1. Example of conceptual error

Question: P and Q represent two fractions on the number line above. P × Q = N. Shows the location of N on the number line?

The student was not able to multiply two fractions and to determine the value of *P* and *Q*. He considered that the way to multiply two fractions is to multiply across that is multiplying the numerator of the first fraction with a denominator of the second fraction and the numerator of the

IOP Conf. Series: Materials Science and Engineering **296** (2018) 012010 doi:10.1088/1757-899X/296/1/012010

second fraction with the denominator of the first fraction, in which he failed to solve the following problem. He made a conceptual error such as multiply two fractions. From the interview, students were not able to understand the way of multiplying two fractions and to determine what value of the number line.

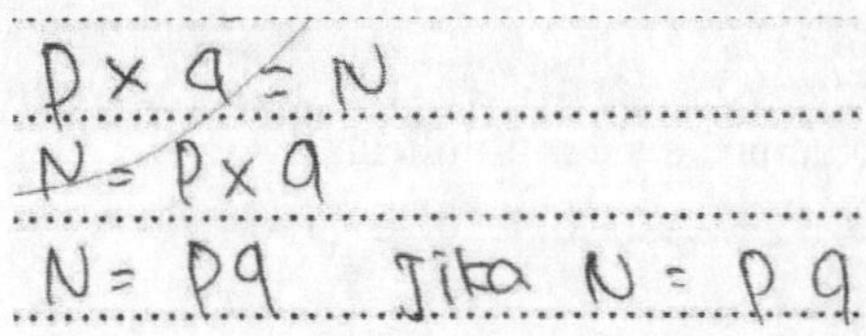

Figure 2. Example of conceptual error

The student was not able to identify the procedure and solve the problem. She considered that item 2 could be solved by algebra operation such as multiplication algebra. She thought whatever problem that contained alphabet must be solved by algebra solution. The interview showed that this error caused by the subject who did not comprehend the mathematical problem given.

3.1.2. Operational Error. Operational error occurs when the students are not able to calculate numbers. This error recorded on *Applying* and *Reasoning level problems*

- For applying level problem

Question: John and Cathy were told to divide a number by 100. By mistake, John multiplied the number by 100 and obtained an answer of 450. Cathy correctly divided the number by 100. What was her answer?

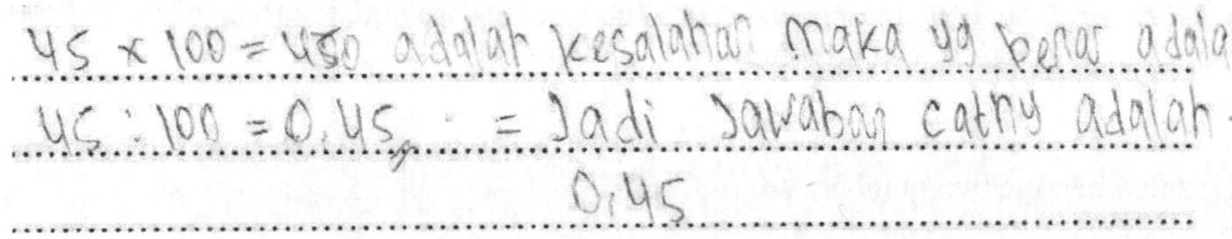

Figure 3. Example of operational error in applying level problem

The student was not able to compute the operation. The subject considered that 45 × 100 = 450, in which he failed to solve the following problem. He made an operational error such as on multiplying. The interview showed that this error caused because the subject was careless.

- For reasoning level problem

Question: P and Q represent two fractions on the number line above. P × Q = N. Shows the location of N on the number line?

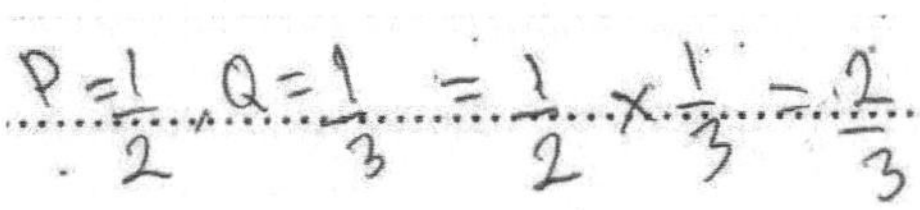

Figure 4. Example of operational error in reasoning level problem

The student was not able to compute the operation. The subject considered that $\frac{1}{2}\times\frac{1}{3}=\frac{2}{3}$ in which he failed to solve the following problem. He made an operational error such as on multiplying two fractions. The interview showed that this error caused because the subject was not able to understand to compute two fractions.

Question: Place the four digits 3, 5, 7, and 9 into the boxes below in the positions that would give the greatest result when the two numbers are multiplied.

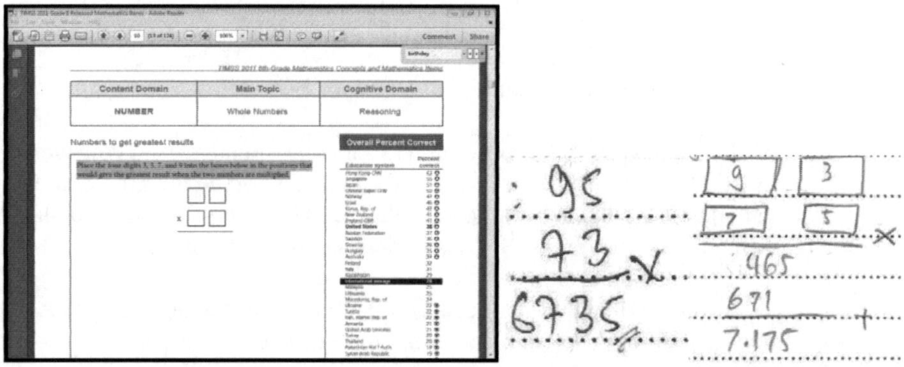

Figure 5. Example of operational error in reasoning level problem

The student was not able to compute the operation. She made an operational error such as on multiplying whole numbers. The interview showed that this error caused because the subject was careless.

3.1.3. Principal Error. Principal error occurs when the students are not able to answer the final answer that is caused by the previous error in the problem-solving. This error recorded on *Reasoning level problems* only.

Question: Place the four digits 4, 6, 8, and 9 into the boxes below in the positions that would give the greatest result when the two numbers are multiplied.

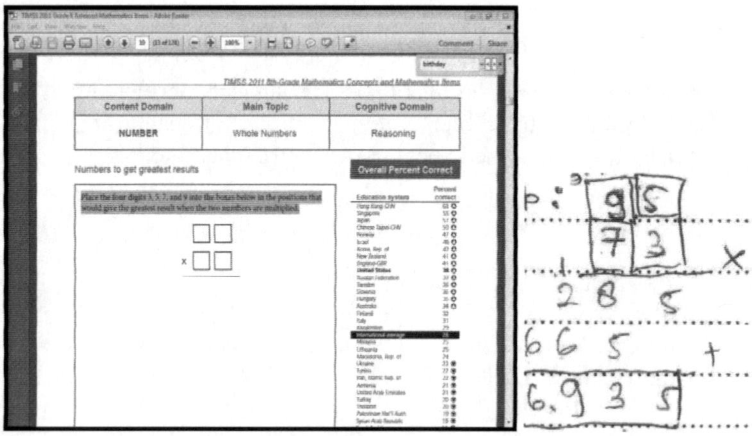

Figure 6. Example of principal error

The student was not able to determine the final answer. She made a major error. The interview showed that this error caused because the subject was not able to determine the suitable numbers that were produced greatest value if they were multiplied.

4. Conclusion

As conclusion, it was found that students tend to make three types of errors; they were conceptual error, operational error, and principal error. Some of conceptual errors were translating word problem to mathematics problem; using multiplication fraction, and determining fraction's order in number line. The operational error that made by students was the subject could not use the operation correctly. For the principal error made by students was determining final solution due to the previous errors.

References

[1] Abdullah A H, Abidin N L Z and Ali M 2015 Analysis of students??? errors in solving Higher Order Thinking Skills (HOTS) problems for the topic of fraction *Asian Soc. Sci.* **11** 21

[2] Lester F K and Kehle P E 2003 From Problem Solving to Modeling: The Evolution of Thinking About Research on Complex Mathematical Activity, eds In R. Lesh, S& H. M. Doerr *Constr. – Model. Model. Perspect. Math. Probl. Solving, Learn. Teach.* 501–17

[3] Polya G 1973 *How To Solve It: A New Aspect of Mathematical Method* (Princeton, New Jersey: Princeton University Press)

[4] Mayer R E 1987 *Learnable aspects of problem-solving: some examples Applications of cognitive psychology: Problem Solving, education and computing, eds D E Berger, K Pezdek and W P Banks* (ERA NJ: Hillsdale)

[5] Widdiharto R 2008 *Diagnosis Kesulitan Belajar Matematika SMP dan Alternatif Proses Remidinya* (Yogyakarta: Pusat Pengembangan dan Pemberdayaan Pendidik dan Tenaga Kependidikan Matematika)

[6] Susanti E, Kusumah Y S and Sabandar J 2014 Computer-assisted realistic mathematics education for enhancing students' higher-order thinking skills (experimental study in junior high school in Palembang, Indonesia) *J. Educ. Pract.* **5** 51–8

[7] Herholdt R and Sapire I 2014 An error analysis in the early grades mathematics: A learning opportunity *South African J. Child. Educ.* **4** 42–60

[8] Yang C W, Sherman H and Murdick N 2011 Error pattern analysis of elementary students with limited english proficiency *Investig. Math. Learn.* **4** 50–67

[9] Olivier A 1996 Handling Pupils' Misconceptions *Pythagoras* **21** 10–9

[10] Ketterlin-Geller, R L and Yovanoff P 2009 Diagnostic Assessments in Mathematics to Support Instructional Decision Making *Pract. Assessment, Res. Eval.* **14** 1–11

[11] Mullis I V S, Martin M O, Foy P and Arora A 2012 *The TIMSS 2011 International Results in Mathematics*

[12] Kementerian Pendidikan Nasional 2011 *Instrumen Penilaian Hasil Belajar Matematika SMP: Belajar dari PISA dan TIMSS* (Yogyakarta)

[13] Utami L 2016 Analisis Kesulitan Siswa SMP Kelas VII Menyelesaikan Soal Operasi Hitung Bilangan dan Solusi Pemecahannya *Konferensi Nasional Penelitian Matematika dan Pembelajarannya (KNPMP I)* (Surakarta: Konferensi Nasional Penelitian Matematika dan Pembelajarannya)

IOP Conf. Series: Materials Science and Engineering **296** (2018) 012011 doi:10.1088/1757-899X/296/1/012011

Analysis of student attitudes towards e-learning using Fishbein Multiattribute approach

Jasuli

Department of Economics Education, Postgraduate School, Universitas Negeri Surabaya

jasulijasuli16070925003@mhs.unesa.ac.id

Abstract. This research aimed to know students' attitudes toward e-learning and to determine what attributes were considered to be dominant by students toward the use of e-learning. The research population was all postgraduate students in the 2016 academic year at Universitas Negeri Surabaya. The sampling technique is using nonprobability sampling and purposive sampling with the sample totaled 100 respondents. The research instrument is using questionnaire with semantic differential scale. The models used to analyze is multi-attribute Fishbein model. The findings indicated that student attitudes toward e-learning are positive and easy accessibility which is considered as the most important attribute by students toward the use of e-learning.

1. Introduction

The rapidly changing globalization followed by the rapidly changing information technology also, resulting in the loss of limits that restricted all human activity. Because of the limitation loss, then all human activity can be done freely. Changes or development of information technology is so rapidly appearing on the internet media. Every year internet users are always improving, everyone knows that. By using the internet, we can exceed the limit. In the sense, we can know everything through the internet, even though it is far from us. Using the internet we can access all the information either from domestic or abroad.

The rapid changes of globalization is followed by the development of information technology. The change of globalization is also followed by the development in the world of education. Previously in the world of education, education can only be done in a class or a specific place. In today's era of globalization, everything has changed. Currently, education can be done either in class or out of class, even from home. Learning can still take place even if there is no meeting between teachers and students.

the outdoor learning and face-to-face learning can be done with the help of fast-growing technology. By utilizing the internet, learning can be done anywhere and anytime. Distance learning by using the internet is called e-learning. E-learning is a process of distance learning by connecting principles in a process of learning with technology [1].

At this time, many educational institutions have used e-learning in the learning process. One of the universities using e-learning as one of the learning processes is Universitas Negeri Surabaya. Vi-learn is one of the forms of e-learning applied in Universitas Negeri Surabaya since 2015.

As an educational institution or college, it is necessary to always improve the quality of all aspects contained in e-learning. To find out what are the deficiencies of e-learning that is used, it is necessary to know the responses or attitudes of e-learning users, i.e. students. An attitude represents a person's general feeling of favorableness toward some stimulus objects [2]. The purpose of knowing the attitude of the student is to know the response to the e-learning which is established whether good or not good. From the student's attitude, the institution can be used as evaluation material to improve the deficiency of e-learning applied.

Thus, the objectives of this research are: (1) to know student attitudes toward e-learning; (2) to determine what attributes are considered dominant by students toward the use of e-learning.

2. Literature review

2.1 Attitude

According to Fishbein and Ajzen, an attitude represents a person's general feeling of favorable or unfavorable toward some stimulus objects [2]. In our conceptual framework, we acquire attitudes toward the object. Kotler and Keller told that an attitude is a person's enduring favorable or unfavorable evaluations, emotional feelings, and action tendencies toward some objects or ideas [3]. Attitudes put us into a frame of mind: liking or disliking an object, moving toward or away from it. It can be concluded that attitude is a response of a person toward an object that favorable or unfavorable. Peter and Olson told that attitude as a person's overall evaluation of a concept [4]. Evaluations are effective responses, usually at relatively low levels of intensity and arousal. Solomon told that attitude is a lasting, general evaluation of people (including oneself), objects, advertisements, or issues. He calls anything toward which one has an attitude object (Ao) [5].

2.2 E-Learning

E-learning is flexible learning using ICT resources, tools, and applications, and focusing on interactions among teachers, learners, and the online environment. E-learning usually refers to structured and managed learning experiences, and may involve the use of the internet, CD-ROMs, software, other media, and telecommunications [6]. Hartley defines e-learning as learning enabled by the internet, intranets, and other electronic networks as well as the development, delivery, and evaluation of content provided to learns through these networks [7].

3. Research method

3.1 Research design

This study used the descriptive analytical method by using survey techniques. Descriptive research is a research that aims to provide an overview of research variables. This study includes data collection, processing, analyzing data obtained and conclusion.

3.2 Population and sample

Population consists of entire research object such as group of people, events or things [8]. The research population was all postgraduate students in 2016 academic year at Universitas Negeri Surabaya.

Sampling is the process of selecting a sufficient number of elements from the population so that a study of the sample and an understanding of its properties or characteristics would make it possible for us to generalize such properties or characteristics to the population elements [9]. Sample is elements or members of population [8]. Sample size larger than 30 and less than 500 are appropriate for most research [10]. For that reason, the sample size is 100 with the sampling technique is using nonprobability sampling and purposive sampling.

IOP Conf. Series: Materials Science and Engineering **296** (2018) 012011 doi:10.1088/1757-899X/296/1/012011

3.3 Operational variable and measurement

Student attitudes toward e-learning attributes are the attributes of student response to the e-learning expressed in seven statements that are divided into two categories: first is very important, important, moderately important, neutral, moderately unimportant, not important, and very unimportant. The second category is strongly agree, agree, moderately agree, neutral, moderately disagree, disagree, strongly disagree. Measurement of positive attitudes and negative attitudes by using semantic differential scale.

Variable attitude toward e-learning in this study are the attributes attached to e-learning [11]:

a. Non-linearity. Users are free to access the learning object and there are facilities to provide requirements depending on the user's knowledge.
b. Self-managing. Lecturers can manage their own learning process by following the structure that has been made.
c. Feedback-Interactivity. Learning can be done interactively and provided feedback on the learning process.
d. Multimedia-Learners style. E-learning provides multimedia facilities. The advantages of using
e. Multimedia, students can understand more clearly and real according to the background of their students.
f. Just in time. E-learning provides whenever the user needs, to solve problems or just want to improve knowledge and skills.
g. Dynamic Updating. Have the ability to update the contents of the material online on the latest changes.
h. Easy Accessibility/Access Ease. Only use the browser (and maybe some devices are installed).
i. Collaborative learning. With learning tools enable interaction, it can communicate directly at the same time (synchronous) or communicate at different times (asynchronous). Users can communicate with creators of materials, other students.

3.4 Data Analysis Method

3.4.1 Validity and Reliability Test. Validity is concerned with whether it already measured the right concept and reliability with stability and consistency of measurement [12]. Validity is a measure of accuracy in measurement. A construct with perfect validity contains no measurement error [13]. In other words, validity is a test to know how well an instrument that is developed to measures the particular concept. Validity is measured from critical values for Pearson's (r). If the r_{count} is more than r_{table} then the question is valid. Reliability is a test of how consistently a measuring instrument measures whatever concept it is measuring [12]. In other words, reliability is a test which will show how far the measure tool can be trusted and reliable. The reliability test in this research is based on Cronbach's alpha. Sekaran and Bougie [14] told that reliabilities less than 0.60 are considered to be poor.

3.4.2 Fishbein Multiattribute Model. In this Fishbein and Ajzen says that a person's attitude toward an object is a function of his belief that the object is associated with certain attributes and evaluative responses that connected to that belief [15]. Mathematical formulation of the multi attributes model of attitude toward the object, by Fishbein, can be formulated as follows:

$$A_o = \sum_{i=1}^{n} b_i e_i \quad (1)$$

Where:
Ao = Attitude toward an object
b = Strength of belief that the object has attribute i

e = The evaluation of attribute i
n = The number of beliefs

4. Result and discussion

4.1 Result

4.1.1 Validity and reliability test

- *Validity test.* The validity test used in this study used with SPSS 21, with the number of respondents in this test amount 30. Therefore r table value with the number of respondents 30 is 0.361 (r table = 0.361). If the value of r count less than r table then not valid, while if value r_{count} more than r_{table} then valid. The results were presented in Table 1, Table 2, and Table 3.

Table 1. Validity Test on Component of Evaluation

Item	Person Correlation	Significant	Status
Non-linearity	0.819	0.000	Valid
Self-managing	0.895	0.000	Valid
Feedback Interactivity	0.581	0.001	Valid
Multimedia Learners	0.819	0.000	Valid
Just In Time	0.720	0.000	Valid
Dynamic Updating	0.625	0.000	Valid
Easy Accessibility	0.781	0.000	Valid
Collaborative Learning	0.843	0.000	Valid

Table 1 shows that all value of r_{count} more than r_{table}. Therefore, it can be stated that all question items on the component of evaluation are valid.

Table 2. Validity Test on Component of Belief

Item	Person Correlation	Significant	Status
Non-linearity	0.841	0.000	Valid
Self-managing	0.845	0.000	Valid
Feedback Interactivity	0.836	0.000	Valid
Multimedia Learners	0.874	0.000	Valid
Just In Time	0.751	0.000	Valid
Dynamic Updating	0.859	0.000	Valid
Easy Accessibility	0.869	0.000	Valid
Collaborative Learning	0.849	0.000	Valid

Table 2 shows that all value of r_{count} more than r_{table} (0.361). Therefore, it can be stated that all question items on the component of belief are valid.

- *Reliability test*

Table 3. Reliability Test

Item	Cronbach's Alpha	Rule of Thumb	Status
Evaluation	0.889	0.60	Reliable
Belief	0.938	0.60	Reliable

Table 3 shows that both items have value of Cronbach's alpha more than 0.06. Therefore, it can be stated that all question items are reliable.

4.1.2 Fishbein Multiattribute Analysis. To answer the research objective, then used Fishbein multiattribute model to measure the overall student attitude. The result is in a form of a score. The measurement is based on the calculation of evaluation (ei) and belief (bi) towards the identified attributes. Table 4 shows the results of attitude analysis of postgraduate students in 2016 academic year at Universitas Negeri Surabaya:

Table 4. Result of Student Attitude Analysis e-learning

No	Attributes	e_i	b_i	$Ao = e_i \cdot b_i$
1	Non-linearity	6.10	6.23	37.94
2	Self-managing	6.02	6.10	36.69
3	Feedback Interactivity	6.06	6.08	36.81
4	Multimedia Learners	5.94	5.96	35.37
5	Just In Time	5.21	5.75	29.90
6	Dynamic Updating	5.22	5.79	30.20
7	Easy Accessibility	6.42	5.97	38.30
8	Collaborative Learning	6.03	6.22	37.51
	$Ao = \sum e_i \cdot b_i$			282.72

The overall score of the student attitudes towards e-learning has been found based on identified attributes from the Fishbein's model. Table 3 shows how the calculation of consumer attitude is computed. The score of consumer attitude towards e-learning, in general, is 282.72. After that, the next step is to determine which score is positive or negative.

To make the interval from very negative to very positive use the following equation:

$$Interval = (m - n)/b \tag{2}$$

Where:
m = the highest score may occur
n = the lowest score may occur
b = the number of intervals will be established

In this study, the interval used is to contain five classes, which are very negative, negative, neutral, positive and very positive. Then, find the highest and lowest score may occur. In this study, the highest score (m) that may occur is 389. The lowest score (n) may occur is 8. Table 5 shows the table of intervals.

Table 5. Attitude assessment interval [2]

$8 \leq Ao \leq 81.8$	Very Negative
$81.6 < Ao \leq 158.6$	Negative
$158.6 < Ao \leq 235.4$	Neutral
$235.4 < Ao \leq 312.2$	Positive
$312.2 < Ao \leq 389$	Very Positive

The attitude score acquired in this research is 282.72. Therefore, based on table 5, the score 282.72 is positive. It means student attitudes toward e-learning is positive. Based on table 4, the most dominant attribute is easy accessibility with the totaled score is 38.30. Thus, the objective of this research is answered.

4.2 Discussion

The students attitude towards e-learning where the Ao is 282.72 is considered as positive. It means that e-learning of Universitas Negeri Surabaya is considered good by the students. Students give the highest score on the attribute of easy accessibility. Where Ao score for easy accessibility is 38.30. It is because e-learning Universitas Negeri Surabaya is easy to access, and easy to access all facilities in e-learning Universitas Negeri Surabaya. Then, the second highest attribute is nonlinearity. Where Ao score for nonlinearity is 37.94. It is because students can freely access all the subject matter or download the material and the tasks from their lecturers, as well as the facilities provided by the Universitas Negeri Surabaya such as wireless to access e-learning. While the attribute that gets the lowest score from the student is just in time. Where Ao score for just in time is 29.29. It is because students assume that the learning material contained in e-learning has not been able to solve the problems of students and not yet able to improve the knowledge and skills of students. Although using e-learning media, face-to-face learning is still needed.

The overall dominant attribute is easy accessibility with the total score is 38.30. However, when viewed on each component (evaluation and belief) on the evaluation component, the attribute gets the highest score is easy accessibility with the total score is 6.42. It is because students expect e-learning that is easy to access. While in the belief component, the attribute that gets the highest score is nonlinearity with the total score is 6.23. It is because students assume that e-learning University-Negeri-Surabaya gives can freely access all the subject matter or download the material and the tasks of lecturers, as well as the facilities provided by the Universitas Negeri Surabaya such as wireless to access e-learning. Then on the evaluation component, the attribute that gets the second highest score is non-linearity with the total score is 6.10. It's because students expect e-learning that gives freedom in accessing e-learning. While on the belief component, attribute that gets the second highest score is collaborative-learning. It is because students assume that e-learning University of Surabaya can enable the communication directly with lecturers or with other students.

5. Conclusion

The conclusions of this research are as follows: (1) Student attitude toward e-learning is considered positive. (2) The dominant e-learning attribute towards student attitudes is easy accessibility.

References

[1] Chandrawati S 2010 pemanfaatan e-learning dalam pembelajaran cakrawala kependidikan *Cakrawala Kependidikan* **8 (2)** pp 1–10

[2] Fishbein M and Ajzen I 1975 Belief, Attitude, Intention, and Behavior (United State: Addison-Wesley Publishing) *An introduction to theory and research. Massachussets: Addison-Wesley* p 216

[3] Kotler P and Keller, K L 2012 Marketing Management (New Jersey: Prentice Hall) *Marketing Management (New Jersey: Prentice Hall)* p 168

[4] Peter J P and Olson J C 2010 Consumer Behavior and Marketing Strategy (New York: Mc-Graw Hill) *Consumer Behavior and Marketing Strategy (New York: Mc-Graw Hill)* p 128

[5] Solomon M R 2013 Consumer Behavior: Buying, Having and Being (United State: Pearson Education) *Consumer Behavior: Buying, Having and Being (United State: Pearson Education)* pp 273–4

[6] Education M O 2010 E-Learning and Implications for New Zealand Schools: A Literature Review (New Zeland: The University of Education) *E-Learning and Implications for New Zealand Schools: A Literature Review (New Zeland: The University of Education)* p 7

[7] Hartley E D 2006 Selling E-Learning, American Society for Training and Development (United

States of America: ASTD) *Selling E-Learning, American Society for Training and Development (United States of America: ASTD)* p 1

[8] Sekaran U and Bougie R 2009 Research Methods for Business: A Skill-Building Approach (New York: West Sussex) *Research Methods for Business: A Skill-Building Approach (New York: West Sussex)* pp 262–3

[9] Sekaran U 2003 Research Methods for Business: A Skill-Building Approach (New York: Wiley & Sons) *Research Methods for Business: A Skill-Building Approach (New York: Wiley & Sons)* p 266

[10] Sekaran U 2003 Research Methods for Business: A Skill-Building Approach (New York: Wiley & Sons) *Research Methods for Business: A Skill-Building Approach (New York: Wiley & Sons)* p 295

[11] Agustina M 2013 Pemanfaatan E-Learning Sebagai Media Pembelajaran Aplikasi Teknologi Informasi *Pemanfaat. E-Learning Sebagai Media Pembelajaran Apl. Teknol. Inf.* pp 8–12

[12] Sekaran U and Bougie R 2009 Research Methods for Business: A Skill-Building Approach (New York: West Sussex) *Research Methods for Business: A Skill-Building Approach (New York: West Sussex)* p 157

[13] Hair J J and Wolfinbarge 2010 Essensial of Marketing Research (New York: Mc-Graw Hill) *Essensial of Marketing Research (New York: Mc-Graw Hill)* p 157

[14] Sekaran U and Bougie R 2009 Research Methods for Business: A Skill-Building Approach (New York: West Sussex) *Research Methods for Business: A Skill-Building Approach (New York: West Sussex)2* p 325

[15] Fishbein M and Ajzen 1975 Belief, Attitude, Intention and Behavior (United State: Addison-Wesley Publishing) *Belief, Attitude, Intention and Behavior (United State: Addison-Wesley Publishing)* p 29

IOP Conf. Series: Materials Science and Engineering **296** (2018) 012012 doi:10.1088/1757-899X/296/1/012012

The implementation of interest-based entrepreneurship curriculum in the Theory of Economics course

J Surjanti[1], D Nugrohoseno[1], Budiono[1] and H Musfidah[2]

[1]Economics Department, Universitas Negeri Surabaya, Ketintang, Surabaya,East Java, Indonesia, 60231

[2]SMAN 5 Surabaya, Jl. KusumaBangsa No.21, Ketabang, Genteng, Surabaya, East Java, Indonesia, 60272

junsurjanti@unesa.ac.id

Abstract. College and university are educational institutions and they should be able to design entrepreneurship program in order to strengthen creativity and innovation. Potential grown-up citizens potential is the main problem of national development solved through the role of universities by reconstructing the curriculum based on 2016 Indonesian National Curriculum Framework. The interest of building entrepreneurship becomes the work of all elements which one of them is a university. Reconstructing interest-based entrepreneurship curriculum is the first step to construct characterized economic. Economics lecturers have a responsibility to develop the curriculum. This study aimed to explore the interest in entrepreneurship used to develop the contextual learning using problem-based learning (PBL) method that refers to Universitas Negeri Surabaya's motto "*Growing with character*" and integrated into 2016 Indonesian National Curriculum Framework. This study was qualitative explorative. The subjects of the study are 86 students in their first semester. The objective of the study was curriculum development interest-based entrepreneurship in the theory of economics course. The results of the study showed that the interest-based entrepreneurship curriculum was able to reach the targeted learning outcome as well as to create a product in the culinary and services project in Wonorejo mangrove forest according to students' interests.

1. Introduction

Creating creative and innovative learning will develop students' creativity and innovation. Universities have important roles to bridge the community needs through entrepreneurship curriculum. Jeffrey [1] stated that lecturers should develop creative teaching and creative learning for the learners/students and universities need to be wiser in adapting in order to grow student's entrepreneurial mindset. Study of entrepreneurship significantly influences the growth of national development in entrepreneurial companies and economic development in a region or country. As Arthur said, "policy-makers should review the current higher education system and change them into entrepreneurial mindset". [2]

Wonorejo mangrove forest in Surabaya is one of the tourist destinations as well as a means of education for the community in Surabaya. Various market activities can be seen in this area and it makes the area becomes a market representative. Observing the area, students are able to see all the possibilities in creating business opportunities. This study focused on mangrove forests as innovative learning media in line with the research conducted by Cruz at all [3] who mentioned that education centers and institutes should provide more facilities and aids for the entrepreneurship program that is lack of support, in order to strengthen and to build the entrepreneurship spirit. Utilizing mangrove forest is expected to give a contextual learning which possibly grows the innovative spirit.

2. Theoretical review

2.1. Previous studies

Entrepreneurship interest cannot directly emerge but it needs to be nurtured. Educational institutions should provide an aid to grow as well as to strengthen the entrepreneur spirit as defined by Cruz [3] at all. Carrying out entrepreneurship material in formal education should be well-designed especially in the curricula. Ertuna & Gurel's [4] study showed that creative and innovative curricula were able to fill the gap in the literature by examining the educational moderation effects of entrepreneurial trait and intention.

Economics is a social science concerned with describing and analyzing production, distribution, and consumption of products and services and focuses on the behavior and agents' interaction as well as their works. Economics cannot be separated from the actors, the entrepreneurs. They are the people who actually run the economy which theoretically debated in the Theory of Economics. Rocha argued that in microeconomics, mainstream economics and Theory of Firms have difficulties in adjusting entrepreneurial practices to theories resulting in a tendency to ignore the existence of entrepreneurs [5]. Reisman [6] referred the word "Schumpeter" as entrepreneurship, although Frank Knight was the one succeeded to instil the word into life which has remained hidden for years [6]. Schumpeter considerably contributed in understanding entrepreneurship mainly from economics lens; he discovered the concept of entrepreneur against the backdrop of economic development by defending it as a dynamic process involving the economic disturbance, especially in remaining the status quo by inviting the entrepreneur to share the responsibility [7].

The importance of entrepreneurs at the student level is significantly affecting the policymakers in developing appropriate educational policies to escalate the entrepreneurial performance [8]. Joyce and Weil [9] mentioned several uses of model in improving entrepreneurship such as a) to identify the procedures followed in the activities; b) to control the covered activities; c) to identify the component used by the administrator; d) to carry out the appropriate changes when discrepancy occurred; e) to help teachers to unify the student tasks.

2.2. Ideal curriculum

The university's ideal curriculum should meet the today's demands. The curriculum is often considered as an official document containing all the necessary information to undertake educational qualification, program, or course. [10] it is also called as curriculum plan or curriculum document. Unesco [11] mentioned that curriculum could ideally be the bridge between education and development. Sarimah Ismail and Dahiru Sale Muhamme [12] concluded that curriculum on Technical and Vocational Education and Training (TVET) in Nigeria did not relate to problem-solving skills and decision-making as well as with long-term learning process. Those resulted in the graduates who did not fit in the work demands. Whether Sitthipon [11] Art-in (2014) study indicated that to develop ideal curriculum, there are six components should be fulfilled, namely: rationale, objective curriculum, content structure, training activity management, measurement and evaluation, and graduation criteria.

2.3. Curriculum of economics study with interest-based entrepreneurship

The study program of management in Universitas Negeri Surabaya designed the curriculum to prepare students to be creative entrepreneurs along with Indonesian National Curriculum Framework. Learning-oriented in students' interest interpreted in social, economic, and cultural context considered as a professional learning activity [13]. Forsythe [14] stated that PBL develops problems, first meeting, research, feedback meeting, and response. Duch [15] stated that in PBL, students are presented with problems, discussion, grades, learning issues, and new integrated knowledge before leading to the core problem. He as well mentioned that students are encouraged to summarize both new and old knowledge as well as to relate them.

3. Methodology

This study was qualitative explorative. The project was to design or create a business opportunity, and analyzed the business, carried out for real. In class activities, students should design and create the project planning at Wonorejo Mangrove Forest. The project was conducted in a group. The subjects of the study were 86 students in the first semester, students of 2014 class who have programmed microeconomics. The objective of the study was to implement interest-based entrepreneurship curriculum in Theory of Economics course. Data analysis was conducted in several steps, as 1) distributing questionnaires to students openly; 2) coding the options, obtained the data of students' interests; 3) collecting data; 4) analyzing data obtained; 5) calculating the frequency of students' interests; 6) analyzing data as learning activity consideration.

4. Results and discussion

This study was qualitative explorative. The project was to design or create a business opportunity, and to analyze the business, carried out for real. In class activities, students should design and create the project planning at Wonorejo Mangrove Forest. The project was conducted in a group. The subjects of the study were students in the first semester of 86 people, students of 2014 class, who have chosen microeconomics. The objective of the study was to implement interest-based entrepreneurship curriculum in Theory of Economics course. Data analysis was conducted in several steps, as 1) distributing questionnaires to students openly; 2) coding the options, obtained the data of students' interests; 3) collecting data; 4) analyzing data obtained; 5) calculating the frequency of students' interests; 6) analyzing data as learning activity consideration.

4.1. Understanding the orientation of students' interest in entrepreneurship

McKimm [14] and Timperley [15] agreed that the first step to analyze the students' interest in entrepreneurship was by collecting data and it can be seen from Table 7.

Table 1. Interest in entrepreneurship

Validity	Frequency	Percentage	Valid Percentage	Cumulative Percentage
No interest	16	18.6	18.6	18.6
Culinary	34	39.5	39.5	58.1
Trading	7	8.1	8.1	66.3
Service	15	17.4	17.4	83.7
Furnitures	4	4.7	4.7	88.4
Garment	5	5.8	5.8	94.2
Farming	3	3.5	3.5	97.7
Fishery	1	1.2	1.2	98.8
Information and Technology	1	1.2	1.2	100.0
Total	86	100.0	100.0	

Based on the data above, it is seen that the highest entrepreneur interest selected by students were culinary as 39%, and followed by service business as 17.4%. The remaining students preferred trading as 8.1%, garment as 5.8%, livestock as 3.5%, fisheries 1.2%, and information and technology as 1.2%, and 18.6% of them did not have any business preferences. This data is very accurate to determine the topic of economics learning as well as it can be used as themes in the project preparation in the assignment. In the end, learning topic agreed was a project study on the mangrove forests for culinary and service business as the learning media. Whether the other students need to adjust to the learning designed was contextual learning using Problem Based Learning (PBL) method.

4.2. Learning of characterized theory of economics using interest-based entrepreneurship
Character Education should be implemented in the study of the theory of economics. The program developed the "Entrepreneur". McKimm [14] defined stages into four, namely: need cycles, assessment, design, and implementation. Moreover, the results were reviewed and evaluated based on the original assessment needs. at Universitas Negeri Surabaya, the Theory of Economics described in the lesson plan should cover all points in the learning outcomes, such as a) mastering the basic concepts of theory of economics and utilizing Internet to know price and policy; b) determining prices and outputs on various types of input and output markets as well as price competition strategy; c) growing the faith, intelligence, independence, honesty, cares and resilience in determining price and output as an economic agent [16].

The micro theory course described the firm of equilibrium output in perfectly competitive markets, pricing and equilibrium outputs. Several textbooks such as the "*Character-Based Economic Learning Edition (Micro-Based Approach)*" and other relevant sources were used in the Theory of Economics course, while Wonorejo Mangrove forest Surabaya became the location selected as learning media. PBL was applied in determining learning design stages including conducting learning experiences, contextual application, literary reading, outclass study, data preparation, presentation, and discussion.

Learning outcomes and evaluation stage were determined using interest-based entrepreneurship curriculum in various culinary and services business as seen in Table 8.

Table 2. Student's work reports curriculum with interest-based entrepreneurship

Team	Material	Project Theme	Criteria
1	Market Competition	Fruits, Leaves, and Roots of Mangrove, JOMA's (Johan & Martin) Seafood	Culinary
2	Perfect Monopoly	Electric bike rents in the area of Mangrove Park Labyrinth	Service
3	Monopolistic	Various products of mangrove chips	Culinary
4	Monopsony	Mangrove leaves fertilizer	
5	Oligopoly	Renting culinary Stand/Booth in Mangrove conservation area, Istana Ice Cream Mangrove, and Café Shop	Service/Culinary
6	Input Market	Syrup, Dodol, and Mangrove jam	Culinary
7	Public properties	Various of public properties In Mangrove Forest	Service

Based on the students' works which were organized at the end of the course, they showed the capability in understanding the Theory of Economics using creativity and producing a product from the project selected in culinary and services as an indicator.

5. Conclusion and suggestion

5.1. Conclusion
Learning activities conducted in the Theory of Economics course using interest-based entrepreneurship was focused on culinary and services located in Mangrove forest. Developing a curriculum of interest-based entrepreneurship was evaluated and targeted on the learning outcomes.

The result of learning activity using contextual learning approach and PBL method showed by the ability of students in writing papers about how to run culinary business by making use of mangrove tree's products (fruits, leaves, and roots), such as by opening JOMA's (Johan & Martin), seafood stall, and café shop, as well as producing mangrove chips, syrup, *dodol* (traditional snack), jam, and ice cream. Aside of culinary, students were able to propose the service business, such as opening electric bike renting in the mangrove forest and in the Labyrinth Park, renting booths and other public facilities in the mangrove forest area. The result of the study indicated that interest-based curriculum was appropriate in growing entrepreneurship interest, yet it was a dynamic curriculum that continuously needs to be improved and developed even it was able to fulfill the targeted learning outcomes.

5.2. Suggestion

Before starting the learning process, the lecturer should perform students' interest mapping in order to obtain targeted learning outcome in the Theory of Economics course. Hereafter, methods, media, learning material, and proper evaluation were decided.

The lecturer should develop the dynamic curriculum through appropriate approached considering the entrepreneurial interest, so the goal of economics learning can be obtained due to meet the community's needs.

References

[1] Jeffrey B and Craft A 2004 Teaching creatively and teaching for creativity *Distinctions Relationships Educ. Stud.* **30** pp 77–87

[2] Arthur S J and Hisrich R D 2012 The importance of education in the entrepreneurial process *A World View J. Small Bus. Enterp. Dev.* **19** pp 500–14

[3] Cruz N M, Escudero A, I R Barahona J H and Leitao F S 2009 The effect of entrepreneurship education programmes on satisfaction with innovation behavior and performance *J. Eur. Ind. Train.* **33** pp 198–214

[4] Ertuna Z I and Gurel E 2011 *The Moderating Role of Higher Education on Entrepreneurship Education + Training* **53** [5] Rocha V C 2012 *The entrepreneur in economic theory: From an Invisible Man Toward a New Research Field- FEP Working Papers School of Economics and Management* (Portugal: University of Porto) [6] Reisman D 2004 *Schumpeter's Market: Enterprise and Evolution* (Cheltenham: Edward Elgar)

[5] Schumpeter J A 1934 *The Theory of Economic Development* (Cambridge MA: Harvard University Press)

[6] Moreno J, Leopoldo J D, Castillo L and Triguero M S 2012 The effect of business and economics education programs on students' entrepreneurial intention *Eur. J. Train. Dev.* **36** pp 409–25

[7] Joyce B and Weil M 1972 *Conceptual Complexity Teaching Style and Models of Teaching A Paper*
Prepared for the National Council for the Social Studies Boston (Boston)

[8] Schugurensky D 2002 The eight curricula of multicultural citizenship education *Multicult. Educ.* **10** pp 2–6

[9] Sitthipon Art-in 2014 The development of teacher training curriculum on learning management to develop students' analytical thinking in Thailand *Procedia - Soc. Behav. Sci.* **116** pp 939– 45

[10] Ismail S and Mohammed D S 2016 Employability Skills in TVET Curriculum in Nigeria Federal
Universities of Technology *Procedia - Soc. Behav. Sci.* **204** pp 73–8

[11] Helen T, Aaron W, Heather B and Irene F 2007 Teacher professional learning and development *BES Iterative Best Evidence Synthesis Program* (Wellington: New Zealand Ministry of Education)

[12] Frank F 2002 *The Handbook for Economics Lecturers Problem-based Learning* ed P Davies

(Staffordshire: University of Staffordshire)
[13] Barbara D, Susan J, Groh E and Allen D E 2006 The power of problem-based learning a practical
"How to" for teaching undergraduate courses in any discipline *Interdiscip. J. Probl. Learn.* **1**
[14] Ibrahim M and Dkk 2016 Naskah Akademik Pengembangan Kurikulum (Surabaya)

IOP Conf. Series: Materials Science and Engineering **296** (2018) 012013 doi:10.1088/1757-899X/296/1/012013

Identifying students' learning performance as a way to determine the admission process in physical education field

J B Prihanto, D C Kartiko and A Wijaya

Department of Physical Education, Faculty of Sports Science, Universitas Negeri Surabaya, Indonesia

junaidibudi@unesa.ac.id

Abstract. The interest in the physical education field has been rising in the past ten years. It can be seen that registrants of the physical education program in several universities increase. This research is meant to analyze students' admission process and its relation to their performance in the learning activities in the department of physical education at Universitas Negeri Surabaya. The design of this study was quantitative data analysis. The research was conducted by collecting students' admission data and their transcripts. The result showed that the most influential factor of admission in physical education program was the student' field of study in high school. In addition, their achievements in sports competitions and family welfare are not likely to be important factors. These results give a recommendation for the next admission process which related to the quality of graduates.

1. Introduction

In this global era, the education quality has an important role in manifesting the excellence and progress of a nation since education will help the human resources to be smart, creative, skilful, competence, moral, and cultured. To increase the competitiveness of the nation, the university has very important roles in developing science and technology which is excellent and relevant to the society needs [1]. Therefore, to ensure the quality of higher education, the government has issued regulations about the standard of every study program [2].

As a program which mission is to develop excellent and professional sports science teachers, sports education, health, and recreation has been working on higher education services through series of activities mentioned before [3]. In order to fulfil that mission, prospective students are expected to have excellent academic and motoric ability. Besides fulfilling the mission, in sports science department, the motoric ability is important since sports teachers are expected to teach movements and sports as well as improve students' fitness, while academic records are needed to teach a healthy lifestyle.

Time management is the challenge for students with achievements in sports who attempt to maintain their achievements at the university [4]. They should be able to organize their time between lecture activities, assignments, independent study and practice to improve their performance. Department of Sports Education of Universitas Negeri Surabaya has attempted to facilitate these athlete-students by providing attendance exemptions as well as independent learning with module and e-learning system [5]. However, the results depend on the students' effort and motivation. The data of the department showed that most of these athlete-students have less satisfactory learning outcomes

IOP Conf. Series: Materials Science and Engineering **296** (2018) 012013 doi:10.1088/1757-899X/296/1/012013

(GPA<3) and spend more time to be graduated (more than 4 years). The worse, some of them are dropouts for failing to follow the lectures well.

Conducting quality assurance for the prospective students' selection and learning process is needed to develop competent graduates to fulfil the vision and mission of the program. For the learning process, the quality assurance mechanism has been manifested on various activities and written on the quality procedure documents at Quality Assurance Unit in the Department of Sports Education. This evaluation is very important because the program will get; 1) information on the distribution of students' school and origin for promotional purposes to attract more qualified students and expand the market, 2) information of which high school whose graduates have good learning achievement and study duration in the Sports Education Department to be priority in the selection (3) Identification of secondary schools whose graduates did not show good performance in the Sports Science Faculty as a consideration of quota reduction 4) Information about students' performance based on school origin that is used as feedback for schools. If the evaluation of students' learning outcomes and study duration based on the region and school can be implemented, the selection process for new students will run more effectively.

2. Method

This research used a descriptive quantitative research. It investigated the status of a group of people or an object. Recording, analyzing, and interpreting the current events were done in conducting this research [6]. The populations were the Sports Science Faculty while the samples were 300 students majoring Sports Education class of 2013 and 2014. The data of the students were analyzed and categorized based on the students' learning outcomes and study duration.

3. Results and Discussion

The managed data was analyzed by using cross tabulation, means, and frequency as follows:

3.1 The Analysis of Students' Origin (Province)

Based on the province, the students were originated from only 5 provinces. 145 students were from East Java; 8 students were from Central Java, and the others were originated from different provinces. The GPA of students coming from East Java was 3.06, students' GPA from Central Java was 3.02, and the others got 3.16 to 3.41.

Based on the data, there was an increase in the number of students' origin distribution ranging from 5 provinces in 2013 to 12 provinces in 2014. In the 2014 academic year, East Java remained the main origin of the students, 162 students. The students from Central Java decreased in the number that was only 2 students. In addition, 2 students were from West Java and only one student was from each of the other 9 provinces.

There was an increase in the students' average GPA score. The students from East Java increased their score from 3.06 to 3.18. The highest increase in GPA score was obtained by students from Central Java, ranging from 3.02 to 3.59. While students from other areas of origin got their average GPA score 2.65 to 3.60. The lowest score of this study was obtained by the student of West Papua and the highest was West Nusa Tenggara.

3.2 The Analysis of Students' Origin (District)

Based on the district, 24 students were from Surabaya, 16 were from Sidoarjo, 11 were from Lamongan, 10 were from Mojokerto and 10 were from Tulungagung. Other districts are from outside of East Java. Based on the district, the average GPA score obtained ranged from 1.89 to 3.58. The students from Blora got the lowest score while students from Jember got the highest one.

After being compared, the data from the 2013 academic year and the 2014 academic year showed the students were originated from the same 5 top districts, except Tulungagung. Sidoarjo was still in the first rank, 24 students. Mojokerto ranked the second position, 17 students. The next ranks were Gresik, 16 students; Surabaya, 14 students; and Lamongan, 10 students sequentially. In terms of average GPA score, Bima placed the first position with 3.6. Then it was followed by Kebumen with 3.59; West Lampung with 3.56; Palembang with 3.52 and Labuhan Batu with 3.49. In this case, the students of Sidoarjo got the average GPA score of 3.06.

3.3 The Analysis of Students' High School

The data showed that students of the Sports Education Department are graduated from UPT SMANOR Jatim (12 students), SMAN 1 Kedungwaru (4 students), SMAN 1 Srengat (3 students), SMAN 2 Lamongan (3 students) and SMAN 1 Bangil (3 students). All schools are located in East Java. If the schools are associated with the students' GPA score, the 5 highest GPA scores were students of SMAN 1 Pulung (3.67), SMAN 1 Kedungpring (3.64), SMAN 1 Giri (3.64), SMAN 1 Babat (3.63) and SMAN 1 Mojosari (3.59). While SMANOR Jatim where the most students came from obtained average GPA of 3.42.

For the class of 2014, most students came from UPT SMANOR Jatim (11 students), SMAN 4 Sidoarjo (6 students), SMAN 1 Mantup (5 Students) and SMAN 1 Manyar, SMAN 1 Gedangan and SMAN 1 Taman with 3 students each. All schools are located in East Java. Associating the schools with GPA, it is found that highest average GPA obtained by the students of SMAN 1 Grati (3.72), SMAN 11 Surabaya (3.71), SMKN 1 Gombong (3.71), SMAN 1 Gondong (3,68) and SMAN 1 Padangan (3.67). While SMANOR Jatim where the most students came from obtained GPA of 2.96.

3.4 The Analysis of GPA based on the Students' Major in High School

Students' major in high school might affect their achievement in higher education. Overall, based on their major in high school, students' average GPA score in the class of 2014 ranging from 2.54 to 3.76. To support the analysis, normality test and ANOVA test were done to complete the descriptive analysis. The normality test showed that all significant value was bigger than alpha (0.05). It means that the distribution was normal.

The results of this normality test were then analyzed by using ANOVA Test. The analysis showed a significant difference in minimum one out of three categories from the high school variable in which the ANOVA significant value (0.022) was smaller than alpha (0.05). ANOVA test then was followed by LSD test to determine which categories had a significant difference. In comparing natural and social science major, it was found that the significant value (0.007) was smaller than alpha (0.05). However, there was no significant difference while comparing other majors with either natural or social science.

3.5 The Analysis of GPA based on Athlete Status and Participation in Organization

Another factor that affects students' learning outcomes was their athlete status and participation in the student organization. The normality test showed that the status and participation of the students in the class of 2014 had bigger significant value than alpha value showing normal distribution. Then the significant value of the ANOVA test (0.29) was greater than the alpha value (0.05) which means that there was no significant difference from the four categories in the students' athlete status and participation in the organization.

3.6 The Analysis of GPA based on the Students' Economic Levels

The students in the class of 2014 got average GPA from five categories ranging from 3.27 to 3.40, in which 2.54 was as the lowest and 3.76 as the highest GPA. The student in the second economic level got the highest average GPA score 3.40 while the student in the first economic level got the lowest average GPA of 3.27. The results of normality test showed normal distribution since all significant value from each category was bigger than alpha value. With this result, the parametric difference test can be applied. Furthermore, the significant value of the ANOVA test (0.29) showed greater value than the alpha value (0.05), meaning that there was no significant difference from the four categories in the students' athlete status and participation in the organization.

Overall, the analysis of provincial data showed that most of the students came from East Java (93%) and the rest (7%) were from other provinces. Students from Central Java consistently existed in the class of 2013 and 2014 even though the quantity decreased. However, the students' origin distribution had significantly increased from 5 to 12 provinces. For the average GPA score among provinces, the highest GPA score of students of the 2013 academic year was obtained by the student from East Kalimantan (3.41), while students from East Java only obtained 3.06. In the 2014 academic year, the highest average GPA score was obtained by the student from Central Java, while the students from East Java obtained 3.18. It can be concluded that the students of East Java have not shown the optimal rate of GPA score because the students' academic performance was varied widely.

From the district data, the students of the 2013 class were 76% while the students of the 2014 class were 68%. This decreasing number showed that the students' origin distribution expands. In the class of 2013, most students came from Sidoarjo (24), Surabaya (16) and Lamongan (11), while the students of the class of 2014 were from Sidoarjo (34), Mojokerto (17) and Gresik (16). In the class of 2013, the highest GPA score was obtained by a student from Jember with the GPA score 3.58, followed by a student from Lumajang with 3.56, and a student from Pemalang with 3.51. on the other hand, the students from Sidoarjo obtained average GPA score 3.30. There were changes in the highest average GPA achiever in the class of 2014. They were students from Bima (3.60), Kebumen (3.59) and West Lampung (3.56).

Based on the data of the students' former high school, in the class of 2013 and 2014, there was an increase in the number of students' high schools from 109 to 12 indicating that the distribution is getting wider. Most of students in the class of 2013 were from SMANOR Jatim (12 students), SMAN 1 Kedungwaru (4 students), and 3 students each from SMAN 2 Lamongan, SMAN 18 Surabaya, SMAN 1 Srengat and SMAN 1 Bangil. In the class of 2014, there were 6 schools in the top five students' origin, namely UPT SMANOR Jatim (11 students), SMAN 4 Sidoarjo (6 students), SMAN 1 Mantup (5 students) and 3 students from SMAN 1 Manyar, SMAN 1 Gedangan and SMAN 1 Taman. All schools are located in East Java.

From the analysis of factors affecting the students' GPA score, it was found that students' major at their high school significantly contributed to the achievement of difference in GPA scores. However, the athlete status and participation in the organization and the economic levels did not show any significant difference.

4. Conclusion and Suggestion

4.1 Conclusion

The students were mainly originated from East Java (93%) in which 68% of them were from Sidoarjo. The determinant factor leading them to get high GPA scores was their high school majors. On the other hand, the athlete status—the participation in the organization and economic levels showed no significant difference.

4.2 Suggestion

The writer suggests other researchers to:

- Conduct evaluation towards students' GPA scores based on the province, district, and previous school in order to determine the potential area.
- Formulate instruments to identify other factors affecting the students' achievement.
- Determine the target area and school in recruiting new students.

References

[1] Michael S L, Merlo C L, Basch C E, Wentzel K R and Wechsler H 2015 Critical Connections: Health and Academics *J. Sch. Health* **85** 740–758

[2] Cimini G, Gabrielli A and Labini F S 2014 The Scientific Competitiveness of Nations *PLoS One* **9**

[3] Tam M 2001 Measuring Quality and Performance in Higher Education *Qual. High. Educ.* **7** pp 47–54
[4] Ennis C D 2013 Reimagining Professional Competence in Physical Education *HHS Public Access. J Strength Cond Res* **19** 662–72
[5] Druckman J N, Gilli M, Klar S and Robison J 2014 The Role of Social Context in Shaping Student-Athlete Opinions *PLoS One* **9** pp 1–14
[6] Ali M 2012 *Metodologi Penelitian dalam Olahraga* (Surabaya: Universitas Negeri Surabaya)

IOP Conf. Series: Materials Science and Engineering **296** (2018) 012014 doi:10.1088/1757-899X/296/1/012014

Hydration status in adolescent runners: *pre and post training*

K Ashadi, D N Mirza and G Siantoro

Department of Sports Coaching Education, Faculty of Sports Science, Universitas Negeri Surabaya, Indonesia.

kunjungashadi@unesa.ac.id

Abstract. The adequacy of body fluids is important for athletes in supporting performance. The purpose of this research was to determine the hydration status of athletes before and after training. The study was a qualitative descriptive by using random sampling. All athletes were trained for approximately 60 minutes. And they were asked to analyze their body fluid pattern routinely. Data were obtained through urine color measurement. The urinary was taken at pre and post training and was immediately assessed in the afternoon. Based on pre-training urine samples, a mean of urine color scale was 3.1 point. It meant that only 31.2% of the athletes were in dehydrated condition. However, after exercising, urine color index showed scale 4.1. And 62.5% of the athletes experienced dehydration. The results showed that there was a significant change in hydration level before and after training. It can be concluded that training for a long time increases the risk of dehydration. It is important for athletes to meet the needs of body fluids in order to avoid functional impairment in the body during sports activities.

1. Introduction

Based on preliminary studies, it is known many athletes are less aware of the importance of hydration for the body. Athletes who do not drink much water and believe the myth that drinking water can cause stomach cramps during exercise are considered as ones who do not have sufficient knowledge and awareness of hydration during exercise. The level of knowledge and education of athletes about hydration affects the hydration pattern during exercise [1]. Information and direction to athletes about their own needs of body fluid will influence them to exercise safely and perform well [2].

Hydration or adequacy of body fluids is an important component that should not be ignored by athletes. Inside the body, water serves to maintain blood flow and pressure, to lubricate the joints and tissues, to facilitate the digestion process and many other functions. Our body consists of 70% fluids. Therefore, it is important to meet the body fluids well [3]. Dehydration may cause people thirsty and decreased urination, damage to physical appearance, muscle pain, increased pulse and respiration, and weaken the body. Dehydration can weaken the limbs, low tension, and, even, make someone fainting [4]. If dehydration lasts for a long time, there is a shift of intracellular fluid out of cells. To overcome that problem, it takes a long time [5]. The body will reach its limit when 20% of water in the body is lost. This condition happens when all organs of the body will not work and can result in death [6].

The purpose of this study was to determine the hydration status of athletes at pre and post-exercise by using urine color chart. Measurement and assessment of urine colors can be used to detect and know the hydration status in hot environments [7,8].

2. Methods

Sixteen running athletes of Pasuruan, consisting of 8 men and 8 women, participated in one-day testing. Data retrieval was conducted in one routine training session that was held in the afternoon. Previously, athlete's urine was taken and after doing exercise for 60 minutes, another sample of the participants was taken for the second time. During the exercise period, the athletes were asked to perform a routine water consumption pattern they did earlier. The sample of urine color was matched to the urine color indices to determine the hydration status of the athlete. To determine the level of athlete's knowledge of hydration, the athlete is asked to fill out a questionnaire which consists of 30 questions related to hydration.

3. Results and Discussion

The mean (± *SD*) physical characteristics of the 16 adolescent athletes were age 16 ± 4.46 years old, height 162 ± 26.96 centimeter, weight 51 ± 24.77 kilogram. Results of urinary indices before and after training presented in Table 1.

Table 1. Adolescent runner results of urinary Indices of hydration status

	Urine Colour	Pre Training	Post Training
Hydrated	1-3	68,75 %	37,5 %
Dehydration	4-5	31,25 %	43,75 %
Severe Dehydration	6-8	-	18,75 %

The majority of the athletes have good body fluid adequacy prior to exercising activities. Urine samples in pre-training were lighter than those of post-training. Most of the samples had hydrated condition in post-training. It is shown by the increasing number of dehydration. In the post-training, few of samples had severe dehydration.

The average of urinary indices of hydration status differ before and after training. It is showed in Table 2.

Table 2. Pre and Post Training Comparison of urinary Indices of hydration status.

	Urine Colour Average	Status
Pre Training	3,06	Hydrated
Post Training	4,06	Dehydration

In pre-training session, some of the samples showed hydrated condition and urine color was light. Dehydration happened in post-training session. The urine color was darker in this session although the majority of the participants have drunk much water during the training session. This shows that many athletes experience dehydration in doing sports activities.

The result of the questionnaire is presented in Table 3

Table 3. Hydration knowledge test result.

Category of knowledge	Amount	Score Scale	Percentage
Very Good	1	25-30	6,25 %
Good	4	19-24	25 %
Fair	7	13-18	43,75 %
Less	4	7-12	25 %
Poor	-	-	-

The majority of the athletes had a fair knowledge of hydration with an average score 16.25 point from total 30 points. The number of athletes had good knowledge of hydration with less category. Only one athlete had very good knowledge of hydration. The data above show that the average athlete has sufficient knowledge of hydration.

Dehydration is the body condition loses too much water and electrolytes (body fluids). The loss of electrolytes with water is often poorly realized. The term dehydration is literally water shortage. The loss of water is always accompanied by loss of electrolyte [9]. Athletes tend to become dehydrated during exercise [10]. Therefore attention should be given to fluid supplementation and individualization of fluid intake for each athlete [11].

Under normal circumstances and activities, the amount of liquid released is very different. An athlete who is practicing or competing at the higher temperature will experience more to lose more fluid and electrolyte. The dehydration consumes water at a rate of 1 - 2 litres in 1 hour. When the body fluid reduced 1 liter, it will cause an increase in internal body temperature of 0.3 degrees Celsius [12]. The evidence suggests that dehydration has a negative impact on physical performance for activities lasting more than 30 minutes [13]. Dehydration can impair distance, accuracy, and distance judgment during sports performance [14-15]. Dehydration can damage physical and mental performance, as well as decrements of greater performance in hot environments and in durable exercise [12]. Finally, it can be concluded that dehydration causes worse cognitive function and motor skills [16]. Therefore, an athlete must keep his / her good condition to achieve optimal sports performance.

Changes in hydration status during exercise can be estimated from changes in body mass. Sweat levels can be estimated if the body fluid intake and urinary loss are measured. Individual drinking strategies will take into account the hydration status of pre, during and after practice [17]. Adequate water intake is a vital issue in sports nutrition. 2 liters loss of water that occurs in 90 minutes of work or heavy exercise in hot environmental conditions takes more than 6 hours to recover [18].

So it can be said that the benefits of water for health is very much. Water has a vital role in human life. One of the benefits of drinking water is to lose weight. It also helps maintain body fluid balance, control calories, and fuel muscle, make brighter skin and improve kidney function. In addition, drinking water can increase productivity, prevent pain and heart disease and improve brain and digestion [6]. Therefore, athletes should drink 400 to 600 ml of water 10 - 15 minutes before exercise [3].

Coaches and coaching staff have important roles to educate athletes about the adequacy of body fluids before exercise. This is due to the high risk of dehydration during exercise [19,20]. Based on the data above, it is known that it is important for athletes to maintain body fluid adequacy before, during and after exercise.

4. Conclusion

Long sports activities increase the risk of dehydration for athletes. Therefore, it is important to consistently maintain body fluid adequacy during exercise.

References

[1] Magee P J Gallagher A M and McCormack J M 2017 High Prevalence of Dehydration and Inadequate Nutritional Knowledge Among University and club level athletes *International Journal of Sports Nutrition and Exercise Metabolism* **27 (2)** pp 158–68 https://doi.org/10.1123/ijsnem2016-0053

[2] Lopez R M 2012 Exercise and Hydration *Strength and Conditioning Journal* **34 (4)** pp 49– 54 https://doi.org/10.1519/SSC.0b013e318262e1d2

[3] Ashadi Kunjung 2015 *Pentingnya Hidrasi Bagi Atlet* (Surabaya: Unipress)

[4] Je'quier and Constant 2009 *Water As an Essential Nutrient: The Physiological Basis of Hydration* http://:.nature.com.ejcn.journal.v64.n2.fu ll.ejcn200 9111a.html//

[5] Syaifuddin 2010 *Anatomi Fisiologi* (Jakarta: Kedokteran EGC Unipress)

[6] Rinzler 2006 *Nutrition for Dummis: 4th Edition* (Indiana: Wiley publishing, Inc)

[7] McKenzie A L Muñoz C X and Armstrong L E 2015 Accuracy of urine color to detect equal to or greater than 2% body mass loss in men *Journal of Athletic Training* **50 (12)** https://doi.org/10.4085/1062-6050-51.1.03

[8] Hahn R G and Waldréus N 2013 An Aggregate Urine Analysis Tool to Detect Acute Dehydration *International Journal of Sports Nutrition and Exercise Metabolism* **23 (4)** pp 303–11 https://doi.org/10.1123/ijsnem.23.4.303

[9] Wiarto 2013 *Fisiologi dan Olahraga* (Yogyakarta: Graha Ilmu)

[10] Arnaoutis G, Kavouras S A, Angelopoulou A, Skoulariki C, Bismpikou S, Mourtakos S, and Sidossis L S 2015 Fluid balance during training in elite young athletes of different sports *HHS Public Access. J Strength Cond Res* **29 (12)** pp 3447–52 https://doi.org/10.1519/JSC.0000000000000400

[11] Webb M C, Salandy S T, and Beckford S E 2016 Monitoring hydration status pre- and post-training among university athletes using urine color and weight loss indicators *Journal of American College Health : J of ACH* p 8481 https://doi.org/10.1080/07448481.2016.1179195

[12] Maughan and Shirreffs 2010 Dehydration and rehydration in competitive sport *Scand J Med Sci Sports* **3** pp 40-47 doi: 10.1111/j.1600-0838.2010.01207.x

[13] Carlton A and Orr R M 2015 The effects of fluid loss on physical performance: A critical review *Journal of Sport and Health Science* https://doi.org/10.1016/j.jshs.2014.09.004

[14] Smith M F, Newell A J, and Baker M R 2012 Effect of Acute Mild Dehydration on Cognitive-Motor Performance in Golf *Journal of Strength and Conditioning Research* **26 (11)** pp 3075–80 https://doi.org/10.1519/JSC.0b013e318245bea7

[15] Bardis C N, Kavouras S A, Arnaoutis G, Panagiotakos D B, and Sidossis L S 2013 Mild dehydration and cycling performance during 5-kilometer hill climbing *Journal of Athletic Training* **48 (6)** pp 741–7 https://doi.org/10.4085/1062-6050-48.5.01

[16] Matthew H Kavitha M and Rabindarjeet S 2015 The Effects of Dehydration on Skill-Based Performance *Internasional Journal of Sports Science* **5 (3)** pp 99-107

[17] Giriwijoyo et al 2007 *Ilmu Kesehatan Olahraga* (Bandung: UPI press)

[18] DuniaFitnes 2015 *Alasan lain mengapa minum air putih itu penting* https:/duniafitnes.com/healthy/alasan-lain-mengapa-minum-air-putih-itu-penting.html.

[19] Thigpen L K, Green J M , and O'Neal E K 2014 Hydration profile and sweat loss perception of male and female division II basketball players during practice *(Journal of Strength and Conditioning Research* **28 (12)** pp 3425–31 https://doi.org/10.1519/JSC.0000000000000549

[20] Williams CA and Blackwell J 2012 Hydration status, fluid intake, and electrolyte losses in youth soccer players *International Journal of Sports Physiology and Performance 7 **(4)*** pp 367–74 https://doi.org/2011-0094 [pii]

IOP Conf. Series: Materials Science and Engineering **296** (2018) 012015 doi:10.1088/1757-899X/296/1/012015

The use of mobile learning application to the fundament of digital electronics course

L Rakhmawati and A Firdha

Department of Electrical Engineering, Universitas Negeri Surabaya, Indonesia

lusiarakhmawati@unesa.ac.id

Abstract. A new trend in e-learning is known as Mobile Learning. Learning through mobile phones have become part of the educative process. Thus, the purposes of this study are to develop a mobile application for the Fundament of Digital Electronics course that consists of number systems operation, logic gates, and Boolean Algebra, and to assess the readiness, perceptions, and effectiveness of students in the use of mobile devices for learning in the classroom. This research uses Research and Development (R&D) method. The design used in this research, by doing treatment in one class and observing by using Android-based mobile application instructional media. The result obtained from this research shows that the test has 80 % validity aspect, 82 % of the user from senior high school students gives a positive response in using the application of mobile learning, and based on the result of post-test, 90, 90% students passed the exam. At last, it can be concluded that the use of the mobile learning application makes the learning process more effective when it is used in the teaching-learning process.

1. Introduction

Mobile learning (*m*-Learning) is a new paradigm in the world concern of erudition. This learning model appears to keep the development of the world of information and communication technology, especially information technology and mobile communication which is very rapid. In addition, it cannot be denied at this time, a mobile communication device is one device that is attached to the daily animation of the actors such as instructor and scholar [1]. M-Learning should have the power to connect to other equipment, especially computers, to present encyclopedias information and to realize bilateral communication between teachers and learners. M-Learning is a unique learning because the user can access code of learning materials, referrals and applications related to learning, whenever and wherever [11]. This will increase attention to learning materials, make learning pervasive, and can encourage learner's motivation to lifelong learning. In addition, compared to conventional learning, m-Learning allows for more chance for collaboration and informal interaction among learners [2]. M-learning is, at the present, more innovative and student-centered, concerning the use of mobile devices, like cell phones, Personal Digital Assistants (PDAs) and smartphones.

M-learning can be defined as "The intersection of mobile computing and e-learning: accessible resources wherever you are, strong search capabilities, rich interaction, powerful support for effective learning, and performance-based assessment. E-Learning is independent of location in time or space." [1]. Based on the definition, mobile learning is a learning model that utilizes information and communication technology. [10] In the concept of learning, mobile learning brings the benefits of the availability of teaching materials that can be accessed at any time and the visualization of interesting materials. It is important to be noted that not every teaching material is suitable for mobile learning. The study of digital electronics is fundamental in electrical engineering courses that need to be taken especially for first to the third year of vocational engineering students. Generally, in these courses, classical teaching uses textbooks as the main element. However, the use of textbooks is very limited, too flat and static [3]. The basic knowledge of digital electronics is a lot of things that the students should memorize and understand. The students do not refer to the books everywhere as it is large, heavy and cannot bring everywhere. The independence of time and place becomes an important factor that is often emphasized. However, in traditional e-Learning, the minimum requirement remains a PC that has the consequence that the independence of time and place is not fully met. This independence cannot yet be fulfilled with the use of notebooks (portable computers) because of the real time and place. It means that a person can learn wherever he can access to learning materials [4]. Therefore, in this new technological context, it is important to develop new m-learning tools for these courses. A new m-learning framework for the digital electronics course was developed at Department of Electrical Engineering of Universitas Negeri Surabaya. The initial course material has been adapted to the *m*-learning environment and new contents were added, enriched with animations and interactive examples. Besides the traditional theoretical lessons and laboratory practices, we propose the use of this m-learning system to reach a wider blended learning methodology.

This paper is organized as follows; in Section II, we introduce related works regarding fundamental digital electronics course and mobile learning projects, while the methodology is presented in Section III. Section IV is dedicated to the presentation of the results and discussion about developed mobile learning. Finally, in Section V, conclusions are presented.

2. Mobile Application Overview

E-Learning or electronic learning was first introduced by the University of Illinois at Urbana-Champaign using a computer-assisted instruction system and a computer called PLATO [5]. Some of the advantages of e-Learning compared with other learning is that it can be used anywhere at any time, most mobile devices have a relatively cheaper price than desktop PC prices, the size of the device is small and lightweight than a desktop PC, it is expected to include more learners because e-Learning utilizes technology commonly used in everyday life [2]. In e-Learning, the independence of time and place becomes an important factor that is often emphasized. However, in traditional e-Learning, the minimum requirement remains a PC that has the consequence that the independence of time and place is not fully met. This independence cannot yet be fulfilled with the use of notebooks (portable computers) because the real time and place independence means that a person can learn wherever he needs access to learning materials.

Mobile learning is one potential alternative to expand access to education [6]. However, there is not much information about the use of mobile devices, especially mobile phones, as a medium of learning. This is unfortunate given the already high level of ownership and usage levels that are underutilized for education [7]. In addition, there is currently very little development effort of mobile-based learning content based on widely accessible devices. Most of the content on the market is still dominated by entertainment content that has less educational aspect and most of it is the result of production from abroad which has different cultural background with our country [8]. This fact raises the need for more mobile, diverse, accessible, mobile-based content development. Factors that limit the utilization of m-learning are mostly related to the limitations of the device. Currently, most mobile devices have limited display screens, storage capacity, and limited power. M-learning also has a somewhat different learning environment with e-learning or conventional learning. M-learning, learners use more m-learning in spare time or idle time (idle time) so that the time to access learning is also limited.

Lack of m-Learning itself is actually gradually will be resolved, especially with the development of increasingly advanced technology [9]. The processor speed of the device is getting better, while the memory capacity, especially the external memory, is currently getting bigger and cheaper. The relatively small display screen can be overcome with the ability of the device to display the output display to TV or to the projector. Limited media input/output issues which consist of few buttons will be resolved by the presence of touchscreen technology (touchscreen) or virtual keyboard. Limitations in the availability of the power supply will be overcome by the utilization of practical, easy-to-obtain and easy-to-carry alternative resources, such as liquid batteries, human motion, solar power and so on.

3. Methodology

This type of research uses Research and Development (R & D). The development procedure follows Brog and Gall procedure [9] that can be done more simply by involving four main steps are 1. conducting an analysis of products to be developed; 2. developing the initial product; 3. expert validation and revision; 4. test the small-scale field and Product revisions. With research, the models are tested systematically, evaluated, improved to obtain specific criteria for the same effectiveness, quality, or standard. This study is the process of implementation of learning steps that are applied in the classroom and the level of creativity to make mobile media applications and student learning achievement as a result of the use of Android-based mobile application media.

Based on the results of needs analysis, the next step of research and development make the design of the product to be developed. The product design that will be developed in this research is designing and preparing the tools and software needed to make the learning media easy to understand. We need Computers/Laptops, Android OS Smartphone, Game Maker Studio Software to implement m-learning.

Validation of this design in the form of learning media android based mobile application taken lecturer by media experts from the electrical engineering department. The validation results are then analyzed and revised according to the validator's suggestion. Next, once the product design is validated through expert judgment or discussion forum, it will know the shortcomings and advantages of the media. The deficiencies will then be revised or revised according to the advice given by media experts. After making a revision of the product design, then the next step researchers conduct product testing in the form of learning media based mobile application android to students of class X Audio Visual Technique of Public Vocational High School (TAV SMK Negeri) 1 Jetis Mojokerto on basic electronic engineering competence.

The steps of development of Android-based mobile application as follows:

1. Prepare the necessary materials such as software, gmx format images, fonts, designs, and audio.
2. Install Game Maker Studio software, CorelDRAW software. Game Maker Studio is used to create mobile apps and CorelDRAW is used to create required images, such as display design drawings, and images that pertain to the material to be explained.
3. Create submenu along with the necessary buttons by using CorelDRAW as supporting software. The submenu view and its sequence scan on the flowchart that has been created. The page is then equipped with the necessary audio and animation.
4. Make forty multiple choice questions about the number system materials, logic gates and Boolean algebra all set to Game Maker and to design drawings, working on the material using CorelDRAW. Problem wrote manually in CorelDRAW then included in Game Maker. How to configure the problem in game maker is to create two objects (right and wrong) create two scripts. The first script to randomize and query in the queue, e.g. (room, go to (choose (Problem 1, Problem 2, Problem 3, Problem 4, Problem 5, Problem 6, Problem 7, Problem 8, Problem 9, Problem 10.) The second script is used to call the queue sequence and delete the one already in Call so that it does not enter the queue of questions; the exercise of the problem is then imported into the evaluation file that has been made before. After that, the file that has been successfully created is stored with the name of the electronic module in one folder in local disc C.
5. Publish mobile application using Game Maker to form the application format .gmx
6. Create designs and images of mobile apps using CorelDRAW.
7. Run the Game Maker Studio program in .gmx format converted into the .apk format using Game Maker program by configuring Game Maker program by configuring Game Maker program then format changed to .apk. Mobile apps that have become .apk format can be directly installed and run on all smartphones with android (gingerbread) operating system upwards.

4. Results and Discussion

The main objective of this project is the implementation of an m-learning tool for the teaching of the fundamental of digital electronics. The purpose of this solution is not to replace the traditional teaching methods, but to create a complementary tool in order to improve the student's interest, motivation, and the resources availability in any place, at any time. The *m*-learning digital electronic course consists of several interactive modules. The contents of the modules cover the following main topics: Number systems, logic gates, and Boolean algebra.

The contents were developed in order to put an emphasis on demonstrating and assisting each theoretical concept with a practical application. Another important feature is the use of images and multimedia animations. As an example, Figure 1a shows a material about the Boolean algebra. With this image, students can see the logic simplification. Several interactive figures have been developed for the understanding and motivation of the students. Figure 1b neither shows an interactive circuit for NOR gate. Students can change the input of the gate to show the output of gate that they want to see.

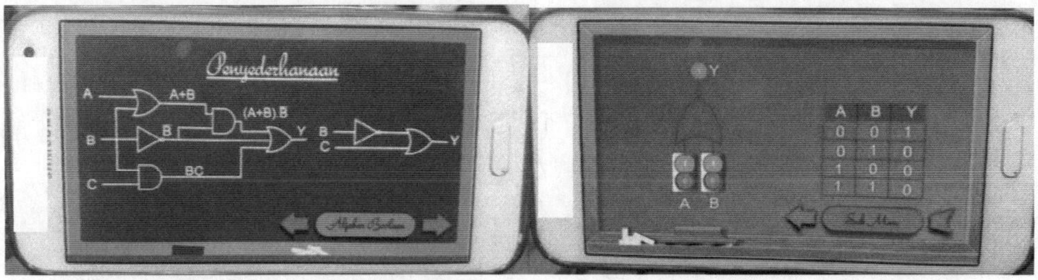

(a) (b)

Figure 1. a).Logic simplification; b). NOR gate's interactive circuit

One important feature associated with the figures and interactive animations is the evaluative quizzes, informing about the results to the questions. In a case of failure, the student must start again the lesson as shown in Figure 2.

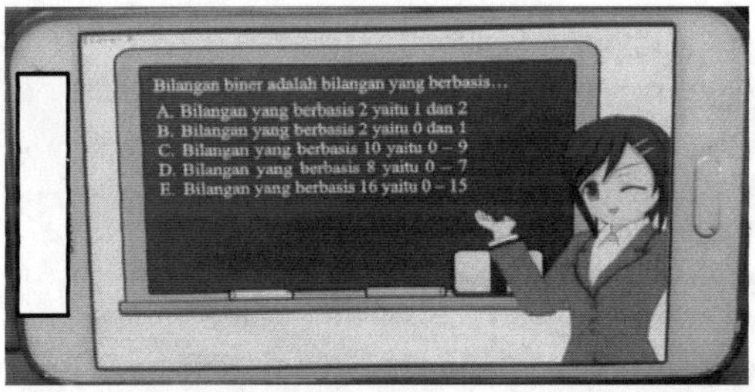

Figure 2. An example of the evaluative quizzes

To assess the readiness, perceptions, and effectiveness of students in the use of mobile devices for learning in the classroom, we have a methodology that mentions before. Based on Table 1, there are four categories of validation: format, design media, material, and language, for general information learning media developed with an average score of 80% are categorized as feasible or at a reasonable level of quality, so that the developed learning media can be used as a medium of learning with little revision. Student response is known through giving questionnaires to students who become research samples. The following data results of student responses to learning media can be seen in Figure 3, with 15 questions, the average is 82%.

IOP Conf. Series: Materials Science and Engineering **296** (2018) 012015 doi:10.1088/1757-899X/296/1/012015

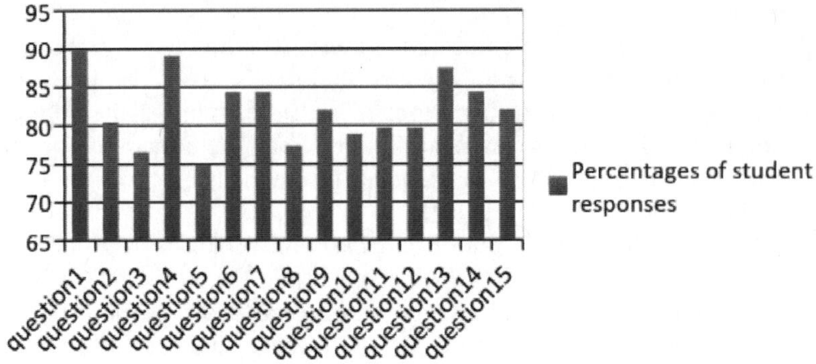

Figure 3. Results Diagram of student responses

Table 1. The results of m-learning validation

	Aspects	Percentages
Format	Systematic presentation	83 %
	Conformity of design	75 %
	The clarity of the use of letters	75 %
	The suitability of color display	75 %
	The ability of simulation in explaining material	91%
Design Media	Display design that served to help the students' understanding	100%
	Simulations are presented clearly, attractively and easily understood	100%
	Media design with a color display presented harmoniously	
	Ease of using the navigation menu	75%
	Giving audio in support of learning	75%
		75%
Material	Material compatibility with indicator	75%
	Truth material simulation with theory	91%
	Material is grouped according to the sequence	75%
	Material suitability with learning objectives	75%
	Material completeness with media	75%
Language	Language is easy to understand	75%
	Grammar	75%
	The language used can clarify the material delivered	75%

To know the extent to which students have attained the competence related to logic gate material students are given a post test in the form of multiple-choice questions. From the data completeness of student learning outcomes during the learning process by using learning media mobile application based on the android basic electronics engineering subjects in class X TAV SMK Negeri 1 Jetis Mojokerto. From the result of recapitulation of student value can be presented diagram as follows: From the results of the data above is known that the level of mastery of students' learning achievement class X TAV SMK Negeri 1 Jetis Mojokerto using learning media mobile application based on Android on basic electronics engineering subjects of 90.90% and at the student level is not complete 9.09%.

5. Conclusions and Future Work

The use of m-learning tools, if correctly conceptualized and built, it constitutes an efficient complementary tool to the traditional teaching methods. So, in this context, it was presented the development of a mobile learning framework for digital electronics course. The theoretical content of the course was structured in several interactive modules. At the end of each module, there are quizzes allowing students to be evaluated by themselves or by course teachers. The result obtained from this research shows that 80 % validity aspect, 82 % of the user from senior high school students give a positive response in using the application of mobile learning, and based on the result of post-test, 90,90% students passed the exam. At last, it can be concluded that the use of the mobile learning application can make the learning process more effective as used in the teaching-learning process. As future work, the teacher can receive automatic SMS from the student's mobile phone. This allows the teacher to have access to the student's progress and comments about the images and multimedia animations.

References

[1] Quinn C 2000 *mLearning. Mobile, Wireless, In-Your-Pocket Learning* (Linezine)

[2] Holla S 2012 Android Based Mobile Application *Int. J. Comput. Trends Technol.*

[3] Madeira R N, Pires V F, Dias O P and Martins J F 2010 Development of a mobile learning framework for an analog electronics course *Proceeding of IEEE Global Engineering Education Conference* pp 561–7

[4] Francisco, Molina J and Castilo 2014 Drivers of Mobile Application *Int. J. E-Services Mob. Appl.*

[5] Shih Y E 2005 Language in Action: Applying Mobile Classroom in Foreign Language Learning-Advanced Learning Technologies *Fifth IEEE International Conference on ICALT*

[6] Borg W and Gall M D 1983 *Educational Research* (Longman)

[7] Hafizul F H and Samsudin K 2012 Mobile Learning Environment System (MLES): The case of Android-Based Learning Application on Undergraduates' Learning *Int. J. Adv. Comput. Sci. Appl* 3 pp 63–6

[8] Liu Z and Zan H 2010 Research Perspective and Development of M-learning *Proceedings of 2nd International Conference on Networking and Digital Society* pp 282–5

[9] Safaat N 2014 *Android Pemrograman Aplikasi Mobile Smartphone dan Tablet PC Berbasis Android* (Bandung: Informatika)

[10] D. Frohberg, C. Göth, and G. Schwabe 2009 *Mobile* Learning projects - a critical analysis of the state of the art *Journal of Computer Assisted Learning* 25 (4) pp 307-331

[11] Yousef M and Hamideh Z, 2013. *Mobile Learning for Education: Benefits and Challenges. International Journal of Computational Engineering Research* 03 (6) pp 93-101

IOP Conf. Series: Materials Science and Engineering **296** (2018) 012016 doi:10.1088/1757-899X/296/1/012016

The implementation of role play in education of pre-service vocational teacher

L Hidayati[1] and P Pardjono[2]

[1] Faculty of Engineering, Universitas Negeri Surabaya, Jalan Ketintang Surabaya, Indonesia.
[2] Faculty of Engineering, Yogyakarta State University, Jalan Karangmalang Yogyakarta, Indonesia.

lutfiyahhidayati@unesa.ac.id

Abstract. This article aims to: 1) explain the enhancement of students' understanding about models of learning through role-play approach, and 2) find out the students' response to the implementation of role-playing. Role-play is a technique that allows students to explore realistic situations by interacting with other people in a managed way in order to develop experience and trial different strategies in a supported environment. This is a Classroom Action Research consisting 2 cycles, held in the Apparel Pre-service Teacher Education Programs, Universitas Negeri Surabaya in 2016, in the subject of Teaching and Learning Strategy. The number of research subject is 54 students. Data collected by giving test and questionnaire. The data analyzed by descriptive analysis. The result states that:1) Role-play succeeds to improve student understanding of learning models; 2) The student's' response shows that the role-play is an exciting activity, provide an opportunity for students to be creative, and easy to be applied. Role-play has flexibility for implemented at the university level, accelerates the student's understanding of learning materials, trains the students' independence, responsibility, awareness to others as important provisions to become a teacher.

1. Introduction

The teacher is the most important variable in school effectiveness [1] and also the most important factor in student learning [2,3]. Future teacher, described as pure water, which clarifies [4]. The main goal of achieving a successful education is the role of teachers and educational institutions in education [5]. Teacher education plays an important role in teaching preparation, not only improving their understanding and skills, but also increasing their chances of surviving in their profession. [6]. Teaching and learning activities should be recognized as a process of continuous reconstruction of experience [7]. Teaching and learning activities conducted at the universities. The universities tend to focus on pedagogical strategies based on learning outcomes of the students. The learning outcomes that are highly valued by the students included creativity, problem-solving skills, decision-making skills, communication skills, team building, and leadership skills [8]. Universities are expected to design and offer courses that can nurture the aforementioned attributes in students. Pedagogical strategies used in university identify a number of essential components that can facilitate the desired learning outcomes. Most of these active learning techniques require the enhanced involvement of the students in comparison to the traditional approaches.

Pre-service teacher education programs aim to prepare graduates to become quality teachers equipped with pedagogical practices that will serve to meet the increasing demands associated with the teaching profession [9]. A vocational education and training (VET) teacher should have (1) cross-linked pedagogical context knowledge; (2) prior theoretical knowledge, which he/she uses to analyze pedagogical situations and to implement appropriate strategies, (3) methods to apply theoretical knowledge and (4) pedagogical competence [3].

Based on the author's 16 year-teaching experience in pre-service teacher education program, some of the following issues are found in the subject of learning strategies: (1) As a prospective teacher, students must understand well with the characteristics of various approaches and models of learning, but in the reality, they have less enthusiastic about educational subjects; (2) The condition is probably caused by teaching materials on learning methods delivered with tutorials by lecturers as well as material presentations by students; (3) The process of understanding teaching materials about learning methods occurs abstractly; and (4) Students are given challenges in small portion. To solve the problem so that the teaching-learning becomes better, the role-play strategy is selected to implement in the subject of the learning strategy. The selection of role-playing as a solution is not without consideration. Based on the study of theory and research results, it proved that role play has many benefits towards the learning process in universities, particularly in pre-service teacher education programs.

Role-play is a pedagogy used in a wide variety of contexts and content areas [10]. The guidelines for the role-play are usually modeled on realistic criteria, so the students can get as close to the real thing as possible. Research on role-play's effectiveness and best practices exist as far back as the 1970s; recently, however, role-play has been touted as a better-suited tool for the needs of today's college student than more traditional teaching methods [11,12].

Role-play has been shown to be effective in reaching learning outcomes in three major learning domains: affective, cognitive, and behavioral [10,13]. By making students take on the role of another person, they practice empathy and perspective taking. It can lead to more self-reflection and awareness on the part of the student [14,15]. When students take the skills, they have learned the theory and put them in practice. It creates a deeper cognitive link to the material, making it easier for students to learn [16]. This pedagogical tool has been used in various fields [14]. Numerous studies have reported that role-play has positive effects for students' learning; for example, the method provided the opportunity to get a deeper understanding of an issue and stimulated further interest in the subject [10,17]. Aspegren's literature review on how medical student best learning communication skills revealed that an experiential training produced much better results than a simple one-way instruction [18]. The use of role-playing has been shown to be better preparation for teachers and construction managers [19]. Not only increasing student engagement, it also increases their knowledge retention [14]. Acting role-plays allow students to practice newly-developed skills by simulating a scenario where that skill may be required. 'Almost Real Life' is a role-play which is as close to the real experience as possible. This type of role-playing enables students applying all their skills in a real situation [10]. The use of role-playing as a learning strategy in higher education has been used in problem-based learning [20]. It is also considered to be a useful strategy in teacher education. The value of role-playing in teacher education lies in a pedagogical-based scenario that results in specific learning designed to create a realistic learning experience for participants [13]. The use of role-playing also facilitates a more comprehensive learning experience for teacher education students compared to more traditional cognitive focus approaches.

2. Method

This study is a Classroom Action Research which is consisting of 2 cycles. Each cycle consists of 4 steps: planning, acting, observing; and reflection. The subject of this research is 54 students of apparel pre-service teacher education programs of Surabaya State University in 2016, in the subject of teaching and learning strategy. The data are collected by testing and giving questionnaire. The data are analyzed by descriptive analysis.

3. Results and Discussion

The role-playing model used in the study consists of nine steps: warm up the group, select participants, set the stage, prepare the observers, enact, discuss and evaluate, reenact, discuss and evaluate, and share experiences and generalize [21]. A unique model was in the fourth stage – preparing the observers. Students who did not participate in the role-playing scenario observed one of the players so that all members of the class became directly involved in the learning process. Then, during the sixth stage - discuss and evaluate - non-participating students were asked to report out on their reaction to the role that was played. In large university classes, students were more likely to become and remain engaged in the role-play if they have been given a direct assignment to observe and to critique one particular player than if they were simply present in the classroom when other students were demonstrating the role-playing activity.

Figure 1 and 2 below showed the students' learning achievement on learning models through role-playing in cycle I and cycle II. The improvement of learning outcomes here including the cognitive level of students in understanding the material about learning strategies, measured by the formative test.

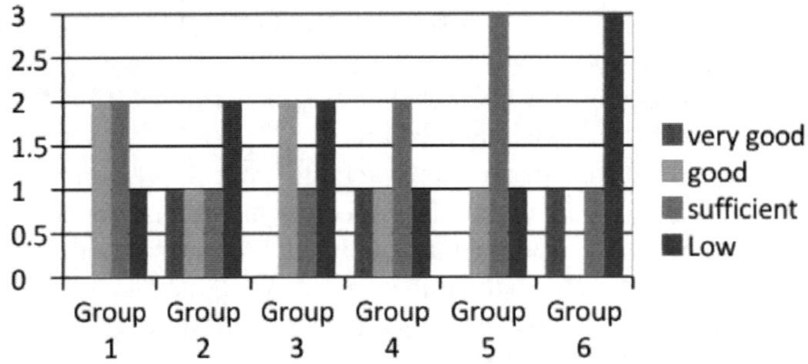

Figure 1. Students' learning achievement through role-playing in cycle I

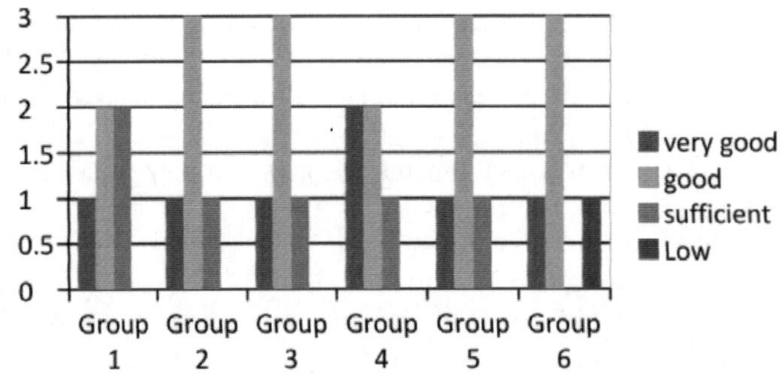

Figure 2. Students' learning achievement through role-playing in cycle II

The comparison of the result from the two cycles identified that there was an enhancement of students' learning achievement in the mastery of learning models through role play. It can be seen in each group that there were changes in the number of low-ability students to be better, from the first cycle to the second cycle. The number of students with excellent qualifications was increased from 10% to 24%. Good qualifications increased from 24% to 53%, or in general, there was an increase from 34% to 77%. Although this enhancement has not been completely able to increase the ability of all students to enter in the category of good and very good, at least it has exceeded the minimum mastery learning standards. In other words, by performing 2 cycles, role-playing succeeded to increase student learning achievement of the learning models. To deal with a number of students who did not achieve mastery learning, it became necessary to follow up in the next cycle.

The result of the student's response to role-playing showed that the learning model:(1) made students easier in understanding learning material than learning by themselves (100%); (2) being able to cultivate the attitude and to train the student's responsibility (90%); (3) was easy to implement in university level (90%); (4) cultivated the attitude and trained student's awareness (100%); (5) being able to cultivate the attitude and to train the student's independence (75%); (6) trained the students as a prospective teacher (100%).

The students stated that during the process of implementing the role-playing: (1) they have responsibility in acting as the scenario requirement (100%); (2) they provided to support their fellow players in the team during role-playing process (80%); (3) they contributed in bringing success to the group (100%); (4) The time available was sufficient to prepare and to perform the role-playing (75%).

Besides its advantages, during the implementation of role play, students encountered some disadvantages. Disadvantages were as follow: (1) Role-play required hard work of all parties; (2) Preparation for the implementation of role-playing would require a relatively more time allocation than any other strategies.

4. Conclusion

The role-playing pedagogy succeeded to improve student learning achievement in the learning models. The students' response to the implementation of role-playing was that it was easy to understand and to be applied in the learning process. It has flexibility for implemented in the university level, It accelerated the student's understanding of learning materials, It trained the students' independence and responsibility as an agent of learning, it trained awareness to others and it trained the students as a prospective teacher.

References

[1] Anderson T J, Barrick R K and Huhges M 1992 Responsibilities of Teacher Education for Vocational Teacher Professional Development Programs *J. Agric. Educ. Summer* pp 43–50

[2] Australian Council of Deans of Education. 2005 *Teaching tomorrow's teachers: ACDE submission to the House of Representatives inquiry into teacher education* (Canberra: Australian Council of Deans of Education)

[3] Barabasch and Watt-Malcolm B 2013 Teacher preparation for vocational education and training in Germany: a potential model for Canada *Comp. A J. Comp. Int. Educ.* **43**

[4] Kemdikbud 2013 *Menyiapkan Guru Masa Depan* (Jakarta: Kementerian Pendidikan dan Kebudayaan)

[5] Zamroni 2000 *Paradigma Pendidikan Masa Depan* (Yogyakarta: Bigraf Publishing)

[6] Darling-Hammons L (Ed) 2000 *Studies of excellence in teacher education: Preparation at the graduate level* (Washington DC: AACTE)

[7] Beck C and Kosnik C 2006 *Innovations in teacher education* (Albany: State University of New York Press)

[8] Biggs J and Tang C 2011 *Teaching for quality learning at university* (England: Open University Press)

[9] Bobbit L M, Inks S A, Kemp K J and Mayo D T 2001 Integrating marketing courses to enhance team-based experiential learning *J. Mark. Educ.* **22** pp 15–24

[10] Rao D and Stupans I 2012 Exploring the potential of role-playing in higher education: Development of a typology and teacher guidelines *Innov. Educ. Teach. Int.* **49** pp 427–36

[11] Rosa J A 2012 Marketing education for the next four billion: Challenges and innovations *J. Mark. Educ.* **34** pp 44–54

[12] Bransford J, Darling-Hammond L and LePage P 2005 *Preparing teachers for a changing world: What teachers should learn and be able to do* (San Francisco: John Wiley and Sons)

[13] Maier H W 2002 Role playing: structures and educational objectives *CYC-online* **36**

IOP Conf. Series: Materials Science and Engineering **296** (2018) 012016 doi:10.1088/1757-899X/296/1/012016

[14] Westrup U and Planander A 2013 Role-play as a pedagogical method to prepare students for practice: The students' voice Ogre utbildning **3** pp 199–210

[15] Sogunro O A 2004 Efficacy of role-playing pedagogy in training leaders: Some reflections *J. Manag. Dev.* **23** pp 355–71

[16] Johnson D W and Johnson F P 1997 *Joining Together: Group Theory and Group Skills* ed Allyn and Bacon (Boston)

[17] Ruhane L 2005 Bridging the divide between theory and practice: Experiential learning approaches for tourism and hospitality management education *J. Teach. Travel Tour.* **5** pp 33–51

[18] Aspegren K 1999 BEME Guide No. 2: Teaching and learning communication skills in medicine-a review with quality grading of articles *Med. Teach.* **21** pp 563–70

[19] Bhattacharjee S 2014 Effectiveness of role-playing as a pedagogical approach in construction education *50th ASC Annual International Conference*

[20] Chan Z C Y 2013 Role-playing in the problem-based learning class *Nurse Educ. Pract.* **12** pp 21–7

[21] Crow M L and Nelson L P 2015 The Effects Of Using Academic Role-Playing In A Teacher Education Service-Learning Course *Int. J. Role-Playing*

IOP Conf. Series: Materials Science and Engineering **296** (2018) 012017 doi:10.1088/1757-899X/296/1/012017

The impact of fashion competence and achievement motivation toward college student's working readiness on "*Cipta Karya*" subject

Marniati and S C Wibawa
Faculty of Engineering, Universitas Negeri Surabaya, Indonesia

marniati@unesa.ac.id

Abstract. This experiment aimed to know the rate of college student's working readiness of fashion's program study to perform 'Cipta Karya' related to cognitive readiness, manner readiness and skill readiness from a variable of fashion's workmanship and achievement motivation. The subject of the experiment was 43 college students who took Cipta Karya subject. Method of collecting data used questionnaire with five alternative answers to Likert ratio model. Data analysis technique used path analysis (double regression). The instrument validity test used product moment correlation while for instrument reliability used Alpha Cronbach's grade. The results showed (1) fashion competence was taking effect significantly on working readiness for 'Cipta Karya'; (2) achievement motivation is taking effect significantly on working readiness for 'cipta karya'; (3) both variables are positive. This means that fashion competence and achievement motivation have a positive effect on working readiness for 'cipta karya' performance.

1. Introduction

College student's success in showing a fashion show reflects their working readiness. They will not be awkward when jumping down to the industrial world because they have discovered their lacks. Slameti [1] stated that working readiness is formed from three aspects: (1) physical, mental and emotional condition; (2) motive and intention needs; and (3) other skills, knowledge, and definition which has been studied before.

College student's readiness in cognitive aspect is in the form of knowledge which is given to them according to the chosen area of expertise, such as design knowledge, pattern construction, sewing technique, textile craft knowledge, and women's clothing manufacturing techniques. While, generally, skill aspect is used to define quickness in applying the knowledge for practice or apprenticeship because this activity is the embodiment of a theory which was done for real. For example, clothing design, pattern making, cutting the materials, apparel sewing and finishing. So, it can be concluded that working readiness in skill aspect is a follow-up from the cognitive aspect of working readiness. This is in line with what Sudijono stated, "the outcomes of learning skill, in fact, is a follow-up from cognitive learning outcomes (understanding something) and effective learning outcomes (which will appear in form of tendency). Cipta Karya is an expertise subject study that can be taken at the end of the semester eight. This subject review concept and skill in apparel design according to the source of inspiration, trend mode, then set them up in a fashion portfolio, manifest them into apparel, and lastly do a fashion show, which fitted with the subject description.

> Cipta Karya Subject Description: assessment both concept and skill in design apparel making according to the source of inspiration and them, then showcasing them in an activity of fashion show, comprise: planning a fashion show, make a fashion portfolio, make and perform fashion show.

This condition demands college students to prepare themselves and their working readiness in order to apply the concept and all the subjects they have got from the first semester to the semester seventh.

Cipta Karya learning outcome is a measurement of achievement learning level with the purpose of teaching that wanted to be achieved. The learning outcome can be seen from practical aspect (the process) and from behavior's change of the students (the result). Pro-productivity or practice aspect refers to the increase of work productivity or college student's accomplishment during and after learning. While for the change of behavior aspect, it indicates an improvement in skill, knowledge, and manner of college students after the learning session done [2]. In addition, according to Mulyasa [3], a learning outcome is an over-all student's learning achievement which becomes basic competence indicator and degree of manner change concerned the students.

2. Method

This experiment uses a quantitative approach. Experiment population is Fashion Education's college students who take Cipta Karya as one of their subjects. The subject of the experiment is 43 college students who take Cipta Karya as one of their subjects. The method of collecting data is questionnaire with five alternative answers of Likert ratio model. Data analysis technique uses correlation analysis and double regression [8]. Instrument's validity test uses product moment correlation while instrument's reliability uses Alpha Cronbach's grade.

3. Results and Discussion

Hypothesis testing is used to know whether the result of data sample processing can be generalized to the population. Regression analysis is used to predict the causal link between a batch of independent variables with the dependent variable. The foundation to answer the problem about the effect between variables used the calculation outcome along with regression analysis and all at once for hypothesis testing.

Decision making for hypothesis testing used p-value with criteria if only $p > 0.05$, then H_0 will be accepted, means the obtained regression coefficient is not significant. And if only $p < 0.05$, then H_0 will be rejected, means the obtained regression coefficient is significant.

3.1 First Hypothesis Testing

Hypothesis H1 declared as the variable presumption of fashion expertise and achievement motivation simultaneously are affecting significantly towards working readiness. F-test result about contribution simultaneously effected from both independent variable towards working readiness is 13.676 ($p < 0,05$) is significant, so, it can be inflicted that fashion's expertise and achievement motivation have a significant effect upon working readiness.

3.2 Second Hypothesis Testing

Hypothesis H2 declared as the variable presumption of fashion expertise and achievement motivation partially affecting towards working readiness. This hypothesis is related to three regression coefficient test result inside regression equation. T-test result towards regression coefficient of fashion expertise to working readiness is 0.386 which is significant ($p < 0,05$). T-test result towards regression coefficient of fashion's expertise to working readiness is 0.338, which is significant ($p < 0.05$). T-test result towards regression coefficient of achievement motivation to working readiness is 0.361 which is significant ($p < 0.05$). The three T-test result is significant, so that can be concluded as data experiment support Hypothesis H2.

3.3 The Relationship of Learning Result of Fashion Expertise to Working Readiness
Subject learning of fashion's expertise practice is suitable with curriculum competence, the subjects in this experiment are mode design, pattern making in the constructive method, sewing technique, textile craft, and women apparel making. Experiment result proves that fashion expertise towards working readiness is significant. This result explains that diversity of working readiness can be explained directly by the amount of ability of fashion's expertise.

Wening [4] stated that fashion skill in practice on pattern making has working steps as (1) design analysis; (2) body measurement; (3) making archetype or deciding pattern making method that will be used to make the archetype; (4) perform archetype manipulation according to the design become pattern that is ready to be used; and (5) decipher the pattern into pieces.

Reducing the obstacle can be done with providing instructor's assistant so that the ratio of college lecturer and college students become smaller. The department also can prepare an interactive and progressive learning module into some level of materials. Another alternative is that college lecturer and department with industry partner's support apply teaching factory subject. This fact is compatible with experiment result by Hidayat [5] that was proven that teaching factory is effective to increase the competence of college student productive skill.

3.4 The Relationship of Achievement Motivation to Working Readiness
Achievement motivation is a driving force or an impetus inside oneself or students to achieve a level of learning achievement as high as possible. The motivation in this experiment is motivation which is inside the college students to try and change their manner to be better in achieving settled goals (learning purpose), i.e. achievement (attained accomplishment), recognition, responsibility, the work itself and advancement.

This experiment's conceptual method is formulated by achievement motivation can be developed through stimulation sourced from an efficient practice activity of Cipta Karya. Santrock [6] stated that motivation can be used as effectively in the class to increase the learning activity, both in cognitive, affective, and psychomotor to accomplish an achievement.

One of the biggest challenges that were faced by college lecturer is deal with a college student with low achievement motivation, as they are often to overlook the applied strategy by the lecturer, including teamwork and motivation to help achieve the targeted purpose. Whereas Schunk [7] stated that program motivation is designed to help student's learning wish and doing task well, because learning method motivated assume that motivation appear before, during, and after learning. As for that, the lecturer can give attention to the college students if they are not spirited. Here, lecturer's role is also needed to be responding or finding out as to why and what happened with her/his students, both from internal or external aspect.

Experiment outcome proves that achievement motivation is affecting positively significant towards working readiness in Cipta Karya practice.

4. Conclusion

According to experiment result as described above, then the core of conclusion can be pulled from experiment findings as follows:

1. Fashion expertise is affecting significantly towards working readiness of Cipta Karya show,
2. Achievement motivation is affecting significantly towards working readiness of Cipta Karya show,
3. Both variables are positive, this can be mean as skill in fashion expertise and achievement motivation of fashion's college students have a positive effect on working readiness for Cipta Karya activity.

IOP Conf. Series: Materials Science and Engineering **296** (2018) 012017 doi:10.1088/1757-899X/296/1/012017

References

[1] Slameto 2010 *Belajar dan Faktor-faktor yang Mempengaruhinya* (Jakarta: Bineka Cipta Sudijono)
[2] Henry S 2008 *Manajemen Sumber Daya Manusia* (Yogyakarta: Penerbit YKPN)
[3] Mulyasa 2009 *Menjadi Guru Profesional* (Bandung: PT Remaja Rosdakarya Offset)
[4] Sri W 2013 *Modul Pembuatan Busana dan Pengembangan Pembelajarannya* (Yogyakarta: Konsorsium Sertifikasi Guru)
[5] Hidayat M D 2011 Model Pembelajaran Teaching Factory untuk Meningkatkan Kompetensi Siswa dalam Mata Pelajaran Produktif *J. Ilmu Pendidik.* **17**
[6] Santrock J W 2011 *Psikologi Pendidikan* (Jakarta: Kencana)
[7] Schunk D H 2012 *Learning Theories An Educational Perspective (Teori-teori Pembelajaran: Perspektif pendidikan)* (Yogyakarta: Pustaka Belajar)
[8] Cohen J, Cohen P, West S G, and Aiken L S 2003 *Applied Multiple Regression/Correlation Analysis for the Behavioral Sciences Third Edition* (New Jersey: Lawrence Erlbaum Associate, Inc.)
[9] Buku Pedomam, 2013/2014. Fakultas Teknik Universitas Negeri Surabaya (UNESA)
[10] Sudijono, Anas.2005. *Pengantar Evaluasi Pendidikan*. Jakarta: PT. Rajagrafindo Persada
[11] Yamin Martinis, 2011. *Paradigma Baru Pembelajaran*. Jakarta: Gaung Persada (GP) Press

IOP Conf. Series: Materials Science and Engineering **296** (2018) 012018 doi:10.1088/1757-899X/296/1/012018

Small-sided games in football as a method to improve high school students' instep passing skills

M Ridwan, G Darmawan and Z Fuadi

Department of Sports Education, Universitas Negeri Surabaya, Indonesia

mochamadridwan@unesa.ac.id

Abstract. This study analyzed the influence of small sided games application toward increasing the learning result of instep passing in football. The research used one group pretest-posttest design. The data were obtained once a week for 135 minutes of small sided games and this activity had been held for four weeks with a final test in the final meeting. According to descriptive data result, there were increases of the mean. The data showed the increase of the application of small sided games resulted in not only the mean of the descriptive data but also the result of T-test. The significant of T-test is 0,000. It means less than 0,05 then the hypothesis Ha received and Ho rejected automatically. The presentation showed that 48,15% data is increasing by small-sided games application. The small-sided games were proven to be the right tool to increase instep passing football technique. We suggested to the apply that kind of games of football learning on physical education subject, especially for pre-university students.

1. Introduction

A good education will not only impact on itself but will also impact on the environment directly or indirectly. This shows that education does not only play roles in knowledge, intelligence, and skill but also for people who believe and pious to The Almighty God. To achieve the national education goals in the 2013 curriculum, there are compulsory subjects (group A) and compulsory subjects (group B), one of which includes subjects of Physical Education, Sport, and Health (PJOK). Physical Education, Sports, and Health are an important part of education as a whole, aiming to maintain students' physical fitness, motion skills, sportive learning, cooperation and many others that are also interconnected with national education goals. By following learning physical education, sports, and health, students are not only able to absorb the values contained in PJOK but also are expected to be fresher in doing activities at school. Students who spend part of their time receiving class lessons are expected to gain new energy and atmosphere when they receive PJOK learning. In the learning of PJOK, there is a variety of sports activities and games that contain lots of values and positive elements and is expected to help to develop not only psychomotor ability but also develop cognitive and affective ability of students. Football is also sport or relatively much-loved by schoolchildren, ranging from elementary to high school level. The students not only play football when classes begin but also in their break time and when there is a race between classes or something sort of this. Small-sided games are games that are played on a smaller field than the big game of 11 against 11. Game 3 versus 3, 4 against 4 is an example of small-sided games.

This review systematizes the SSG use as a training method and highlights the main restrictions of its application in soccer. All factors affecting the SSG are analyzed separately according to studies conducted in order to understand its importance in the response of athletes to training [1]. Even the game can be played even though the number of players is not the same [2]. Small side games have become part of the history of football in the United States. Many of our parents, our grandparents, and our ancestors immigrated to America by playing this beautiful game on the streets of their home country [2]. According to Kamp, small-sided games philosophize on the pleasure and freedom of expression with limited emphasis especially for players at development ages. It was the informality of the game and the various obstacles children had to overcome while playing, such as a number of games being played at once in the playground that bred generations of footballers who really felt the rhythm and heartbeat of the game [3]. Learning model of small-sided games can be done in open areas and, thus, are possible not to use soccer field. In addition to not using soccer field, this model can also be designed with a variety of variations, such as adding or reducing the ball or players in the field. These games are expected to make students react quickly, for example after receiving feedback, students are expected to provide or connect the bait to his team-mates quickly and accurately. In addition, the minimum time is needed to successfully redirect the kick was assessed [4]. Due to the small field and the contact restrictions on the ball, space is narrower. This will make students have to react quickly so that the ball is not taken by opponents and this will be expected to affect the ability of these students.

The advantages of small side models are: a) it's fun, kids practice with the ball better, b) playing soccer to learn soccer, c) learning by trying, d) the most effective technique, insight and communication are developed in situations of the game e) children naturally learn the situation of the game constantly and repetitively and with the frequent contact with the ball; f) small-sided games maximize direct involvement in football situations g) freedom in failure, h) increase creativity and spontaneity [4]. Since the problem occurs at State Senior High School 4 Sidoarjo. It has no field to deliver soccer material, thus, small-sided games model is appropriate to be applied because this model does not need to use field with the size of the soccer field. This research will take football material especially passing with the inner leg because passing with the inner leg is the most commonly used technique in the game.

2. Method

This research uses quasi-experimental research with quantitative approach. Experimental research is a rigorous study to determine the cause-effect relationship among variables [5]. Meanwhile, quasi-experimental research is the closest to the real experiment where it is impossible to control/manipulate all relevant variables. There must be a compromise in determining internal and external validity in accordance with existing constraints [6]. The study design is a design of how a study will be conducted [5]. The research design used in this research is one group pretest-posttest design, which is a research and there is pretest before treatment and posttest after treatment. The time of this research is 4 (four) meetings with duration of 3x45 minutes per meeting. The sample in this research is 37 students of class X in State Senior High School 4 Sidoarjo.

3. Results

After the research data about passing learning results with the foot in soccer is collected, then researchers do a calculation analysis on the results of research data by using the technique of IBM SPSS v20 computer program. There is an increase in pre-test results and post-test passing with a foot inside soccer learning in PJOK in Social class X-2 State Senior High School 4 Sidoarjo in the treatment group. This can be described further in the table 1 and 2:

Table 1. Results of Pre-Test Calculation and Post-Test Passing

	N	*Pre-Test*	*Post-Test*
Mean	35	1,54	2,29
Std. Deviasi	35	1,314	1,582
Varian	35	1,726	2,504
Maks	35	4	5
Min	35	0	0

Table 2 Calculation of Normality Test

	Pre-Test	Post-Test
p value	0,143	0.152
Significant	0,05	0,05
Category	Normal	Normal

The influence of this model of the small-sided game is the result of passing learning with a foot of inner part of a Social student X-2 at State Senior High School 4 Sidoarjo. It is used formula as follows:

$$\text{Percentage} = \frac{MD}{Mpre} \times 100\%$$
$$= \frac{0{,}743}{1{,}543} \times 100\%$$
$$= 0{,}4815 \times 100\%$$
$$= 48{,}15\%$$

4. Conclusion

Based on the results, it can be stated as follows: a). There is the influence of the application of small-sided games model to the learning passing results with a foot in soccer. This is proven by the result of t counted of 4,018> t table of 2.042 with a significant level of 0.05. b). The amount of influence of the implementation of small sided games model on the learning passing with a foot in the soccer is proven by the percentage of influence that is equal to 48.15%. As for suggestions that can be presented according to the results of research, discussion and conclusions are a). PJOK teachers are suggested to provide the right material and lesson plan especially for large soccer ball game material supported by other materials in order to improve the basic soccer techniques. B). Students are expected to develop the ability to play soccer in performing skill passing with a foot inside based on the analysis in this research.

References

[1] Marco A, Botelho G, Lag C, Macas V and Sampaio J 2012 A review on the effects of soccer small-sided games *J. Hum. Kinet.* **33** pp 103–13

[2] Snow S 2011 *US Youth Soccer Coaching Education Department* (Europe: Human Kinetics)

[3] Kamp J Van Der 2005 A field simulation study of the effectiveness of penalty kick strategies in soccer: Late alterations of kick direction increase errors and reduce accuracy *J. Sports Sci.* **24** pp 467–77

[4] Paul C 2006 Small sided games *Learning Through Play* (Sportfacilities & Media BV)

[5] Ali M 2012 *Metodologi Penelitian dalam Olahraga* (Surabaya: Universitas Negeri Surabaya)

[6] Barbara H, Marije G, Wendy P and Christian V 2010 Development of dribbling in talented youth soccer players aged 12-19 years *J. Sports Sci.* **28** pp 689–98

IOP Conf. Series: Materials Science and Engineering **296** (2018) 012019 doi:10.1088/1757-899X/296/1/012019

The characteristics of bioethanol fuel made of vegetable raw materials

Muhaji and D H Sutjahjo

Mechanical Engineering Department Universitas Negeri Surabaya, Indonesia

muhaji61@unesa.ac.id

Abstract. The aim of this research is to identify the most potential vegetable raw as the material to make a bioethanol fuel as the alternative energy for gasoline. This study used experimental method. The high-level bioethanol was obtained through the process of saccharification, fermentation and stratified distillation. ASTM standards were used as the method of testing the chemical element (D 5501, D 1744, D 1688, D 512, D 2622, D 381), and physical test (D 1613, D 240, D 1298-99, D 445, and D 93). The result of the analysis showed that from the seven bioethanols being studied there is one bioethanol from Saccharum of icinarum linn that has physical and chemical properties close to the standard of bioethanol. Meanwhile, the others only meet some of the physical and chemical properties of the standard bioethanol.

1. Introduction

The energy crisis has swept across the world either in a developing country or a developed country. This is caused by the availability of fossil fuels which is very limited and nonrenewable. Thus, it is needed to look for the alternative energy that renewable. As reported by the world energy board that about 82% of the world's energy needs come from petroleum, natural gas and coal [1]. With the current fuel consumption rate, it is predicted that fossil fuel sources will be depleted for petroleum in 50 years, natural gas in 65 years and coal in the next 200 years [2]. However, a team of scientists from UC Davis claims that petroleum will be exhausted 90 years before the availability of technology that replaces petroleum. According to Debbie Niemeier, one of the study participants, and scientists have proven to develop renewable energy production will take longer than expected as total energy consumption. The Niemeier study suggests the possibility of renewable energy will not inflict damage, development, economics, and ecosystems [3]. Bio-renewable fuels can be considered renewable if they can be used within the boundaries of some areas. Renewability is determined by the ratio between reproduction and consumption, not growth rate [4].

In terms of development and exploitation of renewable energy sources, it seems to be articulated by the European renewable energy council. The European EREC Council and government from the European Union agreed that the target of renewable energy usage is 20% by the end of 2020 and 100% by 2050 [5]. While in Indonesia, the use of renewable energy has been regulated by the government through Presidential Regulation no. 5 of 2006 [6] on the target of national and renewable energy utilization policy until 2025, namely: for fuel oil 20%, 30% gas fuel, 33% coal, 5% biofuels, 5% geothermal, others (nuclear, biomass, Water, solar, wind, and liquid coal 7%).

A review of the related literature study showed that the researchers focused more on the use of bioethanol and biodiesel fuel in the gasoline engines and diesel engines. In which, the internal combustion engine needs a fuel with a relatively high energy. Globally, bio-renewable fuels (bioethanol and biodiesel) for the current transportation sector are available to replace gasoline and diesel fuel [7]. Currently, bio-renewable fuels are around 10% of the energy used in the world, that the major contribution of bioethanol production is in North America, South America, and Central America [8]. In America, biofuel production accounts for three-quarters of the world's total production. It has been used as an automotive engine fuel by the automotive industry [9]. The use of this fuels in the gasoline engines and diesel engines can lower the amounts of particles and emissions in the exhaust [10]. The level of octane in the bioethanol is higher and its emissions is lower than gasoline and diesel fuels. Moreover, it also can be directly used in the machine without modifying the machine components [11]. Other studies [12] that investigated the impact of combining bioethanol fuel-biodiesel fuel-and oil on engine emissions, found that the ppm quantity in exhaust emissions decreased, and there was an increase in the amount of carbon dioxide (CO_2). While, the application bioethanol of amorphophallus paenifolius as the mixture of gasoline fuel to the exhaust emissions and engine performance on 4 stroke engine gasoline 1 cylindrical, it resulted that E20 fuel can improve performance and decrease CO and HC emissions [13].

Compared with fossil fuel, bioethanol fuel has the following special quality: 1) It has a high-octane level; 2) It is able to reduce the level of particle emission that endangers health (CO); 3) It is similar to gasoline, so its use does not require engine modification; 4) It is also regarded as renewable fuel that does not have CO_2 emissions; 5) It has low-cost production; 6) It is an alternative solution in facing globalization since it is eco-friendly; 7) It can be made at home and low-cost production than fossil fuel; 8) The physical characteristic of bioethanol in its pure form is soluble in all proportions with water and also with ether, acetone, benzene, and some other organic solvents [14,15]; 8) And its characteristic chemical function are dominated by the OH-group, which can readily help reactions in chemical industries such as dehydration, halogenation, ester formation, and oxidation [15,16].

The energy from vegetable materials has the potential to contribute the future energy scenarios. The production of bioethanol is now widely known to bring a good environmental, economic, strategic, and infrastructure impacts [17]. It is a type of alcohol, which is produced from the fermentation of agricultural products containing sugar starch such as, sugarcane, corn, wheat, sweet, fruit, and so on [18].

Indonesia is a tropical country that is rich with plants (vegetable) as a source of basic materials to make bioethanol and biodiesel, so the opportunity to produce bioethanol and biodiesel in large scale is very possible. Fuel derived from vegetable can be used for fuel, both in the form of bioethanol and biodiesel [19]. While the fuel from vegetables that must be continuously developed and already known its power for raw material bioethanol are *amorphophallus variabillis*, *musaceae*, *solanum licopercium*, *alocasia macrorrhiza*, *maranta arundinacealin*, sugar cane *saccharum officinarum linn*, and *suweg amorphophallus campanulatus*. As for biodiesel raw materials are palm, peanut, jatropha curcas linn, *gassibium herbaceum*, soybean, *ceiba pentandra*, coconut *cocos nucifera,* and *hellianthus annuus* [19,20].

The aim of this study is to analyze the characteristics of renewable fuel capability (bioethanol) from vegetable raw materials. From the results of this study, it is expected that we will get the type of potential material of biofuel, which can be used as a reference for the development of bioethanol fuel production in a larger scale.

2. Method

The high-level bioethanol obtained through the process of saccharification, fermentation, and stratified distillation up to 4 levels. Distillation is carried out by inserting the fermentation broth into the boiler for evaporation. Then, the steam goes into the distillation pipe. After that, the vapor is condensed through the condenser pipe. Finally, bioethanol will come out through the condenser hole.

The characteristics of chemical bioethanol capability that will be analyzed are the levels of its ethanol, methanol, water, Cu, Cl, and gum. While its physical properties include: heating value, density, viscosity, and flash point. The methods of testing the chemical properties are using ASTM standard (D 5501, D 1744, D 1688, D 512, D 2622, D 381) and its physical properties test is using ASTM standard (D 1613, D 240, D 1298-99, D 445, and D 93).

3. Result and Discussion

Laboratory test results of chemical properties of bioethanol from various types of vegetable raw materials are presented in Table 1, while the results of physical properties test are presented in Table 2.

Table 1. The results of chemical properties test of bioethanol from various types of vegetable.

No.	Material	Bioethanol content (%vol)	Methanol content (mg/l)	Water content (%vol)	Cu content (mg/kg)	CI content (mg/l)	Gum content (mg/100ml)
1	*Amorphopallus variabiliti*	94	0.0049	0.202	0.072	17.758	5.2
2	*Mucasea*	93	0.0088	0.653	0.052	39.547	1.4
3	*Solanum lyycopersicum*	94	0.0013	0.684	< 0.01	35.354	6.5
4	*Alocasia macrorrizha*	93	0.0052	0.353	0.095	19.651	11.5
5	*Maranta arundinacealinn*	94	0.0029	0.435	0.056	15.732	9.8
6	*Saccharum officinarum linn*	95	0.0024	1.035	< 001	30.654	1.3
7	*Amorphophallus campanulatus*	93	0.0037	0.253	0.083	21.543	4.6
8	Bioethanol Standard [21]	99.5 min	300 max	1.0 max	0.1 max	40.0 max	5.0 max

Table 2. The results of physical properties test of bioethanol from various types of vegetable.

No	Material	pH value (mg/l)	Heating value (kcal/kg)	Density (g/cm^3)	Viscosity (cSt)	Flash point (°C)
1	*Amorphopallus Variabiliti*	2.5	6786	0.858	2.762	15
2	*Mucasea*	5.8	5524	0.824	2.149	12
3	*Solanum Lyycopersicum*	5.3	5696	0.817	2.843	13
4	*Alocasia Macrorrizha*	3.8	6445	0.836	1.961	16
5	*Maranta Arundinacealinn*	4.6	5969	0.796	3.054	14
6	*Saccharum officinarum linn*	8.5	8756	0.801	1.637	13
7	*Amorphophallus campanulatus*	4.7	5892	0.823	2.376	19
8	Bioethanol Standard [21]	6.5-9.0	6380	0.789	1.525	12

3.1 Chemical Properties of Various Vegetable Materials

Table 1 shows the results of laboratory tests of the chemical properties of bioethanol from various vegetable materials produced by 4-level distillation. The seven vegetable materials have bioethanol content above 90%, and the level of bioethanol that approaching the standard is *saccharum officinarum linn* (95%). Because *saccharum officinarum linn* has sucrose content that reached 19%, so its bioethanol content is higher than the others. Bioethanol is a fuel (*ethyl alcohol* with the chemical formula (C_2H_5OH) produced from biofuels. It is obtained from the fermentation and distillation process. Generally, it is classified into 3 groups, namely as the beverage (alcoholic), the alternative fuel (energy), and for the industrial solvent. In the fuel sector, bioethanol can be used as fuel because it is combustible. However, bioethanol can only be used as a mixture of gasoline fuel to raise the value of octane (*High Octane Mogas Component*) HOMC for gasoline.

The methanol content of organic chemicals in the form of toxic liquid, colorless, odorless, molecular weight 32.04, boiling at 64.7°C, volatile and flammable. The standard of bioethanol

IOP Conf. Series: Materials Science and Engineering **296** (2018) 012019 doi:10.1088/1757-899X/296/1/012019

stipulated that its methanol content should not exceed 300 mg/liter. From the results of the laboratory test, the lowest methanol content is *musacea* (0.0088 mg/l).

The results of laboratory tests showed that the lowest water content of bioethanol is from *amorphophallus variabillis*, it is 0.202% vol. The water content in fuel is one of the benchmarks of the fuel quality, the less water content in bioethanol, the better the bioethanol is. The maximum limit of water content is 1%, it is determined by the standard of ethanol quality. The water content more than 1% in bioethanol can cause damage to the inside of the furnace surface when burning process. Water can also cause a spark of flame at the tip of the burner, which can turn off the flame and lower the temperature of the flame.

Table 1 showed that the lowest copper content of the seven bioethanol is from bioethanol *solanum lycopersicum* and *saccharum officinarum linn* (<0.01). Whereas according to the standard of bioethanol, the maximum copper content is 0.1 mg/kg. Copper is a highly active metal that catalyzing the oxidation of hydrocarbons at low temperatures. Copper with concentrations that is greater than 0.012 mg/kg in gasoline can lead to a significant increase in the rate of gum formation.

Based on the result of chloride ion test (Table 1), all of seven bioethanol in this research are qualified to the standard (the standard is maximum 40 mg/l), and *maranta arundinacealinn* is the lowest one. The effect of chloride content on the fuel is if chloride enters the combustion chamber, it becomes crust since it is corrosive. Moreover, if the fuel has a high content of chloride ions it can make the burning process less efficient and harmful to the engine. Thus, the smaller the chloride ion content in bioethanol, the better the quality of bioethanol is.

The gum content in bioethanol will cause the evaporation process of the fuel to produce a precipitate on the surface of the fuel induction system and the stickiness of the inlet valve. The insoluble gum may also clog fuel filters. Laboratory test results (Table 1) showed that the lowest gum content was obtained in bioethanol *saccharum officinarum linn* (1.3 mg/100 ml) and the highest one is bioethanol *alocasia macrorrizha* (11.5 mg/100 ml), whereas the standard of bioethanol quality is maximum 5 mg/100 ml.

3.2 Physical Properties of Various Vegetable Materials

The physical properties of bioethanol from various vegetable materials produced by 4-level distillation are shown in Table 2. The pH value is a benchmark of acid strength in bioethanol fuel. It is a good indicator for knowing the potential of bioethanol corrosion as fuel. The results of laboratory tests of the seven bioethanol whose pH content meets the standard of bioethanol is *saccharum officinarum linn* that is 8.5 mg/l. Whereas, the standard of bioethanol is should be around 6.5-9.0 mg/l. And if a pH value <6.5 the fuel injector, the engine cylinder and fuel pump may fail to work. Moreover, when the ethanol pH value is > 9.0 parts of the fuel pump plastic will fail to work.

The laboratory test of heating value (Table 2) showed that the highest heating values, which was 8756 kcal/kg, was obtained from the bioethanol *saccharum officarum linn* is 8756 kcal/kg. Whereas, the standart quality of heating value of bioenol is around 6380 kcal/kg. The heating value is the amount of energy released in the combustion process of the union volume or unity of time. The heating value of fuel determines the amount of fuel consumption per time unit. The higher the heating value of the fuel indicates the less the fuel consumes.

Referring to Table 2, the results of density test showed that the bioethanol that approaching the standard is *maranta arundinacealinn* (0.796 g/cm^3) since the standard of bioethanol density is around 0.789 g/cm^3. Density is the ratio of fuel mass to the volume of fuel at a temperature of 150^{0}C. Density is measured by a device called a pycnometer and hydrometer. The pycnometer is a laboratory instrument used to measure the density of fluid type or volume accurately. The hydrometer is a tool used to measure the specific gravity of a liquid with a ratio of fluid density and water density. This density greatly affects the quality of ignition of fuel.

The standard of viscosity on bioethanol is 1.525 cSt. While the Table 2 shows that the lowest viscosity is bioethanol *saccharum officinarum linn* (1.637 cSt). Fluid viscosity is a measurement of material resistance to a stream. Viscosity depends on the temperature. It decreases as temperature increases. It is measured by using a tool called viscometer. Viscosity is essential properties in the storage and use of fuel. If the viscosity of the fuel is too thick, it will make the engine difficult in pumping, igniting the burner, and flowing. In addition, the high-level viscosity in the fuel worsens the atomization, triggering the formation of carbon deposition on the cylinder wall of the engine.

Flashpoint is a flammability of fuel. It is also a parameter to know the impact of hazards during travel or fuel storage. The flash point standard of bioethanol is 12°C, and it is obtained by the bioethanol from *musacea*.

4. Conclusion

Based on the research objectives, the result and discussion of bioethanol performance characteristics of vegetable raw materials, it can be concluded that:

1. The highest ethanol content was obtained 95% vol in *saccharum officinarum linn*, the highest methanol content 0.0088 mg/l in *musacea* peels, the lowest *moisture* content 0.202% vol in *amorphopallus variabillis* tube, the lowest copper content <0,01 mg/kg in *saccharum officinarum linn and solanum lyycopersicum*, the lowest chloride ion content of 15.73 mg/l in the *arundinacealine maranta* tube, and the lowest gum of 1.3 mg/100 ml in the *saccharum officinarum linn.*
2. The best acid/pH level is 8.5 in *saccharum officinarum linn skin*, the highest heating value 8756 kcal/kg in *saccharum officinarum linn* skin, the best density 0.819 g/cm^3 in *arundinaceal maranta*, the best viscosity 1.837 cSt in *saccharum officinarum linn* skin, and the best flash point is 19°C in *amorphophallus campanulatus*.
3. Generally, bioethanol of vegetable raw materials that physical properties and chemical properties approaching standard is *saccharum officinarum linn* bioethanol.

References

[1] Soetaert W and Vandamme E J 2009 *Biofuels. Wiley Series in Renewable Resourses* ed A John Wiley and Sons (Belgium: Ltd. Publications)
[2] Vandamme E J and Soetaert W 2004 *Industrial Biotechnology and Sustainable Cemistry*
[3] Sutt J K 2010 *Kas naftavarud ammenduvad enne alternatiivsete lahenduste leidmist* (Estonian: Novaator)
[4] Muiste P and Kask U 2000 Investigation and Usage of Renewable Energy Sources *Conference proceedings* ed Estonian Agricultural (Estonian: University Publishing Centre) pp 60–1
[5] Adelekan B A 2011 Comparison of Ethanol Fuel Productivity of Some Selected Varieties of Tropical Crops *Proceedings of XXXIV CIOSTA CIGR V Conference 2011*
[6] Peraturan Presiden No. 5 2006 *Kebijakan Pencapaian Target Energi Nasional Terbarukan Sampai Tahun 2025* (Jakarta)
[7] Demirbas A 2008 Biofuels sources, biofuel policy, biofuel economy and global biofuel projections *Energy Convers. Manag* **49** pp 2106–16
[8] Indluj B 2017 BP Statistical Review of World Energy 2017 *Br. Pet.* pp 8–11
[9] Vares V 2008 Biomassi tehnoloogiauuringud ja tehnoloogiate rakendamine Eestis p 176
[10] Yilmaz N and Sanchez T M 2012 Analysis of operating a diesel engine on biodiesel-ethanol and biodiesel-methanol blends *Energy* **46** pp 129–129
[11] Aydogan H 2011 *Investigation of engine performance and exhaust emission effect using bioethanol-diesel fuel blends (E-Diesel)* (Selcuk University)

[12] Shiet X, Pang X, Mu Y, He H, Shuai S, Wang J, Chen H and Li R 2006 Emission reduction potential of using ethanol–biodiesel–diesel fuel blend on a heavy-duty diesel engine *Atmos. Environ* **40** pp 2567–74
[13] Muhaji 2017 Aplikasi Bioetanol Umbi Walur (Amorphophallus Paenifolius) Sebagai Extender Premium Pada Motor Bensin 4 Tak 1 Silinder *Seminar Nasional Art dan Teknologi (SNATS) 1* (Gorontalo)
[14] Ullmann 1990 Ullmann´s Encyclopedia of Industrial Chemistry *Ethanol* pp 588–9
[15] Ritslaid K, Kuut A and Olt J 2010 State of the Art in Bioethanol Production *Agron. Res.* **8** pp 236–54
[16] Ullmann 1990 Ullmann´s Encyclopedia of Industrial Chemistry *Ethanol 1990b* pp 281–2
[17] Wyman C E 2001 Twenty years of trials, tribulations and research progress in bioethanol technology *Appl. Biochem. Biotechnol* **91**–**93** pp 5–21
[18] Park S H and Lee C S 2013 Combustion performance and emission reduction characteristics of automotive DME engine system *Prog. Energy Combust* **39** pp 147–68
[19] Soerawidjaja H T 2011 Peluang, Potensi dan Rintangan Pengembangan Industri Bahan Bakar Nabati di Indonesia *Konggres Ilmu Pengetahuan Nasional (KIPNAS)* (Jakarta)
[20] Shereena K M and Thangarai T 2009 Biodiesel: an Alternative Fuel Produced From Vegetable Oil by Tranesterifikasi *Electron. J. Biol.* **5** pp 67–74
[21] Dirjen Migas 2008 *Keputusan Direktorat Jenderal Minyak dan Gas Bumi No:23204.K/10/DJM S/2008*

IOP Conf. Series: Materials Science and Engineering **296** (2018) 012020 doi:10.1088/1757-899X/296/1/012020

Integrating STEM education through Project-Based Inquiry Learning (PIL) in topic space among year one pupils

Chee Hoe Ng [1], M Adnan [2]
[1] Excellent Teacher Of Mathematics, SJK(C) Kai Chee, Pulau Pinang, Malaysia.
[2] Assoc. Prof. Dr., Deputy Director (Creativity and Innovation), Sultan Idris Education University, Education Research Laboratory (UERL), Malaysia.

ngcheehoe79@gmail.com

Abstract. This research aims to investigate the effect of integrating STEM education through Project-based Inquiry Learning (PIL) and the users of the STEM modules which consists of five projects on topic Space in Year One Mathematics Syllabus in *Kurikulum Standard Sekolah Rendah* (KSSR) of Malaysia. STEM education in primary school focuses on the introduces and awareness of students about the importance of STEM education. The projects in STEM modules are covering the different ethnic cultures in Malaysia. The modules are designed using the four phases in PIL. Concepts and the explanation of STEM education on each project are emphasized and provided in the modules so the teachers able to carry out the projects by using the modules. By using the modules in primary Mathematics, the students and teachers will be more understanding on how to integrate the Mathematics' concepts in STEM education.

1. Introduction

To enhance the quality of education in Malaysia, the Ministry of Education (MOE) has developed and launched the Malaysia Education Blueprint 2013-2025 (PPPM) which aims to transform the education system in Malaysia in order to compete with the growing international education system. To achieve this objective, the MOE has identified 11 shifts that can result in changes in the education system in Malaysia. The first shift in this blueprint has stated: Provide Equal Access to Quality Education of an International Standard. One of the benchmarks in the first shift is to strengthen the quality of Science, Technology, Engineering, and Mathematics (STEM) education.

In order to produce the educated citizens and meet the needs of 21st-century skills, STEM education is an important interdisciplinary subject which fulfilled the needs. Each individual must know and understand the basic concepts of the STEM. STEM education has given the best opportunities for students to understand the world in a holistic manner. According to Hays Blaine Lantz, Jr. (2009), STEM education offers students one of the best opportunity to make sense of the world holistically, rather than in bits and pieces. STEM education removes the traditional barriers erected between the four disciplines, by integrating them into one cohesive teaching and learning paradigm [1].

To equip the students with the skills needed to face the challenges of a changing global with the application of STEM, the MOE is responsible for ensuring an increase in the ratio of students who wish to pursue scientific and technical fields. In the PPPM 2013-2025, the MOE has stated the five factors that cause the decline and the quality of student outcomes in STEM education,

- Lack of awareness in STEM education.
- STEM subjects considered difficult.
- A solid curriculum content.
- Less consistent with the quality of teaching and learning.
- Old and inadequate infrastructure.

These factors or problems must be overcome so that the desire of MOE in PPPM 2013-2025 can be achieved. In 2014, Dr. Suhaidah Tahir, Dominador D. Mangao and Dr. Nur Jahan Ahmad of the Southeast Asian Ministers of Education Organization Regional Center for Education in Science and Mathematics (SEAMEO RECSAM) have designed a conceptual framework for the development of STEM education in Malaysia. In this framework, it divided the development of STEM education into six phases, which is from early childhood to industry or community. In this paper, we focused on the second phase which is on the primary or elementary school, the focus is helping the students to make the connection or build the foundation in STEM education [2].

In the primary or elementary school, the STEM education focuses on the introduction and gives the awareness to students about the importance of STEM education. In other words, it gives a draft idea of the integration of the four fields in STEM education. Initial step to introduce STEM education is to provide the standards-based learning which based on the problem-solving and inquiry-learning method in real situations to link up the four subjects in STEM education. In other words, the goal of STEM education is designed based in the inquiry-learning by used the thinking skills in science and technology to solve the everyday real life problems.

The purpose of this research is to enhance the integration of STEM education in Mathematics *Kurikulum Standard Sekolah Rendah* (KSSR) through an inquiry-based approach which the activities focus on project-based learning or using the Inquiry-Based Learning Project (PIP).

2. Project-based Inquiry Learning (PIL)

Inquiry-Based Learning Project (PIP) is the method that emphasizes on the sciences skills and attitude, high order thinking skills, creative problem solving, design and construction of the object-based technology and encourages the children to communicate and work in teams [3]. There are four phases in PIP, namely Inquiry, Exploration, Experimentation, and Reflection.

In the Inquiry Phase, students ask questions about what they want to know and with the guidance of teachers, the students may decide the issues or topics they want to be investigated.

In the Exploration phase, students are looking the information about issues or topics that have been identified. Search information by using various methods, such as the internet, video, nature walk, visit, study science books, and so on. The students develop the new knowledge and strengthen the existing knowledge on the subject they studied or had been exploring. After receiving all the necessary information, the students will determine the suitable materials for the projects.

In the Experimentation phase, students build or develop their inventions or ideas by creating the model according to what they had planned in the Exploration phase. Students demonstrate the product and answer questions that they ask in the first phase.

In the Reflection phase, students reflect on the learning process which they had gone through and give a view of the product or invention that they have produced. In this phase, students also will be asked to give the opinion on the interest, awareness, appreciation and what they want to do next.

These four phases in PIP are adapted and modified from the implementation of STEM education in the PERMATA curriculum. The aim of modified PIP is hoped that it can be implemented into the Malaysia's primary school curriculum [4].

PIP is the method should use in teaching STEM education which can be represented in Figure 1. The inquiry is the initial phase to stimulate the curiosity of students. 3E in the PIP is Exploring, Experiment and Experience and 3C in the PIP are Collaborate, Create and Communicate.

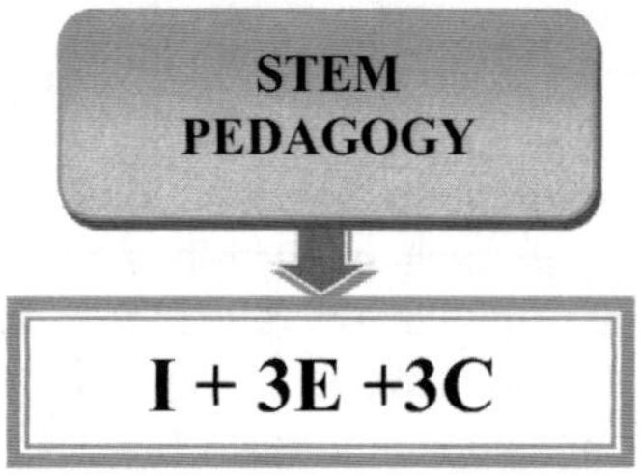

Figure 1. Inquiry-Based Learning Project (PIP) method

3. STEM modules

In Malaysia, the development of STEM education is an important agenda for the transformation of the education system and prepares the students to meet the challenges of the 21st century. To help the teacher to enhance the knowledge and development of STEM education, in this research a STEM modules using the PIP method based on the ADDIE Module has designed.

STEM modules in this research have been characterized by using variety traditional culture of Malaysia. Projects in the STEM modules chose the traditional culture in Malaysia because Malaysia is a country made up from different races and ethnic such as Malay, Chinese, Indian and other which practices different and unique traditional culture. In the modules, students not only understand the STEM concepts involved, at the same time the students also can understand the tradition of different races in Malaysia.

In these modules, five projects that use PIP methods have been designed. The skills in these five projects are integrating the knowledge of science, technology, engineering, and mathematics while producing products. These five projects have been a focus on the topic of space in Revised KSSR Year 1 Mathematics Syllabus. These five projects are:

Table 1. Five STEM Modules

No	Project	STEM Concept Involved
1	Tangling	Shape and size of the Tanglung, a number of materials that needed such as color paper, burning, light, and design of Tangling.
2	Tudung Saji	Measure and counting, shape and size, the usage and design of the Tudung Saji.
3	Kolam	Measure, shape, size, symmetry and how to stick the rice or sands on the paper and the design of Kolam.
4	Wau	Measure, shape, size, symmetry and how to let the Wau fly in the sky.
5	Chinese Yoyo	Measure, shape, size, symmetry and how to balance the Chinese Yoyo while playing.

4. Research methodology

The respondents in this research are a focus on the year one student in the ordinary schools under the Ministry of Education of Malaysia. Two classes of the year, one students were chosen as the population for the study. One class of the student as the treatment group while another class as control class. To get effectiveness on integrating STEM education in Mathematics, this research used a quasi-experimental design for the study group and used the existing year one class. This research will be conducted in six weeks and depend on the progress of the students[5].

In this research, four instruments were built to evaluate the effectiveness of implementing the PIP method based on STEM modules. These instruments are intended to support the validity and reliability of the PIP. These instruments are pre-test and post-test, questionnaires, interviews and the project implementation monitoring report to support the outcome of the research. This study aimed to identify the reliability integration of the STEM education in the field of Mathematics, results of pre-tests and post-tests, questionnaires and interview can only give a rough idea about the integration of STEM education.

Questionnaires to evaluate the understanding, interest, and attitude towards the STEM education was distributed to the treatment group's students. The Cronbach's alpha test was conducted to determine the internal consistency. This method assumes that each item is considered to be a similar test and all correlations between the measured items are the same. The result of the questionnaires shown that confidence coefficient Cronbach's alpha value is 0.76, so the reliability for this questionnaires is accepted.

Reliability Statistics

Cronbach's Alpha	Cronbach's Alpha Based on Standardized Items	N of Items
.756	.775	30

Figure 2. Reliability statistics

5. Pilot Study

A pilot study had been done using the project five in the STEM modules, make a Chinese Yoyo. The project had carried out in two places, for example, the *Expo Mampan* in RTC Gopeng, Perak and the *Majlis Professor Negara* at FELDA UMAS, Tawau, Sabah. During the activity, teacher or the facilitator guides the students on how to make a Chinese Yoyo. The teacher or the facilitator used the instructions in module five to carry the project. While the students are making the Chinese Yoyo, the facilitator asked the students questions about the shape, the size and how to make a yo-yo balance while playing. The questions asked by facilitator as below

- What is the shape of Chinese Yoyo?
- What is the design of Chinese Yoyo to make it spin?
- How to make a Chinese Yoyo?

These questions are asked in the Inquiry Phase, the purpose asked these questions is to cause the curiosity among the students. From the curiosity, the students are interested to find out more about the Chinese Yoyo. In the same time, the facilitators guide the students followed the four phases in Project-based Inquiry Learning (PIL) and let the students understand the integrated STEM knowledge in this project. The Chinese Yoyo in STEM modules can consider special and interesting because the materials that used in this project are recycled materials, such as mineral water bottles, chopstick, screw, and nylon string to make this project.

From the responses and feedback from the students, noticed that the students are enjoying themselves while making the Chinese Yoyo, they also feel excited because the used the recycled bottles to make a Chinese Yoyo. The Chinese Yoyo in this project not only because it is a Chinese traditional game but it also consists of STEM knowledge, such as how to make a balance Yoyo, how to keep its balancing while played the Chinese Yoyo and what is the design can use for Chinese Yoyo.

3.1. Students make the Chinese Yoyo

3.2. Facilitator gives instructions

3.3. Students play the Chinese Yoyo

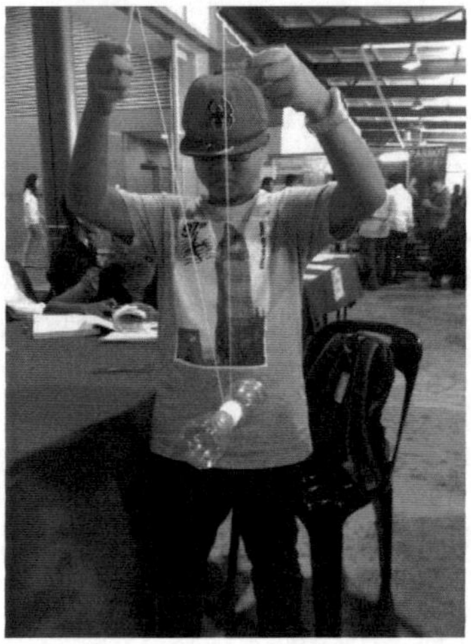

3.4. Students play the Chinese Yoyo

Figure 3. Pilot Study at *Expo Mampan* in RTC Gopeng, Perak

This two pilot study was facilitated by main author and co-author. The project made by students has followed the four phases of PIL in the STEM modules. There was no boredom among the students, from the participated of students, they had employed the hands-on and minds-on while making the Chinese Yoyo.

4.1 Students sit in a group

4.2 Students sit in a group

4.3 Facilitator gives instructions

4.4 Facilitator gives instructions

Figure 4. Pilot Study at *Majlis Professor Negara* at FELDA UMAS, Tawau, Sabah

6. Conclusion and Discussion

The purposes of this study are to improve the integration of STEM education in Mathematics through PIP. The stem is a teaching and learning approach that involves the application of knowledge, skills, and values of STEM to solve the real-life problem in the context of daily life, society, and the environment. This approach encourages students to ask questions and explore the environment through inquiry and solve the issues related to the real world situation to cultivate the practice of STEM. By using the STEM modules based on the PIP, the teacher will be more confident and able to carry out the STEM education in school. To achieve the goals of PPPM 2013-2015 of MOE, the teacher themselves must change their mind and accept the challenges of the teaching and learning in this new era, so they can produce the students fulfilled the need of 21st century [6].

References

[1] Ong Eng Tek, Aminah Ayob, Md Nasir Ibrahim, Mazlini Adnan, Jameyah Shariff, Noriah Ishak. 2016 The Effectiveness of an In-Service Training of Early Childhood Teachers on STEM Integration through Project-Based Inquiry Learning (PIL) *Journal of Turkish Science Education* **13** (Special Issue) pp 44-58

[2] Hays Blaine Lantz 2009 *Science, Technology, Engineering, and Mathematics (STEM) Education, What Form? What Function?*

[3] Malaysia Education Blueprint 2013-2025 (Preschool to Post-Secondary Education) 2013 Kementerian Pendidikan Malaysia

[4] Mazlini Adnan, Aminah Ayob, Ong Eng Tek, Mohd Nasir Ibrahim, Noriah Ishak, Jameyah Sheriff 2016 Memperkasa pembangunan modal insan Malaysia di peringkat kanak-kanak: Kajian kebolehlaksanaan dan kebolehintegrasian pendidikan STEM dalam kurikulum PERMATA Negara

[5] Panduan Pelaksanaan Sains, Teknologi, Kejuruteraan dan Matematik (STEM) dalam Pengajaran dan Pembelajaran 2016 *Bahagian Pembangunan Kurikulum, Kementerian Pendidikan Malaysia*

[6] Suhaidah Tahir, Dominador D Mangao, and Nur Jahan Ahmad 2014 *Making Stem Education Work: Exploring Partnership Between Seameo Recsam And Moe Malaysia* SEAMEO RECSAM

High-capability students' difficulties in problem solving on the concept of ratio

N F Kusuma[1], S Subanti[2] and B Usodo[2]

[1]Students of Magister Mathematics Education, Universitas Sebelas Maret Jl. Ir. Sutami No. 36 A, Surakarta 57126, Indonesia
[2]Lecturer of Magister Mathematics Education, Universitas Sebelas Maret Jl. Ir. Sutami No. 36 A, Surakarta 57126, Indonesia

niyankusuma@gmail.com

Abstract. This research is descriptive qualitative study which aimed to describe students' difficulties in problem-solving ratio with high-capability. Two students were chosen as students with high-capability in mathematics. This study used qualitative method. The result of this study showed that there were two difficulties in problem-solving ratio, including the difficulty in conducting cross-product and the difficulty in determining the ratio. The factor that causes students feeling difficulty in solving ratio is the students who accustomed to using the formula contained in the book. The solution offered to reduce that difficult, teachers should give exercises by using the concept of the ratio as the solution so thatt the students do not just memorize the formula from their books.

1. Introduction

The multiplicity of students' intellectual ability, particularly, in the mathematics field in Junior High School is highly varied. This ability includes the capability to recollecting, comprehending, interpreting information, solving a certain problem, and much more. Those various issues concerning students would be developed along with the adjustment of the learning environment, which is both directly and indirectly felt by the students. On the other hand, the learning difficulties are not only faced by below average students but also encountered by students with any ability from any faction with level and source of impediment that multiple diverse [1].

Ratio and proportion are used in a variety of other student areas; for instances, geography student in senior high school. The ratio and proportion are used for the concept of population density. The concepts of ratio and proportion are important for some other students, such as in cartography (map-making); in the science to define various natural phenomena; in economics and statistic to calculate profit-and-loss and probability; and in technological studies for calculations in engineering, mechanics, robotics, computer science, among other [2]. By seeing the importance of ratio, it is clear that these concepts are the materials that are required to be taught in junior high schools. That is because the ability of students about the concept of the ratio in junior high school became the basis for students studying the concept at a higher level and applying it in their life.

According to the interview result with mathematics teacher of class IX junior high school 2 Ampel, it was identified that there were a lot of mistakes which made up by the students in answering the question regarding basic ratio student. The simplest faulty that done by the students was changing the units; furthermore, students often made ratio reversely. There was also found the students' mistake in multiplying operation. The teacher asked the students not to directly multiply the ratio number so that the resulting number was not big, but to simplify the ratio first. However, there were errors of students' works in simplifying the ratio. These errors occurred generally due to the unwillingness of students in finding other ratio problems references and only relied on problems which were given by their teacher, so the students lacked practice and their knowledge regarding ration problems were not varied.

The ability of grade 5 students to solve ratio and proportion problem is still low. Most of students experienced obstacles. Students used some strategies that given by the teacher so that they get difficulties in completing the problems which have different contexts. The learning material of ratio and proportion that presented in textbooks is still partial (not depth). In addition, the lack of ability of the teacher in creating the problems also makes student confuse in understanding the main purpose of the problems [3]. Other research in 6th-grade students indicated they most frequently used cross-multiplication algorithm strategy during the solution of these problems [4].

Problem-solving is one of a major aspect of mathematics curriculum which required students to apply and to integrate many mathematical concepts and skills as well as making a decision [5]. In mathematics, solving problems represents the most effective concept to contextualization and re-contextualization of concepts, to operational and basic mathematical knowledge transfer to ensure a sustainable and meaningful learning [6]. Therefore, it is important to know students' difficulties in problem-solving of the ratio in junior high school and what factors are causing it. This study aimed to find students' difficulties in ratio and to analyze what factors are causing it.

2. Method

This research was conducted at one of junior high school in Boyolali Central Java Province, Indonesia. There are 2 students as participants. They are students of grade 9, 1 male and 1 female with high-capability and with good communication skills were chosen to be subjects of the study. The male student is coded by SF and the female student is coded by SL.

This qualitative study was carried out in the following four procedures. Firstly, researchers collected data from the individual written test. Secondly, we collected data from the individual interview. Thirdly, we analyzed data by comparing data on collected from the individual written test with data on collected from the individual interview. Fourthly, we made conclusions.

3. Result and Discussion

Based on data analysis, we classify the results into two sections. The sections are seen from student test result and students interview data results. There were three problems given by researcher. It aimed for identifying students' difficulties in solving a problem that given by researcher. Students' difficulties will be described in the following section.

> Problem 1:
> "A map was made with a scale 1: 1.200.000. If the actual distance between two cities is 95 kilometers, the distance that was showed on the map is supposed to be ..."

Based on the test result of student SF, there was an error in the problem 1, as shown in Figure 1.

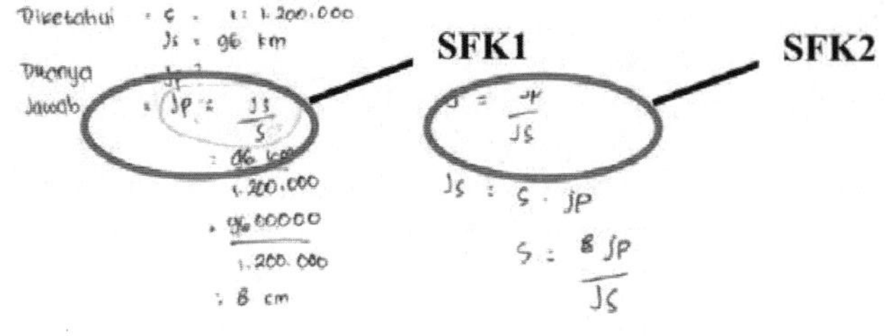

Figure 1. The student SF answer to question problem 1

The outcome of student's works with code SKF 1 showed that student SF encountered mistake in writing the formula, in spite of the correct answer was made. On problems with code SKF 2, it was expected that student SF had known the formula to determine the scale, but then student SF did mistake in defining distance formula on the map based on scale formula that the student had.

According to the interview result with student SF, it was identified that student had memorized the scale formula. Student SF was doing error in writing the formula to determine the distances that should be shown on the map. The cause of the error was the confusion of student SF in cross multiplying. Based on scale formulation that the student had to acquire the formula to find the distance that should be shown on the map. Based on the test result of student SL, there was also error on problem 1, as shown in Figure 2.

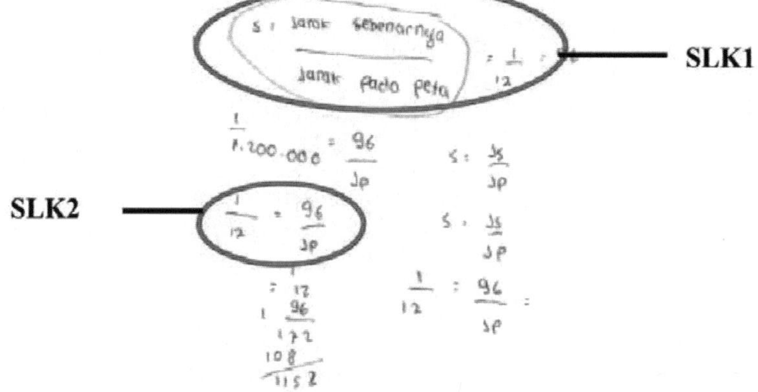

Figure 2. The student SL answer to question problem 1

The outcome of student's work with code SLK 1 was expected that student SL faced difficulties in finding the correct formulation. Student SL reversed the scale formula that supposed to have the on-the-map distance on the upper side and substantive distance on the bottom side. Then, student SL started to continue finding the answer, but on the problem code, SKL 2 showed the student SL only did the works up to that step. Student SL did not try to cross multiplying the result from the previous step. Furthermore, student SL re-write the scale formulation although the formula that the student wrote was still wrong, continue to find the answer, and stop at the very same step.

Based on the interview result with student SL, it was known that the student had memorized the formula. On the other hand, student SL did the mistake in writing the formula because the student forgot the formula that had been memorized before. Student SL was nervous at that time so that the student felt that the atmosphere became tense and he forgot the formula. Furthermore, student SL admitted that he was lack of carefulness in reading the problem. It was found that student SL did

IOP Conf. Series: Materials Science and Engineering **296** (2018) 012021 doi:10.1088/1757-899X/296/1/012021

errors in writing measurement unit that should be in centimeters to be in kilometers. Figure 3 presents the formulation of Problem 2.

Problem 2:

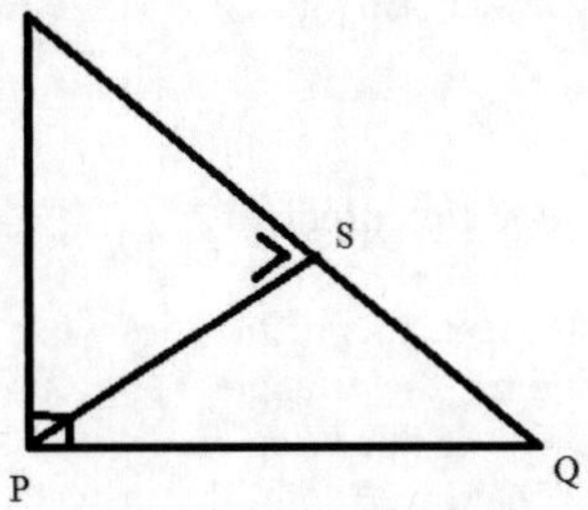

"Based on the picture, triangle PQR, beveling in P. The length of QS = 6 cm, SR = 18 cm, the length of PQ is…"

Figure 3. Problem 2

Based on the test result which was given to student SF, there was an error on problem 2 as shown in Figure 4.

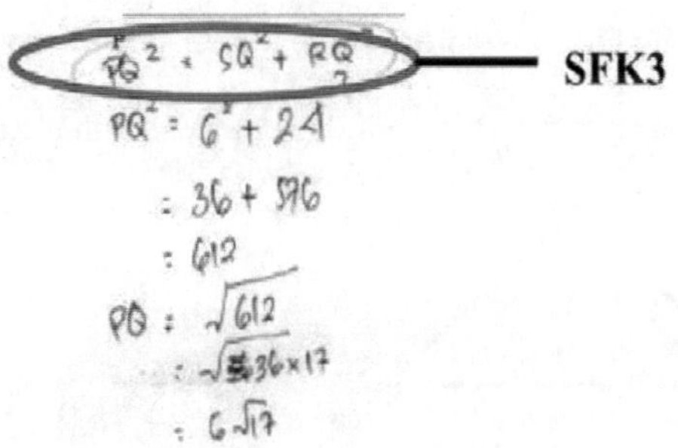

Figure 4. The student SF answer to question problem 2

The student's works with code SKF 3 showed that students SF encountered mistake in writing the formula. Student SF did not do rationing first. It was expected that student SF wrote the formula directly although the formula was wrong.

According to interview result with student SF, it was known that student memorized the formula. Yet, student SF did error when he wrote the mistake in finding the length of PQ. Student SF forgot the formula that had memorized before. Student SF considered that the problem used Pythagoras theorem.

Based on the test result of student SL, it was found that there was also a mistake as shown in Figure 5.

IOP Conf. Series: Materials Science and Engineering **296** (2018) 012021 doi:10.1088/1757-899X/296/1/012021

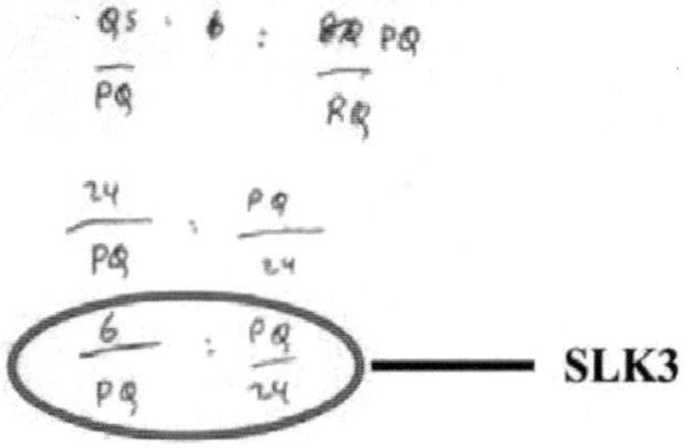

Figure 5. The student SL answer to question problem 2

According to the student's works, it was expected that the students met difficulties when trying to find out the unknown variables. Students SL had done rationing the triangle based on the problem statement, but then, student SL was confused when he reached the third step on the problem code SLK 3. Student SL did not do cross multiplying from the result that he had got that it was presumed that student SL did not found the answer and stuck in that step.

Based on the interview result with student SL, it was identified that the student had memorized the formula. Yet, when the student SL was asked to solve the problem by using ratio, student SL encountered difficulties when he wanted to determine bottom side length if there were only one side length known. Student SL found difficulties in solving the related problem due to lack explanation from the teacher regarding ration and he only looked at the formula from LKS.

The following Figure 6 represents the formulation of Problem 3.
Problem 3:

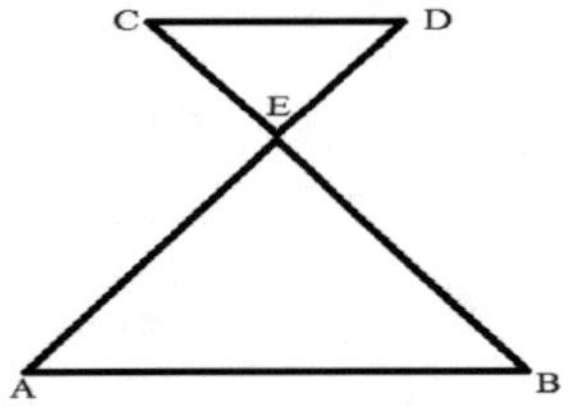

"Based on the picture, triangle ABE was congruent with triangle CDE. If the length of BE = 10 cm, DE = 8 cm, CE = 5 cm and CD = 10 cm, then, the length of AB is…"

Figure 6. Problem 3

Based on the test result of student SL, there was an error in solving the problem 3, as shown in Figure 7.

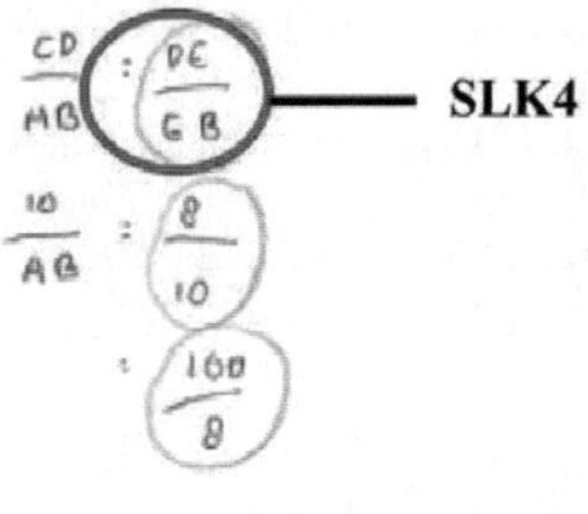

Figure 7. The student SL answer to question problem 3

According to the student's work, it was presumed that student SL faced difficulties in writing the ratio. Student SL thought that side BE would be comparable to side DC, and side CE would be comparable with side AE so that CD was comparable with AB. On this matter, student SL was wrong when she wrote the ratio of that triangle above.

Based on the interview result with student SL, it was known that student SL did mistake in determining the ratio. Student SL previously thought that the triangle was isosceles, but then, student SL called the triangle figure as an irregular triangle. Student SL determined the ratio without paying attention to the scale of formula that as large as on the triangle. Student SL thought that the triangle was isosceles and directly defined the ration without notice the same angle which has the same number.

According to the result analysis of students' answer an interview which had been done, students encountered many obstacles in solving ratio problems. Both of the students, that is SF and SL face the same difficulties, although the student SF could solve more problem with correct answer compared with student SF. Bot students, could not do cross multiplying. Student SF found difficulties in changing formula

$$s = \frac{jp}{js} \text{ into } jp = s \times js \text{ or } js = \frac{jp}{s} \qquad (1)$$

Student SL found an obstacle in determining the result of the ratio. Student SL felt confused when was given ration in the following form:

$$\frac{6}{PQ} = \frac{PQ}{24}, \qquad (2)$$

so that, student SL did not try to do cross multiplying. The cause of the obstacles that occurred is that students lack the practice to use cross multiplying in solving problems, thus when the students faced problems regarding cross multiplying, the would find troubles.

Student SL could not found the ration of two triangles. Student SL only looked at the ratio on the upper side with the bottom side without paying attention towards the number of angles. The cause of the difficulties is because the student SL used to apply the formula that he knew from LKS, so that, when he was asked to determine the ratio, student SL could not do all the steps to find the correct answer. Student SF in solving the problem rarely used ratio and only implement the formula known. This situation made student SF found difficulties in determining the ratio.

Based on the analysis of the difficulties which were encountered by the students in solving mathematics problem regarding ratio student, it is highly advised to find alternative solutions to decrease those obstacles that are; teachers should often give practice by using ratio concept in solving the problem thus the students not only memorized the formula written in LKS. Teachers also need to explain previous matters which related to the matters which are taught in the current situation. As an example, when determining the ratio, it would be better for the students if the teachers re-explain the position of angles which has same numbers so that the students will define the ratio easier.

4. Conclusion

Based on the results, there were two difficulties in problem-solving ratio, including the difficulty in conducting cross-product and the difficulty in determining the ratio. Students are accustomed to use the formula contained in the book. The solution offered to reduce that difficult, teachers should give exercises by using the concept of the ratio as the solution and explain previous matters which related to the matters which are taught in the current situation.

References

[1] Widdiharto R 2008 *Diagnosis Kesulitan Belajar Matematika SMP dan Alternatif Proses Remidinya* (Yogyakarta: Pusat Pengembangan dan Pemberdayaan Pendidik dan Tenaga Kependidikan Matematika)
[2] Chaim D B, Keret Y and Ilany B S 2012 *Ratio and Proportion: Research and Teaching in Mathematics Teachers' Education (Pre- and In-Service Mathematics Teachers of Elementary and Middle School Classes)* (Rotterdam: Sense Publishers)
[3] Andini W and Jupri A 2017 *Student Obstacles in Ratio and Proportion Learning* (United

Kingdom: IOP Publishing)
[4] Avcu C and Avcu S 2010 *Proc. Social and Behaviorial Sciences* (Elseiver LTD)
[5] Tambychik T and Subahan M M T 2010 Proc. Social and Behavioral Sciences (Elseiver LTD) pp 142–51
[6] Caprioara D 2014 *Proceeding Social and Behaviorial Sciences* (Elseiver LTD)

Mathematics Pedagogical Standards: A Suggested Model of Instruction in Enhancing the Mathematics Teacher's Quality of Instruction

N S Saad[1], M Jemali[1], Z Hj Zakaria[1] and Q Yusof[2]

[1]Kolej University Perguruan Ugama Seri Begawan, Brunei Darussalam
[2]University Pendidikan Sultan Idris

noorshah5156@gmail.com

Abstract. The paper aims at identifying the standards for teaching and learning of mathematics based on National Council of Teacher of Mathematics (NCTM, 2000), The Australian Association of Mathematics Teachers (AAMT, 2006) and Training and Development Agency for School (TDA, 2007). These known standards were used as a guide in identifying the constructs of the mathematics teacher's instruction in the classroom. The survey method used in which a questionnaire instrument encompassed on the four identified constructs on the standards for teaching and learning of mathematics, namely *professional practices*, *professional attributes*, *professional knowledge,* and *professional instructional processes*. The instrument was tested during a pilot study and a Cronbach's Alpha reliability index of greater than 0.85 was obtained. The actual research was carried out in Peninsular Malaysia involving 224 secondary schools with 1.120 mathematics teachers and 108 primary schools with 540 mathematics teachers. From the selected schools, only 820 secondary mathematics teachers (73.2%) and 361 primary teachers (66.9%) gave a response to the mailed questionnaires. The findings of the study revealed that the secondary and primary mathematics teachers strongly agreed on three constructs; *professional practices*, *professional attributes* and *professional instructional processes*.

1. Introduction

Shulman highlighted the notion that, *'teaching necessarily begins with a teacher's understanding of what is to be learned and how it is to be taught'* [1]. As such, decisions made by the teacher before, during and after the instruction are crucial not only to the teacher but to the learners as well in enhancing the quality of the instruction. The decisions made must be based on a credible professional guidance. Hence, to help the teachers, NCTM [2], introduced '*The Professional Standards for Teaching Mathematics'*. The standards proposed by NCTM were to guide American mathematics teachers on how the teaching and learning of mathematics should be practiced in the classrooms. NCTM also described two types of standard namely the content standard for teaching specific mathematical topics and the process standard that involves pedagogy and the methods of instruction. This emphasized through the statement, '*Teachers need to know and use mathematics for teaching that combines mathematical knowledge and pedagogical knowledge'* [2]. In the *'Teaching Principle'* NCTM [2] put forward three main aspects of effective instruction. These aspects are the needs of a sound knowledge and understanding of mathematics, the students as learners and sound pedagogical strategies.

The standards for teaching mathematics also introduced in Australia for teachers to practice and this known as the *'Standard for Excellent in Teaching Mathematics in Australian School'* [3]. The proposed standards would serve as a guide for mathematics teachers about the quality required to be an effective mathematics teacher, ways to implement his or her roles and responsibilities competently and creating his or her own teaching philosophy.

The situation in Malaysia is a bit different. Here, the various divisions in the Ministry of Education have the responsibilities towards the development of education. In addition to producing quality teachers, the Ministry also wishes that these quality teachers remain within the orbit of the national education system and sustain their quality until they retire from the educational service [4]. To achieve these aims the Ministry introduced the 'Malaysian Standards for A Quality Education' [5] to evaluate the quality of teaching in the classroom and 'Malaysian Standards for Teachers' [6] to enhance the quality and knowledge of pre-service teachers. These two standards are apparently different from the perspective of their respective aims. This paper discusses the constructs that have been identified as the bases for teaching mathematics and that have been evaluated by using the *Structure Equation Modelling (SEM).* SEM is used to determine the constructs that contributed greatly to enhancing the quality of teaching and learning of mathematics and hence forming a standard mathematics pedagogical model that can be practiced to achieve an effective mathematics instruction.

2. Methods

Four constructs or components of the pedagogical standards for teaching mathematics were identified and they were *Professional Knowledge*, *Professional Practices, Professional Instructional Processes* and *Professional Attributes.* Overall, 17 elements were identified from the literature review as in Table 1. The items for the questionnaire were constructed based on the four components or construct to gain a degree of approval from the respondents using a five-point Likert scale. Each item was then analyzed based on the frequencies and percentages of the degree approval towards the pedagogical standards for teaching mathematics. This was followed by the analysis of the average for each component of the standards. The degree of perception and the mathematics teachers' knowledge category were measured using the scales as proposed by Nugent, Sieppert dan Hudson [7] as in Table 2. In addition, Cronbach's Alpha Reliability Index of the instrument was determined during the pilot study with an index of 0.85 [8] and 0.95 [9].

Table 1. Mathematics Pedagogical Standards components

Element	Component
PA	**Professional Attributes**
SA	Self Attribute
PD	Professional Development
RSC	Responsibility for School & Community
PP	**Professional Practices**
PMI	Planning of Mathematics Instruction
PIMI	Pedagogical Implementation of the Mathematics Instruction
ME	Monitoring & Evaluation

Element	Component
PK	**Professional Knowledge**
MCK	Mathematics Content Knowledge
KICTAMI	Knowledge of ICT Application in Mathematics Instruction
KIQ	Knowledge of Intellectual Quality
KL	Knowledge of Learners
KMC	Knowledge of Mathematics Curriculum
PIP	**Professional Instructional Processes**
UTEMI	Usage of Tools in Enhancing Mathematics Instruction
ACLE	A Conducive Learning Environment
MMT	Meaningful Mathematical Tasks
TR	Teacher's role
LR	Learner's role
AMI	Analysis of the Mathematics Instruction

Table 2. Degree of Perception and Pedagogical Knowledge Category

Scale	Perception	Category
1.00 – 1.99	Very Low	Weak
2.00 – 2.99	Low	Moderate
3.00 – 3.99	High	Good
4.00 – 5.00	Very High	Excellent

In the second stage, the survey method was employed using the constructed questionnaire about the pedagogical standards for teaching mathematics. This was administered by post to 224 secondary schools and 108 primaries selected randomly throughout Malaysia. A total of five mathematics teachers from each selected school were asked to respond to the mailed questionnaire. However, only 73.2% or 820 secondary mathematics teachers and 66.8% or 361 primary mathematics teachers responded to the questionnaires.

3. Findings of the Study

The findings of this study were analyzed statistically (descriptive) on the four components of the pedagogical standards for teaching mathematics. Further, an exploration of the findings using the Structured Modelling Equation employing the Partial Least Squares Method was carried out to obtain the 'Pedagogical Standards for Teaching Mathematics Model' (Diagram 1).

3.1. *Profiles of the Respondents*

Table 3 shows the detailed profiles of the mathematics teachers who took part in this study. Based on gender, there were 888 female respondents (75.2%) and 293 male respondents (24.8%) involved in this study. Ethnically, there were 842 Malay respondents (71.3%), 224 Chinese respondents (18.9%), 104 Indian respondents (8.8%) and 11 respondents from other ethnicities (1.0%). Almost 90% of the secondary mathematics teachers and 36.3% of the primary mathematics teachers were first degree holders with education. Among the secondary mathematics teachers, 528 (64.4%) were graduates who majored in mathematics while the rest were graduates who majored in non-mathematics fields. Among the primary mathematics teachers, 183 (50.7%) graduated with an option in mathematics while 178 (49.3%) graduated with a non-mathematics option. It was also discovered that 586 (71.5%) secondary mathematics teachers and 202 (55.9%) primary mathematics teachers were between 31 and 50 years old. As far as the teaching experience was concerned, 516 (62.0%) secondary mathematics have between one to ten years of experience in teaching KBSM Mathematics while 747 (91.1%) teachers claimed to have between one to ten years of teaching Additional Mathematics. Among the primary mathematics teachers, 209 (57.9%) of them claimed to have between one to ten years of teaching KBSR mathematics while 152 (42.9%) teachers have teaching experiences of more than 10 years.

Table 3. Profiles of Secondary and Primary Mathematics Teachers

Profile	Secondary No.	Secondary Percentage	Primary No.	Primary Percentage	Total
Gender					
Male	188	22.7	113	31.3	293
Female	640	77.3	248	68.7	880
Ethnicity					
Malay	651	77.3	191	22.7	842
Chinese	136	60.7	88	39.3	224
Indian	27	26.0	77	74.0	104
Others	6	54.5	5	45.5	11
Major					
Mathematics	528	64.4	183	50.7	711
Non-mathematics	292	62.1	178	37.9	470
Age					
< 31 years	234	59.5	159	40.5	393
> 31 years	586	74.4	202	25.6	788
Teaching Experience					
KBSM (1-10 years)	516	62.9	-	-	820
KBSM > 10 years	304	37.1	-	-	
KBSR (1-10 years)	-		209	57.9	361
KBSR > 10 years	-		152	42.1	

3.2 *Standards for Teaching Mathematics Components*

3.2.1. *Professional Attributes Component.* Table 4 shows the mathematics teachers' perception of the professional attributes component. There are three elements in this component which are self-attribute, professional development and school and community's responsibility. The overall mean score for this component is 4.06 (SD= 0.51 for the secondary teachers and 4.12 (SD = 0.46) for the primary teachers. This indicates that the level of perception of the respondents on this component is high and the teachers' knowledge is categorized as excellent. For the elements of self-attribute, professional development and school & community's responsibility the mean scores and standard deviations for secondary/primary teachers are (Mean = 4.29/4.20; SD = 0.50/0.45), (Mean = 4.08/4.10; SD = 0.54/0.49) and (Mean =

3.80/4.05; SD = 0.67/0.56) respectively. Therefore, it could be said that the respondents' levels of perception of the elements of self-attributed and professional development are very high and mathematics teachers' knowledge can be categorized as excellent.

Table 4. Secondary and Primary Mathematics Teachers Perception on the Professional Attribute Component

Component	Secondary		Primary	
	Mean	SD	Mean	SD
Self-attribute	4.29	0.50	4.20	0.45
Professional Development	4.08	0.54	4.10	0.49
School & Community's Responsibility	3.80	0.67	4.05	0.56
Overall	4.06		4.12	

3.2.2 *Professional Practices Component.* Table 5 shows the three elements of the professional practices component which are mathematics instructional planning, monitoring and evaluation and pedagogical implementation in the teaching and learning of mathematics. Overall, the mean scores for this component are 4.04 (SD = 0.53) and 4.15 (SD = 0.47) for the secondary and primary teachers respectively. This shows that the mathematics teachers' level of perception is very high and their knowledge can be categorized as excellent. For each of the element or construct, the mean scores and standard deviations for secondary/primary teachers are 4.08/4.16 (SD = 0.54, 049), 3.99/4.12 (SD = 0.59, 0.47), 4.06/4.14 (SD = 0.56, 0.52) respectively. Among the secondary mathematics teachers, the respondents show a high level of perception concerning the elements of mathematics instructional planning and pedagogical implementation in the teaching and learning of mathematics while the teachers' knowledge can be categorized as excellent.

Table 5. Secondary and Primary Mathematics Teachers' Perception of the Professional Practice Component

Component	Secondary		Primary	
	Mean	SD	Mean	SD
Mathematics Instructional Planning	4.08	0.54	4.16	0.49
Monitoring and Evaluation	3.99	0.59	4.12	0.47
Pedagogical Implementation in the Teaching and Learning of Mathematics	4.06	0.56	4.14	0.52
Overall	4.04		4.15	

3.2.3 *Professional Knowledge Component.* Table 6 shows the five elements associated with the professional knowledge components which are mathematics content knowledge, knowledge on the application of ICT in mathematics instruction, knowledge of learners, knowledge of the curriculum and knowledge on intellectual quality. Overall the mean scores for this component are 3.94 (SD = 0.54) for the secondary teachers and 4.04 (SD = 0.46) for the primary teachers. This indicates that level of perception for this component is high and the secondary teachers' knowledge is deemed good while the primary teachers' knowledge is categorized as excellent. Only the element of content knowledge is reported to be at a very high level with a mean score of 4.11 (SD = 0.59) while the teachers' knowledge is reckoned to be excellent. For the secondary teachers, the other four elements are said to have a high level of perception with mean scores of 3.70 (SD = 0.69), 3.92 (SD = 0.58), 3.98 (SD = 0.70), and 3.99 (SD = 0.61) respectively while the teachers' knowledge is deemed good. For the primary teachers, only the element of ICT with a mean score of 3.73 (SD = 0.64) can be said to have a very level of perception and the teachers' knowledge can be categorized as good. The other four elements; mathematics content knowledge (mean = 4.12, SD = 0.49), knowledge of learners (mean = 4.07, SD = 0.50), knowledge of the curriculum (mean =

4.17, SD = 0.55), and knowledge on intellectual quality (mean = 4.06, SD = 0.51) too have a very high level of perception among the primary teachers while their knowledge on the four elements can be described as excellent.

Table 6. Secondary and Primary Mathematics Teachers' Perception of the Professional Knowledge Component

	Secondary		Primary	
Component	Mean	SD	Mean	SD
Mathematics content knowledge	4.11	0.59	4.12	0.49
Knowledge of the application of ICT in mathematics instruction	3.70	0.69	3.73	0.64
Knowledge of learners	3.92	0.58	4.07	0.50
Knowledge of the curriculum	3.98	0.70	4.17	0.55
Knowledge of intellectual quality	3.99	0.61	4.06	0.51
Overall	3.94		4.04	

3.2.4 *Professional Instructional Processes Component.* Table 7 shows the six elements associated with the component of professional instructional processes namely knowledge on the application of tools in enhancing the mathematics instruction, knowledge of a conducive environment, knowledge of meaningful mathematical tasks, knowledge of the teacher's roles, knowledge of the learner's roles and knowledge of how to analyze the mathematics instruction. Overall, the mean scores for this component are 4.07 (SD = 0.53) for the secondary teachers and 4.05 (SD = 0.43) for the primary teachers. This indicates that for this component, the degree of approval for both sets of teachers is very high indeed, while their knowledge can be categorized as excellent. As for the element of application of tools in enhancing the mathematics teachers is concerned, the level of perception can be considered high for both secondary (mean = 3.76, SD = 0.62) and primary (mean = 3.72, SD = 0.55) teachers. It also revealed that their knowledge concerning this element deemed good. For the element of knowledge concerning the conducive environment, the mean scores for the secondary and primary teachers are 4.02 (SD = 0.62) and 4.03 (SD = 0.62) respectively. Similarly, the mean scores for the element of knowledge of meaningful mathematical tasks are 4.09 (SD = 0.59) and 4.10 (SD = 0.48). As for the knowledge about the teacher's roles, the mean scores are 4.2 (SD = 0.55) and 4.19 (SD = 0.48). Meanwhile, the mean scores for the element of the learner's roles, are 4.22 (SD = 0.58) and 4.17 (SD = 0.50) respectively. Lastly, the mean scores for the element of knowledge on the analysis of the mathematics instruction are 4.13 (SD = 0.62) and 4.12 (SD = 0.51) respectively. Moreover, the teachers' knowledge regarding this component is excellent.

Table 7. Secondary and Primary Mathematics Teachers' Perception on the Professional Instructional Processes Component

	Secondary		Primary	
Component	Mean	SD	Mean	SD
Knowledge of the application of tools in enhancing the mathematics instruction	3.76	0.62	3.72	0.55
Knowledge of a conducive environment	4.02	0.62	4.03	0.50
Knowledge of meaningful mathematical tasks	4.09	0.59	4.10	0.48
Knowledge of the teacher's roles in the mathematics instruction	4.20	0.55	4.19	0.48
Knowledge of the leaner's roles during the mathematics instruction	4.22	0.58	4.17	0.50
Knowledge about the analysis of the	4.13	0.62	4.12	0.51

mathematics instruction		
Overall	4.07	4.05

The overall mean scores for the four components are 4.06 and 4.09 respectively for the secondary and primary teachers. This indicates that both sets of mathematics teachers have a very high degree of approval for these components of the pedagogical standards for teaching mathematics. In tandem, their knowledge too can be categorized as excellent.

3.2.5. *Structural Equation Modelling for the Pedagogical Standards for Teaching Mathematics.* In the actual research involving 820 secondary mathematics teachers and 361 primary mathematics teachers, *Kuder-Richardson* and *Alpha Cronbach* coefficients are used to determine the reliability of the questionnaire items relating to all constructs: Professional Practices (PP), Professional Attributes (PA), Professional Knowledge (PK), Professional Instructional Processes (PIP) and Pedagogical Standards for Teaching Mathematics (PSTM). For the secondary teachers, the coefficient value is between 0.859 and 0.934 while for the primary teachers, the coefficient value is between 0.734 and 0.959. This range of values exceeds the range of values between 0.70 and 0.80 which is considered good as proposed by Kaplan & Saccuzzo [10]. Thus, all the items deemed reliable.

This study used the reliability value of the Dillon-Goldstein's composite variables indicator and Fornell & Larcker's extracted mean variance to test the reliability and fitness of the external structural equation modeling as suggested by Lohmoller et al. [11]. The reliability coefficient values for the five constructs of the tested external model are as in Table 8.

Table 8. Reliability coefficient values for the external model of PSTM

Construct	Dillon-Goldstein (≥0.70) Secondary	Dillon-Goldstein ((≥0.70) Primary
Professional Practices (PP)	0.9478	0.8929
Professional Attributes (PA)	0.9205	0.8599
Professional Knowledge (PK)	0.9280	0.7481
Professional Instructional Processes (PIP)	0.7499	0.7381
Pedagogical Standards for Teaching Mathematics (PSTM)	0.9835	0.8196

The Dillon-Goldstein's coefficient values obtained from the study on the primary and secondary mathematics teachers exceed 0.70. Fornell & Larcker's value of 0.9013 also exceeds 0.70. Hence, based on the suggestion made by Lohmoller [11] concerning the acceptable values to be reckoned as significant, the developed pedagogical standards for teaching mathematics model in this study can be accepted. As for the external model that includes PP, PA, PK, PIP and PSTM, the Fitted Index (FI) is used in determining the acceptance of the model. The calculation of FI depends on the value of the extracted mean-variance and the determination coefficient. In this study, the FI values for the secondary and primary mathematics teachers are 0.8216 and 0.8666 respectively. According to Tenenhaus et al. [12], FI represents the overall fitted degree of a developed model. If the FI is more than 0.75, the model is considered fit. Based on the values of the FI obtained from this study that is 0.8216 and 0.8666, the internal model of the developed Pedagogical Standards for Teaching Mathematics (PSTM) can be accepted.

The Pedagogical Standards for Teaching Mathematics Model that involve PP, PA, PK PPI and PSTM is shown in Diagram 1.

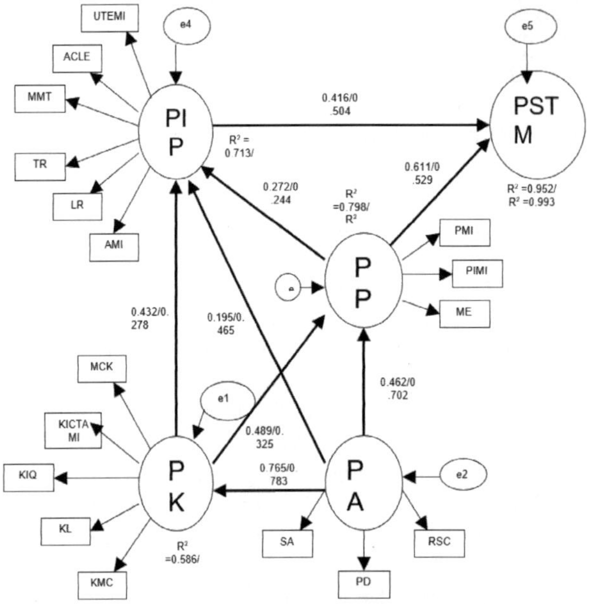

Diagram 1. Pedagogical Standards for Teaching Mathematics (PSTM) Model

The external model equations that have been successfully developed for secondary and primary schools' are as follows:

Secondary mathematics teachers:		Primary mathematics teachers:	
PSTM = 0.611 PP + 0.416 PPI	(i)	PSTM = 0.529 PP + 0.544 PPI	(i)
PP = 0.462 PA+ 0.489 PK	(ii)	PP = 0.454 PA + 0.465 PK	(ii)
PK = 0.765 PA	(iii)	PK = 0.783 PA	(iii)
PPI = 0.272 PP + 0.195 PA + 0.432 PK	(iv)	PPI = 0.244 PP + 0.465 PA + 0.278 PK	(iv)

These two sets of external model equations demonstrate that a causal relationship between the five latent variables that were developed in this model does exist. The structural equation modeling developed shows a direct relationship between PP against PSTM and as well as between PPI against PSTM. All of the regression coefficients for the exogenous and endogenous variables in the model display significant statistical testing values. The causal relationship mentioned earlier refers to the direct effect of the latent variable PP on PSTM and PPI on PSTM. The estimated correlational and regression coefficients values obtained between PP and PPI against PSTM are 0.611;0.529 and 0.416;0.544 respectively. Each of PP and PPI correlated significantly with PSTM. This explained vividly that PP and PPI are the main components in developing the pedagogical standards for teaching mathematics. The regression coefficient value indicates that a unit change in PP would cause an increase of 0.416;0.529 unit change in PSTM while a unit change in PPI would cause an increase of 0.611;0.504 unit change in PSTM.

IOP Conf. Series: Materials Science and Engineering **296** (2018) 012022 doi:10.1088/1757-899X/296/1/012022

The indirect effect shown in the PSTM model is the effect of PK on PSTM through the endogenous variables PP and PPI. The indirect relationship between PA and PSTM can also be found through the mediator variables PP and PPI.

4. Conclusion

The study was carried out with reference to the standards for teaching mathematics put forward by NCTM (USA), AAMT (Australia), and TDA (United Kingdom). However, these standards were not tested statistically as far as the significance and the suitability of the standards' constructs was concerned. As an example, the six standards for teaching mathematics espoused by NCTM came about from discussions among a group of experts that comprised teachers, teacher educators, administrators, researchers, and mathematicians. From these discussions, 30,000 drafts of the standards for teaching mathematics constructed were distributed throughout the United States. This was followed by presentations at various mathematics education conferences. In all, 650 individuals and 70 groups had voiced out their opinions on the standards for teaching mathematics. The feedback was given and the issues raised were used to improve the standards for teaching mathematics to be used in America.

AAMT is an Australian professional body of about 5,500 members that comprises mathematics teachers, teacher educators, and administrators at every level. The Standards for Excellence in Teaching Mathematics was constructed jointly by AAMT and University of Monash through a research project financed by the *'Industry Research Grant'*. This project utilized the *'grounded research methodology'* and was carried out continuously to obtain the mathematics teachers' feedbacks. A 'Teacher Focus Group' which comprised 50 members was formed in each province. This group held discussions for three years in order to produce the *'Standards for Excellence in Teaching Mathematics in Australian Schools* [3]. Three components of the standards were identified but were not tested statistically in determining the suitability of the components.

The model that is produced from this study is known as the Pedagogical Standards for Teaching Mathematics or in short PSTM. This model refers to the level of knowledge about professional knowledge (PK), professional practices (PP), professional attributed (PA) and professional instructional processes (PIP) that a mathematics teacher must have in order to make a transformation in carrying out these standards before, during and after the mathematics instruction. The finding from the study revealed that the secondary and primary mathematics concurred highly with PA, PP, and PPI. However, the respondents' level of knowledge on PK was only satisfactory. Based on the mathematics teachers' perception, the mean scores for the overall mathematics pedagogical standards were 4.09 for the secondary teachers and 4.06 for the primary teachers. This shows that the respondents displayed a very high degree of approval towards the PSTM. They concurred that the four knowledge components of PK, PP, PA, and PPI formed a crucial knowledge for mathematics teachers to achieve excellence in teaching.

The instrument for the Standards for Teaching Mathematics's Model Instrument that had been developed was successful in exploring and testing statistically the model that was based on the four components namely, professional knowledge, professional attribute, professional practice and professional process of instruction. The composite reliability indices of the indicator variables that consisted of constructed items were tested and the results indicated that the items fitted nicely into the Pedagogical Standards for Teaching Mathematics (PSTM) model. Testing on the five constructs of PP, PA, PK, PPI, and PSTM showed that values of the tested reliability coefficients satisfied the condition to be accepted as the constructs in the Standards for Teaching Mathematics Model. The results of the statistical test using the Structural Equation Modelling (SEM) based on the Partial Least Squares (PLS) method showed that the model could be fitted significantly. Consequently, the four components plus the items constructed could be used as indicators for the mathematics teachers' standards for teaching and in determining the excellent level of mathematics instruction in Malaysia.

The component of 'Professional Practices' (PP) is seen as a direct contributor towards PSTM while 'Professional Knowledge' (PK) gives an indirect contribution via 'Professional Attributes' (PA) and 'Professional Processes of Instruction' (PPI). This means that PA gives a direct contribution towards PPI and PPI, in turn, gives a direct contribution towards PSTM.

To achieve 'Professional Practice' (PP) a teacher needs to have 'Professional Attribute' (PA) and 'Professional Knowledge' (PK). These practices of PA and PK give indirect contributions towards PSTM. 'Professional Knowledge' is seen as a direct contributor towards 'Professional Process of Instruction' (PPI) and 'Professional Practice' (PP). Nevertheless, PI also contributes indirectly towards PSTM but through PA and PPI.

Furthermore, 'Professional Knowledge' (PK) and Professional Attribute' (PA) are looked upon as the pillars for 'Professional Practice' (PP) and 'Professional Process of Instruction' (PPI) in improving the quality of mathematics instruction. For these two components to be implemented successfully, the mathematics teacher needs to have knowledge of mathematical content, mathematics curriculum, tools of technology, the learners, the learners' intellectual qualities while delivering the instruction as well as displaying concerns for them, enhancing his or her professional development from time to time and contributing significantly to school and community in improving the level of mathematics achievement.

References

[1] Shulman L S 1986 Those who understand: Knowledge growth in teaching *Educ. Res.* **15** pp 4–14
[2] National Council of Teachers of Mathematics 2000 *Principles and Standards for School Mathematics* ed V A Reston (New York: Guilford Press)
[3] The Australian Association of Mathematics Teachers 2006 *Standards for Excellence in Teaching Mathematics in Australian Schools* (Adelaide: The Australian Association of Mathematics Teachers Inc.)
[4] Kementerian Pelajaran Malaysia 2007 Pelan Induk Pembangunan Kementerian Pelajaran *Learn. Sp. Websites*
[5] Kementerian Pelajaran Malaysia 2003 *Standard Kualiti Pendidikan Malaysia* (Jemaah Nazir Sekolah)
[6] Bahagian Pendidikan Guru 2007 *Draft Standard Guru Malaysia* (Malaysia: Kementerian Pelajaran Malaysia)
[7] Nugent W, Sieppert J and Hudson W 2000 *Practice evaluation for the 21st Century* (Pacific Grove, CA: Brooks/Cole)
[8] Saad N S, Tajudin N M, Mohamed Z, Musa M, and Boon L K 2009 Mathematics teaching standard: Raising the quality of mathematics teacher in Malaysian school *Prosiding 1st International Conference on Educational Research and Practice (ICERP 2009)* (Putrajaya: Marriot Hotel)
[9] Pazin A H 2012 *Standard Pengajaran Matematik Dalam Kalangan Guru Matematik Sekolah Rendah* (Malaysia: Universitas Sultan Idris)
[10] Kaplan R M and Saccuzzo D P 1997 *Psychological testing: Principles, applications, and issues* (Belmont: Brooks)
[11] Lohmoller J B 1989 *Latent variable path modeling with partial least squares* (Heidelberg: Physica Verlag)
[12] Tenenhaus M, Vinzi V E, Chatelin Y M and Lauro C 2005 PLS path modeling *Comput. Stat. Data Anal.* **48** pp 159–205

An early start to STEM education among year 1 primary students through project-based inquiry learning in the context of a magnet

N Safiee, Z M Jusoh, A M H M Noor, O E Tek and S M Salleh

Sultan Idris Education University, Malaysia

norabunga@gmail.com

Abstract. For the needs of the 21st century, the Government of Malaysia has conceptualized the Malaysia Education Blueprint 2013-2025 which embodies 11 strategic and operational shifts. In Shift 1, it is emphasized that the quality of Science, Technology, Engineering, and Mathematics (STEM) Education will be enhanced. This study employed the mixed-method approach using the "one-group pre-test and post-test design". Accordingly, this paper describes the pedagogical practice of Project-based Inquiry Learning (PIL) which promotes STEM Education among Year 1 students in the move to progress in tandem with Shift 1. Specifically, using the context of a magnet which has been stipulated in the Primary School Standard Curriculum, Year 1 students experienced the STEM Education through the STEM Pedagogy in which they raised questions upon the presentation of a relevant stimulus (Inquiry Phase), explored the ways in which a train carriage or coach could be assembled by means of recycled materials and magnets (Exploration Phase), designed a train carriage (Design Phase), and ultimately reflected on their inventions (Reflection Phase). The cognitive and affective impacts through the use of this Project-based Inquiry Learning are presented. Implications for the teaching and learning of science are discussed within the context of STEM Education.

1. Introduction

In the quest of becoming a developed nation by 2020, Malaysia has given great emphases on education. Moreover, given that science, technology, engineering and mathematics education plays a big role as the catalyst in meeting the challenges and demands of our present and future economy, Malaysian government has instituted a number of related policies such as the 60:40 Science/Technical: Arts (60:40) Policy in 1967. Statistics indicate that, as of 2014, only about 45% of students graduated from the higher secondary schools were from the Science stream, including technical and vocational programmes. Additionally, the percentage of secondary school students who chose not to pursue the Science stream despite meeting the requirement based on their Form 3 National Standardised Examination (PMR) had increased to approximately 15% [1].

Meanwhile, in the context of rising international education standards and the aspiration of better preparing Malaysia's children for the needs of the 21st century, the Government of Malaysia has conceptualized the Malaysia Education Blueprint 2013-2025 [2] which embodies 11 strategic and operational shifts. Essentially, there are five outcomes that the Blueprint aspires to achieve for the

Malaysian education system as a whole: access (i.e. 100% enrolment across all levels from preschool to upper secondary by 2020), quality (i.e. top third of countries in international assessments such as PISA and TIMSS in 15 years), equity (50% reduction in achievement gaps: urban-rural, socioeconomic and gender, by 2020), unity (an education system that gives children shared values and experiences by embracing diversity), and efficiency (i.e. a system which maximises student outcomes within current budget).

Research indicates that when children engage in projects in which they conduct investigations and/or involve in creation or invention around their personal questions, their intellectual capacities are very likely to be provoked and utilized [3–5]. Furthermore, engagement in projects places children in an active and interactive role where they take responsibility and initiative in inquiring by means of generating questions which they would like to seek answers to; exploring by means of collecting relevant data or information culminating in a suitable design or procedures; investigating based on the procedures or inventing based on the design; and showcasing or reporting their work [4], and ultimately yields better school participation and achievement in the long term [5].

Ong et al. [6] concocted a STEM pedagogical approach termed as Project-based Inquiry Learning or PIL which entails 4 phases, namely Inquiry, Exploration, Invention, and Reflection. Using the context of a ship, Ong et al. [7] illustrate that the inquiry-based learning centres on "Inquiry" regarding a ship, culminating in a question on "how to build a ship?" which in turn, leads to the "Exploration" of the phenomenon on floating and sinking, and of information pertaining to ships of different shapes and sizes which concludes with a design of a ship. Such design will then be accomplished and materialized through project-based "Invention", and this is then followed by showcasing and reflection of the completed projects/inventions/investigations, and appreciation of the projects by peers. The STEM integration in this Project-Based Inquiry Learning encompasses the *science* concepts on buoyancy and energy, the *technology* which is reflected in building ships, the *mathematics* which is manifested in size and symmetry in ships, and *engineering* which involves designing of a ship.

2. Methods

2.1 Research Design

This study employed the mixed-method approach using the "one-group pre-test and post-test design" in which quantitative attitudinal data were collected and then triangulated by interview data which mainly solicit children's view on their self-perceived understanding [8]. In the qualitative data, interview approach was used by means of interviewing seven randomly selected students from the single group students after they have undergone the learning of magnet using PIL approach to ascertain their interest towards the learning of magnet based on Project-based Inquiry Learning and equally, their understanding of the concept of magnet and their reflection. The reflection of a science teacher (the first author) in conducting a lesson on magnet through the using of a Project-based Inquiry Learning was shared.

2.2 Sampling

A purposive sampling was used as selection criteria that only involves a school namely Primary School A (name withheld for anonymity purpose) in the Ipoh district where students were in Year 1 class. Table 1 provides a breakdown of the sample by gender. Overall, there were 27 years 1 students who participated in this study.

Table 1. *Breakdown of sample based on gender*

Gender	Total
Boys	15
Girls	12

2.3 Instrumentation
The questionnaire contains two parts. Part 1, consisting of three-point Likert-scale items, measures students' attitudes towards learning of science through the Project-based Inquiry Learning approach, while Part 2, consisting of 3 interview questions: (a) Do you enjoy making train carriage? (2) What have you learned about magnets from the train project? (3) How can you improve your train (invention)? The reliability of Part 1 of the Questionnaire was established using the data set of 27 students of which the Cronbach's alpha was measured at 0.917, indicating that the questionnaire has high internal reliability [9].

2.4 Lesson Plan on Project-Based Inquiry Learning Model
The PIL-based lesson plan on Magnet, carried out for the duration of 3 weeks, entails four major steps/phases. In the first phase (inquiry), the teacher draws the attention of students to the various pictures and video clips of a train as a stimulus, and students are asked to state what they knew about the train and subsequently, what they want to know more about the train. They also explore how magnet could be used to connect the carriages of the train. In the third phase (design), students build and design train carriages (i.e., engineering inaction) based on their earlier sketches produced in the second phase.

2.5 Data Gathering and Analysis Procedures
Part 1 of the questionnaire was administered to the students before and after the lesson so as to obtain feedback on their attitudes towards learning science through Project-based Inquiry Learning. The data collected were analyzed using a paired samples t-test. Meanwhile, seven students were interviewed based on the 3 interview questions in Part 2 so as to obtain feedback on their interest towards learning science and the knowledge acquired, and their reflection on the aspects their invention could be improved. The interview data will be transcribed and analyzed thematically.

3. Results
Research Question 1: To what extent does the Project-Based Inquiry learning increase Year 1 students' attitudes towards learning science?

Table 2. *Results Obtained from t-Test for Paired Samples (Students' interest in learning Science)*

Attitudes Before PIL			Attitudes After PIL				
N	Mean	SD	N	Mean	SD	T	p
27	11.04	4.49	27	24.00	0.00	14.99	.000

A paired-samples t-test was conducted to compare the self-perceived attitudes towards science before and after the magnet science lesson using Project-based Inquiry Learning (PIL). There was a significant difference in the scores for the attitudes towards science before magnet lesson using PIL ($M=11.04$, $SD=4.49$) and the attitudes towards science after the magnet lesson using PIL ($M=24.00$, $SD=0.00$) conditions; $t(26)=14.99$, $p < .001$. These results suggest that Project-based Inquiry Learning does have an effect on students' attitudes towards science. Specifically, our results suggest that when Project-based Inquiry Learning was employed, the attitudes towards science increases.
Research Question 2: To what extent does the Project-Based Inquiry learning increase Year 1 students' students' interest, understanding about magnet and reflective ability?
Based on the responses of seven students to Question 1, resounding major theme emerged, which has students enjoyed science. The enjoyment was expressed in different forms such as they enjoyed making and creating their own train carriage models, and enjoying the lesson as a whole. This theme is supported by the following transcripts of the interview data:-

Yes, I enjoy making the train carriage. (Student A)
I like to make train carriage. (Student B)
Yes, I want to make train carriage again. (Student C)
Yes, I was very excited making train carriage. (Student D)
Yes, I enjoyed [the lesson]. (Student E)
Oh, I enjoyed making the train carriage. (Students F & G)

When asked about what they have learned which was posted as Question 2, the responses revealed students' understanding of magnets which could be categorized into 3 general themes:

- ***The use of magnet***

Students A, D and G seem to show the understanding that magnets are able to connect the carriages, although the way in which the connection may occur was not explicated specifically.

Yes, I understand how to use magnets to connect train carriage. (Student A)
The magnet can connect the carriages. (Student D)
The magnet can attract the train carriage. (Student G)

- **The things that a magnet attracts**

One of the understandings which emerged from the interview data is that of the things which a magnet attracts. During the exploration phase, students explored about the magnets, and it was specifically explicated in the interview data that …

The magnet can attract iron. (Student B)

- **Differing poles of magnets attract**

Students did learn the basic facts of a magnet such as same poles repel while opposing poles attract each other. This is supported by the following transcript:

Magnets can attract each other if [they are of] different poles. (Student C)

When asked how the students could improve their train (invention) which was posted as Question 3 as a means to solicit their reflection, the responses could be categorized into three aspects in which the students would like to see improvement of their designs. The three aspects are:

- **Loads**

Students would like to see that improvement could be made in terms of load. They would like to see an improvement is made to their design so that their invention would be able to withstand and carry more loads as succinctly supported by the following transcription:-
My train carriage will carry more goods. (Students A & F)

- **Speed**

Meanwhile, the majority of the students interviewed expressed an improvement in the speed that the train could move. They have not have learned, let alone explored in the science classrooms about friction, axles, and wheels although they have encountered them in the real-life situation, the following expressions should someday be realized:-
My train carriage will move faster. (Student B)
My train carriage will be much … faster. (Student C)
My train can move faster. (Student D)
My train carriage will move faster. (Student E)
My train magnet carriage moves faster. (Student G).

- **Size**

Finally, the third aspect of improvement to the invention is in terms of the size of the train. They would like to see that, in future, they are able to build a bigger train as depicted in one of the transcriptions:-
My train carriage will be much bigger and faster. (Student C)

4. Reflection and Conclusion

The findings of this study indicated that students perceived that their attitudes towards learning Science have been elevated through the Project-based Inquiry Learning, evident in the significant differences between self-perceived attitudes towards science before and after learning science through PIL. Additionally, students' understanding of the concept at hand has also been elevated and the understanding was categorized into three themes, namely the use of a magnet, the things that a magnet attracts, and differing poles of a magnet attract. Students also indicated that should they asked to improve on their designs, improvements to the loads that a train may carry, the speed that a train travels, and the size of a train would be their priority.

There are some pedagogical implications for teachers based on the experience of conducting the current study. Firstly, the teacher should allow students the opportunity to raise questions upon the presentation of a relevant stimulus train. Project-based Inquiry Learning is one of the ways teachers could do so. Besides, the teacher needs to be able to handle student questioning and be tactful in convincing students which question should be explored and investigated. In this case, how to make a train using magnets was the question being explored and investigated to which students did enjoy themselves and learn the concept at hand (i.e., magnets). As evident in this research, teachers should be convinced on the beneficial use of Project-based Inquiry-based Learning as it not only enhances students' understanding of the concept at hand, it also increases creativity students when Project-based Inquiry Learning method is maximally utilized due to the higher level of student engagement and on-task behavior.

References

[1] Azian T S A 2015 STEM Education: Policies and prospects towards achieving the International standard and meeting national development needs *STEM Education: Policies and prospects towards achieving the International standard and meeting national development needs* (Kuala Lumpur: STEM Education)
[2] Kementerian Pelajaran Malaysia 2012 *Pelan Pembangunan Pendidikan Malaysia 2013- 2025* (Putrajaya: Kementerian Pelajaran Malaysia)
[3] Katz L G 2010 STEM in early years *SEED (STEM Early Educ. Dev. Conf.*
[4] Katz L G and Chard S C 2000 *Engaging children's minds: The project approach* (Stamford: Ablex)
[5] Helm J H and Katz L G 2001 *Young investigators: The project approach in the early years* (New York: Teacher College Press)
[6] Ong E T and Ong Eng Tek , Aminah Ayob, Md Nasir Ibrahim , Mazlini Adnan, Jameyah Shariff N I The effectiveness of an in-service training of early childhood teachers on STEM integration through Project-Based Inquiry Learning (PBIL) *Turkish J. Sci. Educ.* **13** 44–58
[7] Ong Eng Tek , Aminah Ayob, Md Nasir Ibrahim , Mazlini Adnan, Jameyah Shariff N I 2016 Integrating STEM into Early Childhood Education: Is it feasible? *Proceedings of the International Conference of Education in Mathematics, Science, and Technology* (Bodrum, Turkey: ISRE (International Society of Research in Education and Science)) pp 385–90
[8] Gay L R and Airasian P 2009 *Educational Research* (Ohio: Prentice Hall)
[9] Nunnally J C 1978 *Psychometric theory* (New York: McGraw-Hill)

IOP Conf. Series: Materials Science and Engineering **296** (2018) 012024 doi:10.1088/1757-899X/296/1/012024

The development of learning material using learning goal orientation approach in digital electronics

P Puspitaningayu, L Anifah and N Kholis

Department of Electrical Engineering, Faculty of Engineering Universitas Negeri Surabaya, Indonesia

pradinip@unesa.ac.id

Abstract. Mastery of digital electronics principles is essential for future engineers in the digital era. This article describes the use of simulations in an undergraduate electrical engineering course to promote the adoption of a learning-goal orientation. This study used experimental method. This was done by providing students with a simulation environment which students freely use to experiment with various circuit models. Students were then invited to reflect on how the simulation results compare with results from lab experiments. The module got 82% of positive rating from 28 students and all of them passed in the examination with 81.8 as the average score. Those majority students were motivated by the combination of two learning goals written in the module. Moreover, they also gain the ability to design more complex systems because of their combined experience. Additionally, the module also has been validated and got 83% of reliability. The final product of this research hereafter can be recommended to be used as teaching material.

1. Introduction

Digital technology in today's information age contributes tremendously to the development of the overall human lifestyle. Beginning with the invention of transistors in 1947, human civilization has shifted due to technological developments ranging from vacuum tubes to semiconductor devices [1]. The presence of semiconductor technology then continues to evolve with the introduction of integrated circuit (IC) which enables the processing of digital data automatically by a tiny component.

With the development of digital technology, binary information processing should also be done according to the logic gate principle [2]. The need for encoders and decoders is also vital for converting human acceptable analog signals into digital signals that can be processed by the computer. Data processing will be more easily done digitally than using the more intrusive and inefficient analog signal principle.

Because of the importance of knowledge about the principles of digital electronics, it is one of the compulsory subjects in Electrical Engineering. The need for instructional media that can effectively meet the learning achievements then emerges, especially for this subject [3-5]. In cognitive theory, there are two goal orientations: performance goal and learning goal. Furthermore, students with learning goal orientation likely have a better motivation and achieve better result compared to students with performance goal [6].

Laboratory modules are commonly structured as properties, goals, and a sequence of steps to reach the main purpose [7]. This is a classic direct instruction which known as the best method to attain procedural knowledge or skills [8]. However, this method is lack of creativity and analytical stimulation for the students because they are only following a single set of instructions. In engineering, students need the ability to solve problems using different approaches. They need to have a constructive mind on any issues an engineer might be encountered. By adding another approach to achieve the learning outcome, students are required to learn more by experimenting on the various way for the same target.

Based on the background that has been mentioned above, this study is intended to develop a laboratory module for Digital Electronics based on learning goal orientation approach. The goal is derived from the subject's learning outcomes then applied in the module design which is designed to have two instructions: simulation practice and laboratory work. By using this approach, the learning process is expected to be effectively directed and students may have a variety of the learning experience.

2. Methodology

The development of digital electronics learning the material with learning goal orientation is organized using Four-D model proposed by Thiagarajan and Semmel [9] which consist of: define, design, develop, and disseminate process which shown in figure 1 below:

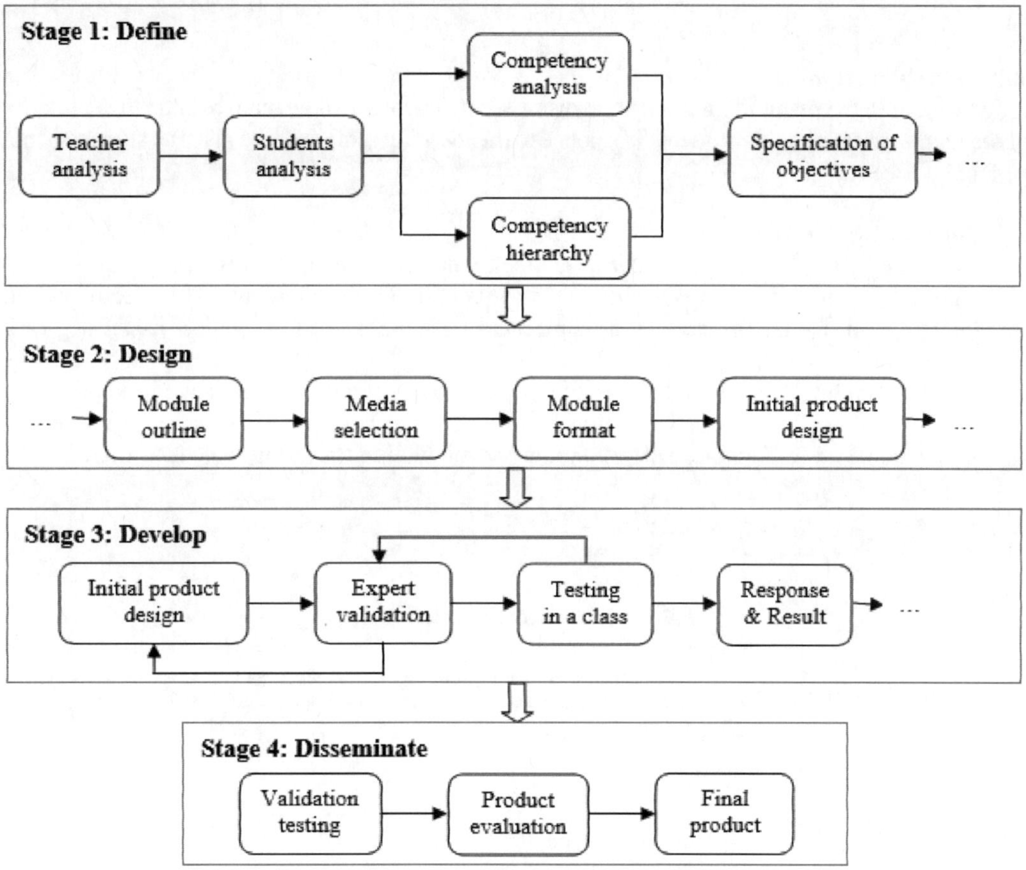

Figure 1. Four-D model for learning material development

The stages are started by defining the problems by analyzing both teacher's and student's current situation, considering the targeted competencies of the subject and its concept, then finally result in a specification of objectives. To specifying the objectives, there's a need for defining instruments. This study is using a group discussion between teachers and students to evaluate the current situation and its potential problems.

Secondly, after determining the specification, a concept of the module can be designed including its format and media selection. The initial design then needs to be evaluated by experts, tested in the developmental stage, and get feedback from those processes.

The third process is the development stages where the material is developed based on the basic needs and also feedbacks from other parties. In this research, the module is developed in a different way where it combined the usual lab work instructions with simulation steps.

The last part is the disseminate stage where the produced module is tested, validated, and evaluated until it can be launched as the final product. Along with the development of the final learning material, students' response and their learning outcome also been monitored.

Furthermore, in conducting the research, it is necessary to have an instrument related to the evaluation process of the whole research process. A research instrument is a tool or facility used by researchers in collecting data to make the work easier and the results better, in a more accurate, complete, and systematic so that more easily processed [10, 11]. Instruments in this study include validation sheet of teaching module and student response questionnaire. Validation of teaching materials will be done by colleagues in Department of Electrical Engineering Universitas Negeri Surabaya, while the questionnaire will be given to the Diploma-3 Electrical Engineering students as the subject of the research.

3. Results and discussion

This section describes the result of this study and discusses it. There are three main results: the validation result of the learning material, the students' response to the learning material, and also the students' final score on Digital Electronics.

3.1. The expert validation result

This research produces the Digital Electronics Practice module. The module was further validated by four lecturers of Electrical Engineering Department at Universitas Negeri Surabaya. Three lecturers with a master degree in electrical engineering act as subject experts and one lecturer with a master degree in education act as a media expert.

Assessment is given by the validator by checking (√) on the list of indicators indicated and processed by the researcher's rating. Instrument validation criteria include: (a) Module Structure, (b) Material Writing Organization, and (c) Language. The formula for calculating the rating scale is as table 1:

$$\text{Score} = \frac{\sum respondents'answer}{\sum respondents'highest\ score} \times 100\% \qquad (1)$$

Table 1. Assessment criteria

Quantitative Assessment	Qualitative Assessment	Score
20% – 35%	Very Invalid	1
36% – 51%	Invalid	2
52% – 67%	Quite Valid	3
68% – 83%	Valid	4
84% – 100%	Highly Valid	5

Table 2 below presents the module validation result data provided by the validator and has been calculated using the above formula.

Table 2. Result of module validation

No.	Aspect	Score 1	2	3	4	5	∑ Validator	Rating
1.	Structure							
	a. Organization for the presentation in general				4		4	80 %
	b. General presentation is attractive			1	3		4	75 %
	c. The consistent linkage between the subject matter				2	2	4	90 %
Average of structure rating								81.25 %
2.	Organization of Materials							
	a. Material coverage				2	2	4	90 %
	b. Clarity and material order			1	3		4	75 %
	c. Material precision with lesson plan					4	4	100 %
	d. The relationship between the problem and the student's life/cognition context contained in the module			1	3		4	75 %
Average of materials rating								85.00 %
3.	Language							
	a. Use of good and appropriate language				4		4	80 %
	b. The language is communicative				3	1	4	85 %
	c. The sentences' structures are effective				4		4	80%
Average language rating								81.67 %

Based on those results, the Digital Electronics Practice module can be categorized valid with the final result rating of 82.64% so it is worthy to be used as a learning instrument in the Department of Electrical Engineering

3.1.1. Validation results of students' response questionnaire

Students' response questionnaire of the Digital Electronics Lab Module is also validated by the validators. Criteria and response validation results are shown in table 3 below:

Table 3. Result of students' response questionnaire validation

No.	Aspect	Score 1	2	3	4	5	∑ Validator	Rating
1.	The module presentation							
	a. The module is well presented so it's easy to understand				3	1	4	85%
	b. The module is presented in graphs and tables					4	4	100%
	c. The module has bibliography			2	2		4	70%
The average of presentation rating								85%
2.	Material							
	a. The material covered in the module is well organized				4		4	80%
	b. Module is easy to understand				4		4	80%

The average of material rating						80%
3	Language					
	a. Use of good and appropriate language		4		4	80%
	b. The language is communicative		3	1	4	85%
	c. The sentences' structures are effective	1	3		4	75%
The average language rating						80%

Based on Table 3, the rating of the student's validation response is 81.67% so that the module is declared eligible to be used to take the student's response in the Electrical Engineering Department.

3.1.2. Result of student response questionnaire

After the questionnaire of the validated student response, this questionnaire was given to 20 D3 Electrical Engineering students class of 2015. The questionnaire results were used to find out the student's response regarding the content of Digital Electronics lab module. Table 4 below shows the questionnaires that were distributed to students. The criteria for assessment of student responses refer to table 4 concerning the validated questionnaire.

Table 4. Result of students' response questionnaire

No	Highly Agree	Agree	Not Sure	Disagree	Very Disagree	$\sum$ Score in each point	Criteria
1.a	8	10	2	-	-	87 %	Very Strong
1.b	2	16	2	-	-	81 %	Strong
1.c	3	15	2	-	-	82 %	Strong
2.a	4	14	2	-	-	83 %	Strong
2.b	2	14	4	-	-	80 %	Strong
3.a	2	14	4	-	-	83 %	Strong
3.b.	5	12	3	-	-	83 %	Strong
3.c	3	12	5	-	-	80 %	Strong

The result of table 4 can also be described in figure 2 below:

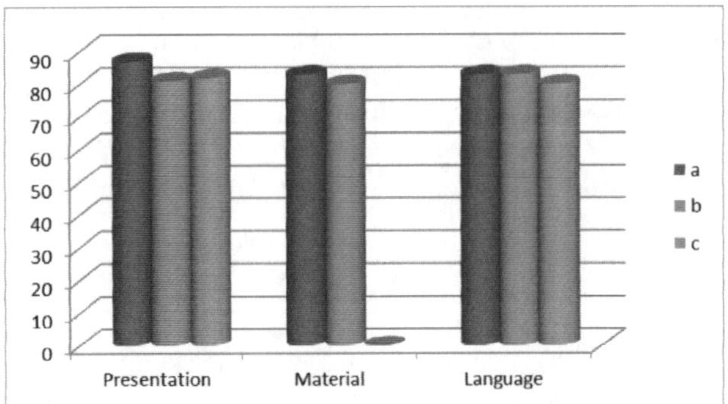

Figure 2. Diagram of students' response questionnaire

Based on the result of a questionnaire of student response, the result that the total score of student response using the following formula:

Σtotal response score = (overall score) / 8

Σtotal response score = 659/8

Σtotal response score = 82.37%

Thus, in Digital Electronics Practice learning, the use of this module gives excellent results with a score of 82.37%.

4. Summary

Based on the formulation of the problem and the results of the data obtained in the study and has been poured on the discussion, it can be concluded that:

- Based on the module validation results, the Digital Electronics lab module obtained 81.25% score on Structure Module criteria, 85% on the criteria of Writing Material Organization, and 81.67% on the Language criterion. Thus, the overall value is 82.64% and the module can be declared eligible to be used as a learning instrument at the Department of Electrical Engineering, Universitas Negeri Surabaya.
- The response result from 20 D3 Electrical Engineering students stated that in three categories: Presentation of Module, Material, and Language contained in Digital Electronics Practice module get the result of rating equal to 82.37%.

References

[1] Boylestad R L and L N 2013 *Electronics Devices and Circuit Theory* (New Jersey: Pearson Education)
[2] Tocci R J 1991 *Digital System Principles and Applications* (London: Prentice-Hall)
[3] Hamalik O 1994 *No TitleMedia Pendidikan. (cetakan ke-7)* (Bandung: Penerbit Citra Aditya Bakti)
[4] Arsyad A 2011 *Media Pembelajaran* (Jakarta: Rajawali Pers)
[5] Musfiqon 2012 *Pengembangan Media dan Sumber Pengajaran* (Jakarta: Prestasi Pustaka)
[6] Maeher M L and Zusho A 2009 *Achievement Goal Theory: The past, present, and future* (Routledge)
[7] Tollefson N 2000 Classroom Applications of Cognitive Theories of Motivation *Educ. Psychol. Rev.* **12** pp 63–83
[8] Arends R I 2012 *Learning to Teach Ninth edition* (New York: McGraw Hill)
[9] Thiagarajan S, S S D and Semmel M I 1974 *Instructional Development for Training Teachers of Exceptional Children: A Sourcebook* (Minnesota: Leadership Training Institute/Special Education)
[10] Arikunto S 1997 *Prosedur Penilaian Suatu Pendekatan Praktek* (Jakarta: Rineka Cipta)
[11] Riduwan *Skala Pengukuran Variabel-variabel Penelitian* (Bandung: Alfabeta)

IOP Conf. Series: Materials Science and Engineering **296** (2018) 012025 doi:10.1088/1757-899X/296/1/012025

The use of Facebook and WhatsApp application in learning process of physics to train students' critical thinking skills

R Kustijono[1] **and F Zuhri**[2]

[1]Department of Physics, Universitas Negeri Surabaya
[2]Department of English, Universitas Negeri Surabaya

rudykustijono@unesa.ac.id

Abstract. The purpose of this research is to describe the learning process by using Facebook and WhatsApp to train students' critical thinking skills. The research steps are: 1) analysis; 2) design; 3) development; 4) implementation; 5) evaluation. The research subjects are 40 students of Physics Department of Universitas Negeri Surabaya. This research used descriptive qualitative approach. The study The validation point, practicality, effectiveness, and critical thinking skills of students assessment use Likert scale. Learning process criteria are eligible if $\geq$ 60% is rated good or excellent. The results are: 1) the use of Facebook and WhatsApp can be implemented in the learning process, and the existing constraints can be overcome; 2) the assessment of students' critical thinking skills is categorized as good and excellent. These results suggest that learning by using Facebook and WhatsApp can be used to train students' critical thinking skills.

1. Introduction

Nowadays, the use of smartphones has threatened the existence of teachers in conventional classes when students are already bored with the learning process in the classroom [1]. Students frequently use smartphones to access today's popular social networks like Facebook and WhatsApp just for fun. WhatsApp is a free messenger application that works on various platforms like iPhone and Android phones, and this app is widely used among students to send multimedia messages like photos, videos, audio along with short messages [2]. It is unwise if teachers forbid students to bring smartphones to school because basically a result of technology is neutral.

To empower the use of smartphones by students, learning with digital communication between groups of students and between students and teachers has become popular over the past decade through various networks including Facebook and WhatsApp [3]. Basic physics which is the basic idea that arises from the application of the scientific method that examines the most basic ideas about the physical properties [4] requires a learning process that can train critical thinking skills is an attempt to ask and answer questions systematically to produce an explanation which is coherent and credible [5]. The previous study of the use of WhatsApp in learning has four main purposes: communicating with students, building a social atmosphere, creating dialogue, and encouraging students to share [6], and with a similar purpose, it is also applied in learning by using Facebook. The delivery of messages, pictures, and videos related to the explanation of physical symptoms can be a means to train students' critical thinking skills.

Research on the use of WhatsApp in the South African University class got positive responses from students stating that it is an easy way to communicate with their teachers and all classmates, and generate useful discourse on relevant issues. In an informal environment, students can learn something authentically intimate and fun [12].

2. Methods

This Research uses Research & Development which encompassed design, development or production, implementation or delivery, and evaluations [20]. Stage analysis is the process of analyzing the need for product development, analyzing the feasibility of the product, and the terms of its development. Product development begins with the gaps that occur due to existing learning which has not been correlated to the needs, learning environment, technology, and characteristics of students. The design stage is a systematic process that starts with setting and learning objectives, designing learning scenarios, designing learning tools, designing learning materials and evaluating learning outcomes.

3. Result of analysis

3.1. Learning design validity

A product is called valid if the product can reflect the soul of knowledge (state of the art of knowledge), which can also be called the validity of the content. The validity of the use of Facebook and WhatsApp in the learning process in terms of instructional aspects and technical aspects. The instructional aspects set include learning objectives, learning scenarios, instructional materials, instructional media, learning environment, and evaluation of learning outcomes. Technical aspects set include: easy to operate, accessible, and cheap. The validation result from 3 peer lecturers towards the use of Facebook and WhatsApp in the learning process as shown in Table 1.

Table 1. Learning design validity

	Facebook				WhatsApp			
	1	2	3	4	1	2	3	4
Instructional:								
• learning objectives	0	0	2	1	0	0	2	1
• learning scenarios	0	0	3	0	0	0	3	0
• learning materials	0	0	3	0	0	0	3	0
• learning Media	0	0	1	2	0	0	1	2
• learning environment	0	0	0	3	0	0	0	3
• evaluation of learning 0 outcomes		0	3	0	0	0	3	0
Technical:								
• easy to operate	0	0	0	3	0	0	0	3
• easy to access	0	0	0	3	0	0	0	3
• inexpensive	0	0	0	3	0	0	0	3

Based on Table 1, the validator stated the instructional aspect, as well as the technical aspects, are 33,3% good and 66,7% very good. Results are shown concurrently in conceptual usage. Therefore, the developed learning design can be applied in the learning process in the classroom.

IOP Conf. Series: Materials Science and Engineering **296** (2018) 012025 doi:10.1088/1757-899X/296/1/012025

3.2. Learning design practicality

A product is said to be practically available to people when it is usable. The practicality of using Facebook and WhatsApp in the learning process is in terms of the implementation and technical. Technical aspect in question: easy to operate, easy to access, cheap, commonly used, no memory consuming, non-time consuming, and non-eyestrain. The instructional aspects set out: access to material search, be able to share material, full of facilitators, study anytime and anywhere, and provide a pleasant environment. Practical data are collected by using questionnaires to the Facebook class (20 students) and WhatsApp class (20 students). The researcher uses the Facebook and WhatsApp in the learning process. The result is shown in Table 2.

Table 2. Learning design practicality

	Facebook				WhatsApp			
	1	2	3	4	1	2	3	4
Technique:								
easy to operate	1	3	9	6	1	3	8	8
Easy to access	2	2	10	6	2	3	9	6
Inexpensive	2	3	8	7	2	2	9	7
get used to	1	2	10	7	1	2	9	8
non-memory consuming	3	4	8	5	6	5	5	4
non-time consuming	7	6	5	2	6	6	5	3
non-eyestrain	8	6	4	2	7	6	4	3
Instructional:								
easy access in searching material	3	4	8	5	2	5	7	5
material is shareable	2	2	9	7	2	4	8	6
facilitator's availability	3	4	9	4	3	4	8	5
learning anytime and anywhere	1	2	10	7	2	3	9	6
providing fun learning	2	2	9	7	1	2	10	7

Based on Table 2, the number of students who stated the practicality of using Facebook in the learning process is: very less 15%, less = 17%, good = 41%, and very good = 27%. While the number of students who stated the practicality level of use of WhatsApp in the learning process is: very less 15%, less = 19%, good = 38%, and very good = 28%. These results indicate that in general the use of Facebook and Whatshap in the learning process can be declared practical (Facebook = 68% and WhatsApp = 66% practicality in both good and excellent category). Technical obstacles in the use of Facebook in the learning process is little and can be overcome. Those obstacles are time-consuming (65%) and eyestrain (70%). While on the use of WhatsApp, there are technical constraints also occur due to consuming memory (55%), time-consuming (60%), and eyestrain (65%).

3.3. Learning design effectiveness

The explanations of each of these elements are [17] : 1) analyzing, the process of learning something to identify elements or relationships between elements; 2) Evaluating, the process of critically reviewing and responding to materials, procedures, or ideas, and assessing them against other objectives, standards, or criteria; 3) applying, the process of using ideas, processes, or skills in a new situation; 4) Generating ideas, a process of expressing thoughts that express originality, speculation, imagination, a personal perspective, flexibility in thinking, discovery or creativity; 5) expressing ideas, the process of presenting initial and logical ideas while using appropriate language for the audience. The rating category uses the rubric as follows: very good if it meets all indicators, good if it meets only 4 indicators, less if it meets only 3 indicators, is very less if it meets less than 3 indicators. Data were obtained from observation of student activity in the group for the Facebook class (20 students) and WhatsApp class (20 students) by 2 observers. The results obtained are shown in Table 3.

Table 3. Student's Critical Thinking Skills

	Facebook				WhatsApp			
	1	2	3	4	1	2	3	4
Analyzing	2	7	30	1	3	6	28	3
Evaluating	2	6	29	3	2	5	31	2
Applying	3	5	31	1	5	8	25	2
Generating ideas	5	5	28	2	6	7	27	0
Expressing ideas	6	7	27	0	6	8	25	1

Based on Table 3, the use of Facebook in the learning process produces critical thinking skills of students as follows: very less 9%, less = 15%, good = 72.5%, and very good = 3.5%. While the use of WhatsApp in the learning process produces critical thinking skills of students as follows: very less 11%, less = 17%, good = 68%, and very good = 4%. These results show that in general, the use of Facebook and Whatshap in the learning process can effectively train critical thinking skills (Facebook = 76% and WhatsApp = 70% in both good and excellent category). Effectiveness is also reviewed from how students respond to learning (looking for a deeper explanation, finding information from multiple sources, sharing experiences, expressing ideas, new experiences of social media in learning). The results obtained are shown in Table 4.

Table 4. Student's Response to Learning Process

	Facebook				WhatsApp			
	1	2	3	4	1	2	3	4
Looking for a deeper explanation of the physical phenomena	0	3	11	6	0	2	10	8
Seeking information from various sources	0	2	12	6	0	1	13	6
Sharing experiences about physical phenomena in everyday life	0	1	10	9	0	0	13	7
Expressing an easy-to-understand idea	0	2	9	9	0	3	11	7
New experiences of social media in learning	0	3	11	6	0	2	12	8

Based on Table 4, the number of students responding to Facebook usage in learning process is: very less 0%, less = 11%, good = 53%, and very good = 36%. While the number of students who respond to the use of WhatsApp in the learning process is: very less 0%, less = 9%, good = 59%, and very good = 36%. The results show that in general, the use of Facebook and WhatsApp in the learning process get a positive response from students (Facebook = 89% and WhatsApp = 91% of students respond which is in good and excellent category).

4. Discussion

Validation results indicate that both instructional and technical aspects are 44% in good category and 56% in the excellent category. It shows that conceptually, the use of Facebook and Whatshap in the learning process is stated valid which means that the developed learning design can be applied in the learning process in the classroom. Learning scenarios, learning materials, instructional media, and learning environments can be applied well because the features make it possible to do so. That is why we can choose mobile learning based on this innovative technology [23]. Technically, it is also compatible with the previous research such as low cost, simple, high accessibility, and efficient [24].

The results showed that in general the use of Facebook and Whatshap in the learning process can be stated practical (Facebook = 68% and WhatsApp = 66% practicality in both good and excellent categories). Technical obstacles in the use of Facebook in the learning process are considered little and can be overcome (those are time-consuming (65%) and tiring (70%)). While on the use of WhatsApp, technical constraints also occur due to memory consuming (55%), time-consuming (60%), and eyestrain (65%). Obstacles experienced are overcome by limiting the duration of the learning process per session for only 100 minutes. Learning using Facebook and WhatsApp overcomes the availability of facilitators and learning can be done anytime and anywhere such as the results of previous research [25].

This research examines the effectiveness of using Facebook and WhatsApp in terms of its success in training critical thinking skills with the following elements: analyzing, evaluating, applying, generating ideas, and expressing ideas. Assessment of each element using indicators: clarity, accuracy, relevance, depth, and logic. The results showed that in general, the use of Facebook and Whatshap in the learning process effectively trained critical thinking skills (Facebook = 76% and WhatsApp = 70% in both good and excellent categories). Effectiveness is also reviewed from how students respond to learning (looking for a deeper explanation, finding information from multiple sources, sharing experiences, expressing ideas, new experiences of social media in learning). The results showed that in general, the use of Facebook and WhatsApp in the learning process get positive responses from students (Facebook = 89% and WhatsApp = 91% of students respond in good and excellent categories).

5. Conclusions

The use of Facebook and WhatsApp in the process of physics learning to train students' critical thinking skills conceptually can be declared worthy of being applied in the classroom. Technical obstacles arise due to the use of Facebook and WhatsApp i.e. time consuming and eyestrain. Besides, the use of WhatsApp also takes a huge memory of the smartphone. The constraint is overcome by limiting the duration of learning time to just 100 minutes. The use of Facebook and WhatsApp is effective because it trains students' critical thinking skills through the elements of analyzing, evaluating, applying, generating ideas, and expressing ideas, with which the categories are measured by clarity, accuracy, relevance, depth, and logic indicators. Moreover, it also gets a positive response from students. Learning by using Facebook and WhatsApp can be used to train students' critical thinking skills.

References

[1] Yavuz F 2016 *Do Smartphones Spur or Deter Learning: WhatsApp Case Study* Int J Edu Sci **15** **(3)** pp 408—15

[2] Lenhart A, Madden M, Macgill A, Smith A. *Teens and social media* York Press

[3] Calvo R, Arbiol A, & Iglesias A 2014 Are all chats suitable for learning purposes? A study of the
required characteristics Procedia Computer Science **27** pp 251—60

[4] Feynman R. 2010 Basic Physics, The *Feynman Lectures on Physics **1** (02)*

[5] Willingham D T 2007 *CriticalThinking Why Is It So Hard to Teach?*American Educator American Federation of Teachers

[6] Bouhnik D & Deshen M 2014 *WhatsApp goes to school: Mobile instant messaging between teachers and students* Journal of Information Technology Education: Research Vol **13** pp 217—31

[7] Bere A 2012 *A comparative study of student experiences of ubiquitous learning via mobile devices and learner management systems at a South African university*. Proceedings of the 14th Annual Conference on World Wide Web Applications; Durban

[8] Facebook 2011 *Facebook press room: Statistics* Retrieved from http://www.facebook.com/ press/info.php?statistics

[9] Fewkes A M & McCabe M 2012 *Facebook: Learning Tool or Distraction?* Journal of Digital Learning in Teacher Education Vol **28** No 3 pp 92—8

[10] Bowers & Campbell J 2008 *Cyber "pokes": Motivational antidote for developmental college readers* Journal of College Reading and Learning **39 (1)** pp 74—87

[11] Church K & de Oliveira R 2013 *What's up with WhatsApp? Comparing mobile instant messaging behaviors with traditional SMS* Proceedings of the 15th International Conference on Human-computer Interaction with Mobile Devices and Services pp 352—61

[12] Bere A 2013 Using mobile instant messaging to leverage learner participation and transform pedagogy at a South African University of Technology British Journal of Educational Technology Vol **44** No 4 pp 544—61

[13] Chipunza P R C 2013 Using mobile devices to leverage student access to collaborativelygenerated resources: A case of WhatsApp instant messaging at a South African University International Conference on Advanced Information and Communication Technology for Education (ICAICTE 2013)

[14] Plana M G C, Escofet M I G, Figueras I T, Gimeno A, Appel C, & Hopkins J 2013 *Improving learners' reading skills through instant short messages: A sample study using WhatsApp* 4th World-CALL Conference, Glasgow, 10-13 July 2013

[15] Rozakis L 1998 *81 Fresh & Fun Critical-Thinking Activities* Scholastic Teaching Resources

[16] Paul R & Elder L 2006 *The Miniature Guide to Critical Thinking Concepts and Tools* Foundation for Critical Thinking

[17] Valentino & Catherine 2000 *Developing Science Skills* Houghton Mifflin Company

[18] Pence H E 2007 *Preparing for the real Web generation* Journal of Educational Technology Systems Vol **35** No 3 pp 347—56

[19] Sayan H 2016 *Affecting Higher Students learning Activity by Using WhatsApp* European Journal of Research and Reflection in Educational Sciences Vol **4** No 3 Progressive Academic Publishing UK p 88-93

[20] Dick W, Carey L, & Carey J 2009 *The Systematic Design of Instruction* (7th ed) Boston MA: Pearson A&B

[21] Fraenkel J R & Wallen N E 2003 *How To Design And Evaluate Research in Education* (Fith Edition) McGraw-Hill Higher Education: New York

[22] Chan L 2005 *WebCT revolutionized e-learning* UBC Reports **51(7)**

[23] Amry A B 2014 The Impact of WhatsApp Mobile Social Learning on The Achievement and Attitudes of Female Students Compared With Face to Face Learning in The Classroom European Scientific Journal **10 (22)** pp 116—36

IOP Conf. Series: Materials Science and Engineering **296** (2018) 012025 doi:10.1088/1757-899X/296/1/012025

[24] Gon S & Raweka A 2017 Effectivity of E-Learning through Whatsapp as a Teaching-Learning Tool MVP Journal of Medical Sciences **4 (1)** pp 40–6
[25] Bansal T, Joshi D 2014 *A study of students' experiences of mobile learning* Global Journal of Human-Social Science **14 (4)**

IOP Conf. Series: Materials Science and Engineering **296** (2018) 012026 doi:10.1088/1757-899X/296/1/012026

How can teachers assess reading skills of generation z learners in German language class?

R N Dirga and P H Wijayati

State University of Malang

ryandirga@gmail.com

Abstract. Generation Z is a digital native generation who has unique characteristics on the daily basis includes reading. In order to assess their reading skills correctly, teachers need to take these characteristics into consideration. This paper aims to describe the process in developing an instrument to assess reading skills of Generation Z learners in German language class. This study used developmental method. The developed instrument has not only good quality but also consists of texts that are suitable for German learners of Generation Z. This instrument can be used as an example in assessing German learners' reading skills in the 21st century.

1. Introduction

Reading is considered one of the most important skills for the learners in both academic and interpersonal context [1, 2]. Through reading, learners should be able to understand texts' topic in order to get ideas, knowledge, and information [3, 4]. In German as a foreign language subject, reading skill is elaborated in an international standard for the German language.

The standard is known as CEFR (Common European Framework of Reference for Language). This standard consists of six levels, such as basic user (A1 & A2), the independent user (B1 & B2) and proficient user (C1 & C2) [5]. Each level has different reading complexity. Learners need to acquire the reading competences based on their German language level in order to achieve successful reading. In achieving successful reading, there are two main supporting factors: text and learners [6, 7].

Text is a meaningful written expression that consists of language elements as a communication medium between writer and readers or learners [8]. This input needs to be comprehensible for the learners, so they can get new knowledge [9]. Especially in learning foreign language context, such as German as a foreign language, the texts are authentic materials [10]. Moreover, each text has different text readability. Readability is a complex interaction between all language elements and content in a written material that affects learners' ability to understand the text [11, 12].

In German language reading, learners or readers need to use their cognitive and affective ability. Learners factors that support reading comprehension are cognitive ability [13, 14]), motivation [15, 16], reading strategy [17, 18], learning style [19], personality [20] and prior knowledge [21, 22]. Furthermore, learners in German language class can be divided into some group based on their generation. There are some generations named silent generation, baby boomers, generation X, generation Y and generation Z. Learners at a university that take German as foreign language right now are generation Z because they were born on 1995 [23]. Generation Z is also called digital native generation. They have been using internet since they were a baby. They have good visualization ability. It means that reading with visual media or visualization can help them to understand German texts [24].

Learners prefer online reading with gadget rather than paper-based-material, then they like colorful reading materials with pictures [24]. These characteristics are only found in generation Z learners. In reality, it's difficult for teachers to make good test instrument [25]. This paper aims to describe the development process of a test instrument for reading skill in German as foreign language class of generation Z learners.

2. Method

This research is a development research. The development model in this research is a test combination of ALTE [24] and Grotjahn & Kleppin [25]. In this research, there are ten stages that have been done. The stages are determining the test objectives, gathering relevant theories, collecting texts, creating a blueprint, item writing, expert reviewing, doing revision, doing trial and doing try out.

3. Results

The result of this research is an instrument test for German as a foreign language subject in level B2 of generation Z learners. This test is developed to measure German learners' reading skill. The test items consist of subjective and objective items. There are texts on the test. Those texts have different readability. Text readability is important in foreign language class, where texts are authentic materials [10]. Readability is a complex interaction between all language elements and content in a written material that affects readers understanding [26]. Besides trial, the readability of texts in the test was calculated with a formula. This Flesch formula was accessed online at www.leichlesbar.ch. This website and formula can measure readability index of a German text [27]. Table 1 below shows the readability index of each text and table 2 shows the Flesch Readability Criteria as follows:

Table 1. Text Readability

No	Text	Readability	No	Text	Readability
1	1	85	16	16	49
2	2	64	17	17	51
3	3	61	18	18	52
4	4	34	19	19	74
5	5	20	20	20	47
6	6	58	21	21	73
7	7	48	22	22	8
8	8	80	23	23	46
9	9	70	24	24	67
10	10	77	25	25	46
11	11	83	26	26	77
12	12	65	27	27	44
13	13	45	28	28	86
14	14	61	29	29	57
15	15	39			

Table 2. Flesch Readability Criteria

No	Index	Criteria
1	81-100	extremely easy (*extrem leicht*)
2	71-80	Very easy (*sehr leicht*)
3	61-70	Easy (*leicht*)
4	41-60	Moderate (*durchschnittlich*)
5	31-40	Litte bit difficult (*etwas schwierig*)
6	21-30	Difficult (*schwierig*)
7	0-20	Very difficult (*sehr schwierig*)

The texts on this test have different readability from easy until very difficult. The purpose is giving students with low reading achievement chance to answer the test and differentiating students' reading ability effectively.

The example of text with very easy readability (85) and its items:

Liebe Katrin, } *(11)* *Berlin, 21. 10. 2016 (12)*

wie geht's? Mir geht es nicht so besonders gut. Ich habe schon wieder Streit mit meinen Eltern. Es ist wirklich nicht mehr zum Aushalten: Meine Noten sind ihnen nicht gut genug, meinen Freund mögen sie nicht und neuerdings stört sie auch mein Aussehen. (13)

Weißt du, letzte Woche habe ich die Haare blau färben lassen. In der Bravo gab es Fotos von Frauen mit dieser Haarfarbe, und ich fand es ganz toll. Jetzt habe ich also blaue Augen und blaue Haare, das sieht richtig gut aus. Nur meinen Eltern gefällt es nicht. Sie lassen mich nun erst recht nicht in Ruhe. Entweder ich färbe sie mir wieder blond oder ich habe den ganzen Tag lang Ärger mit ihnen.
Kannst du mir sagen, was ich machen soll? Am liebsten würde ich ja einfach zu Florian ziehen, aber seine Eltern sind auch nicht besser. Vielleicht tue ich einfach so, als würde ich sie nicht hören. Irgendwann werden sie schon daran gewöhnen. } (14)

So, jetzt muss ich aber Schluss machen.
Viele liebe Grüße, } *(15)*
deine Silke

Struktur eines Briefes
Wie heißt das Teil des Briefes?
(11). ____________
(12). ____________
(13). ____________
(14). ____________
(15). ____________

The example of text with moderate readability (48) and its items:

25
Die Angst als Erwachsener keinen Job zu bekommen, rangiert bei den 12- bis 24-Jährigen bereits an erster Stelle ihrer Sorgen. Schon Schüler fürchten heute, später einmal arbeitslos zu sein. Jeder Zweite – ganz gleich, ob in Ost oder West, ob männlich oder weiblich -, der nach den Hauptproblemen der Jugendlichen gefragt wurde, nannte das Thema Arbeitslosigkeit.

A. Schlechtere Karrierechancen für Frauen
B. Bundesregierung gibt Ratschläge zur persönlichen Sicherheit
C. Unerfüllte Berufsträume
D. Maschinen übernehmen unsere Arbeit
E. Bewerben aber richtig

IOP Conf. Series: Materials Science and Engineering **296** (2018) 012026 doi:10.1088/1757-899X/296/1/012026

F. Typisch weibliche Eigenschaften
G. Wie sicher ist mein Arbeitsplatz?
H. Gute Einfälle sind gefragt
I. Angst vor der Zukunft
J. Ratgeber für eine gute Ausbildung

Table 3. The example of text

Text	***Theme***
21	
22	
23	
24	
25	

The example of text with very difficult readability (8) and its items:

Lesen Sie zuerst die 10 Situationen und dann die Anzeigen. Welche Anzeige passt zu welcher Situation? Sie können jede Anzeige nur einmal verwenden. Es ist auch möglich, dass Sie keine passende Anzeige nicht finden können. Wenn es keine richtige Antwort gibt, schreiben Sie einfach "x".

56. Zum Geburtstag möchten Sie Ihrem Vater eine schicke Jacke kaufen.
57. Sie singen gern und möchten am Samstagabend Ihr Können vor Publikum zeigen.
58. Eine Freundin will in Weiß heiraten, aber sie möchte das Kleid nicht kaufen.
59. Sie möchten mitten in der Stadt im Freien etwas trinken.
60. Ein Freund, 20, sucht eine neue Hose zum Ausgehen. Er hat nicht viel Geld.
61. Ihre Tochter braucht eine neue Kinderbrille. Sie soll billig sein, weil sie ihre Brille ständing verliert.
62. Eine nicht ganz schlanke Freundin sucht einen preiswerten Rock.
63. Sie möchten am Sonntag in einem Biergarten zu Mittagessen gehen.
64. Sie brauchen gute Sportbrille. Die Qualität muss stimmen, dafür darf sie auch ruhig etwas kosten.
65. Sie möchten gleichzeitig gute Musik hören und etwas für Menschen in Not tun.

Table 4. Entry

Situation	***56***	***57***	***58***	***59***	***60***	***61***	***62***	***63***	***64***	***65***
Anzeige										

After all of the items were developed, try out has been held. Then the results were analyzed in order to know test's validity, reliability, item difficulty, item discrimination and distractors' function as shown figure 1.

IOP Conf. Series: Materials Science and Engineering **296** (2018) 012026 doi:10.1088/1757-899X/296/1/012026

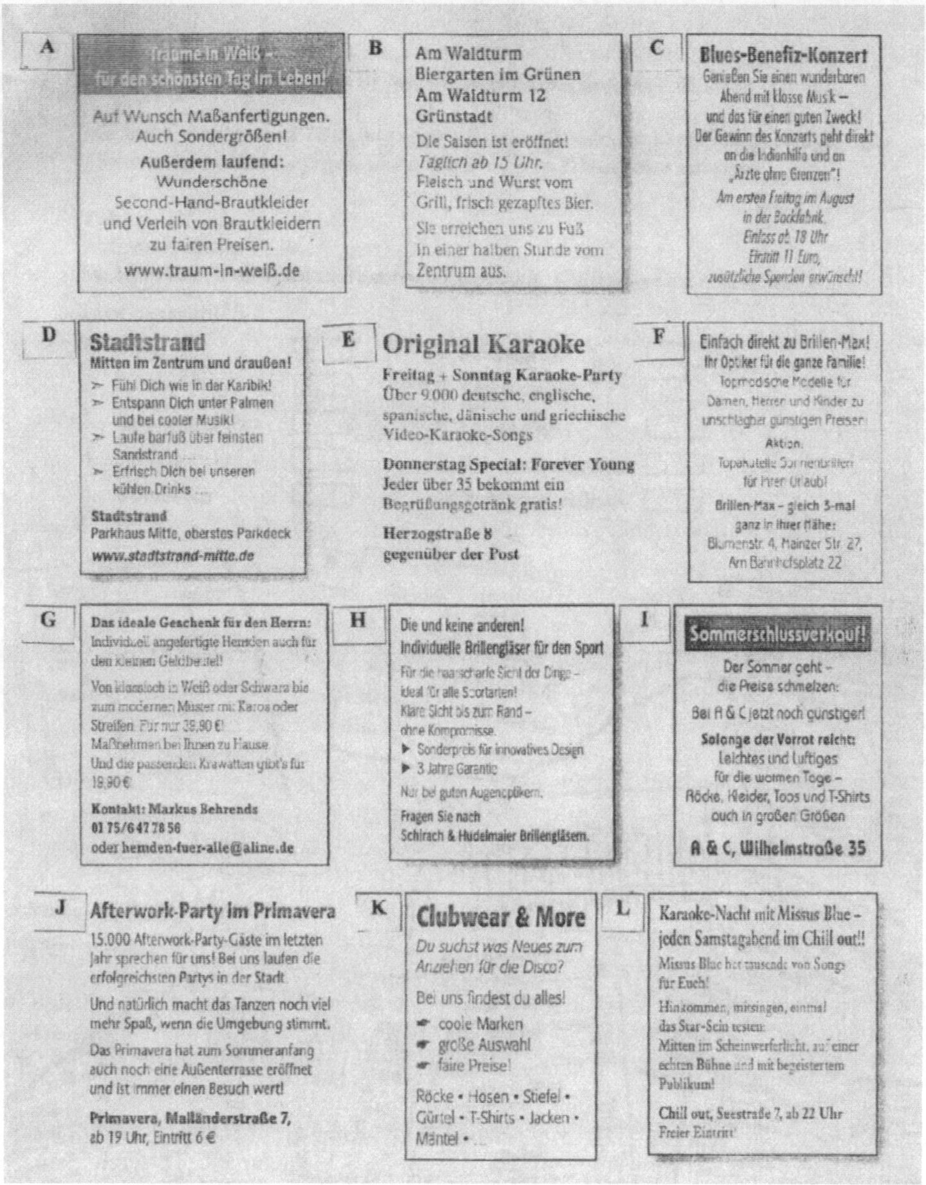

Figure 1. Function Analysis

The validity was measured with point biserial [28]. The analysis result shows that the developed test is valid. Then, test reliability was assessed with KR 20, because the items are dichotomous [29]. The reliability index is 0,920. Furthermore, each item was also analyzed. The distractors also can do its function properly.

4. Conclusion

The test that has been developed in this research is a test for German as a foreign language subject in level B2 with target learners teenager that are considered as generation Z. Moreover, this test is intended to measure learners' reading skill. The development model is a combination of test developmental model of ALTE and Grotjahn & Kleppin. The items were developed based on the generation Z learners' characteristics. There are 65 items with subjective and objective types.

This study concludes that the test has good readability. It can be seen from trials result and readability measuring formula (Flesch formula). Moreover, the test also has good validity, reliability and items. Expert judgment and point biserial were conducted in order to test validity. The reliability index is 0,920. The developed test's strengths are: appropriate with CEFR, suitable with generation Z characteristics, the test has good texts (readability), the test has good items, the test has good validity, reliability, distractors.

References

[1] McKee S 2012 Reading Comprehension, What We Know: A Review of Research 1995 to 2011 *Language Testing in Asia*, **2 (1)** pp 45–58

[2] Khaki N 2014 Improving Reading Comprehension in a Foreign Language: Strategic Reader *The Reading Matrix* **14 (2)** pp 186 – 200

[3] Bilikozen N & Akyel A 2014 EFL Reading Comprehension, Individual Differences, and Text Difficulty *The Reading Matrix* **14 (2)** pp 263–96

[4] Collins-Thomson K 2014 *Computational Assessment of Text Readability: A Survey of Current and Future Research* (Michigan: University of Michigan)

[5] *CEFR (Common European Framework of Reference for Languages)* 2005 (Strasbourg: Council of Europe)

[6] Yang X 2016 Study on Factors Affecting Learning Strategies in Reading Comprehension *Journal of Language Teaching and Research* **7(3)** pp 586–90

[7] Antoniou F & Souvignier E 2007 Strategy Instruction in Reading Comprehension: An Intervention Study for Students with Learning Disabilities *Learning Disabilities: A Contemporary Journal* (Online) (http://researchgate.net/publication/215757766), accessed 8 Oktober 2016

[8] Woolley G 2011 Reading Comprehension: Assisting Children with Learning Difficulties *Springer Reading Journal* (Online) (http://springer.com), accessed 27 September 2016

[9] Krashen S 2013 The Case for Non-Targeted, Comprehensible Input *Journal of Bilingual Education Research & Instruction* **15(1)** pp 102–10

[10] Sekiziyivu S & Mugimu C 2015 The relationship between Learners' German Language Communicative Abilities and Their Prior Performance in a National Ugandan Certificate Examination. *Journal of Language Teaching and Research* **6(1)** pp 43–52

[11] Revesz A & Brunfaut T 2013 Text Characteristics of Task Input and Difficulty in Second Language Listening Comprehension *Studies in Second Language Acquisition* **35** pp 31–65

[12] Crossley S, Greenfield J & McNamara 2008 Assessing Text Readability Using Cognitively Based Indices *TESOL Quarterly* **42 (3)** pp 475–93

[13] Guthrie J, Wigfield A, Barbosa P, Perencevich K, Taboada A, Davis M, Scafiddi N & Tonks S 2004 Increasing Reading Comprehension and Engagement Through Concept Oriented Reading Instruction *Journal of Educational Psychology* **96 (3)** pp 403–423

[16] Cain K & Oakhill J 2006 Profiles of Children with Specific Reading Comprehension Difficulties *British Journal of Educational Psychology* **76** pp 683–696

[17] Schwabee F, McElvany N & Trendtel M 2014 The School Age Gender Gap in Reading Achievement: Examining the Influences of Item Format and Intrinsic Reading Motivation *Research Reading Querterly* **50 (2)** pp 219–232

[18] Gurses M & Bouvet E 2016 Investigating Reading Comprehension and Learning Styles in Relation to Reading Strategies in L2 *Reading in a Foreign Language* **28 (1)** 20–42

[19] Alharbi M 2015 Reading Strategies, Learning Styles, and Reading Comprehension: A Correlation Study *Journal of Language Teaching and Research* **6 (6)** pp 1257–68

[20] Arabsarhangi M & Noroozi I 2014 The Relationship between Self-awareness and Learners' Performance on Different Reading Comprehension Test Types among Iranian EFL Elementary Learners *Theory and Practice in Language Studies* **4 (4)** pp 675–85

[21] Sanchez E, Garcia J & Gonzalez A 2007 Can Differences in the Ability to Recognize Words

Cease to Have an Effect Under Certain Reading Conditions? *Journal of Learning Disabilities* **40 (4)** pp 290–305

[22] Taboada A, Tonks S, Wigfield A, & Guthrie J 2008 Effects of Motivational and Cognitive Variables on Reading Comprehension *Springer Science+Business Media B.V.* (Online), (http://springer.com) accessed 8 September 2016

[23] Nicholas A 2008 *Preferred Learning Methods of the Millennial Generation*. Salve Regina University: Digital Commons, (Online), (http://salve.edu) accessed 27 Oktober 2016

[24] Rothman D 2015 *A Tsunami of Learners Called Generation Z* (Online), accessed 27 Oktober 2016

[25] Dirga R N & Wijayati P H 2016 *Is The German Language Text Too Short For The Senior High School Students?* Makalah disajikan dalam ICERE, Program Studi Penelitian dan Evaluasi Pendidikan, Pascasarjana UNY, Yogyakarta, 30 – 31 Mei 2016

[26] ALTE 2012 *Handbuch zur Entwicklung und Durchführung von Sprachtests, Zur Verwendung mit dem GER* (Frankurt am Main: telc GmbH)

[27] Grotjahn, R. & Kleppin, K. 2015. *Prüfen, Testen, Evaluieren* (München: Langenscheidt)

[28] Smith M & Taffler R 1992 Readability and Understandability: Different Measures of The Textual Complexity of Accounting Narrative *Accounting, Auditing & Accountability Journal*, **5 (4)** pp 84–101

[29] Bachman C 2014 *Testen Sie Ihren Text* (Online), (http://leichtlesbar.ch), accessed 19 Desember 2016

[30] Azwar S 2012 *Reliabilitas dan Validitas Edisi 4* (Yogyakarta: Pustaka Pelajar)

[31] Sulistyo G 2015 *Assessment At Schools: An Introduction to Its Basic Concepts and Principles* (Malang: CV Bintang Sejahtera)

IOP Conf. Series: Materials Science and Engineering **296** (2018) 012027 doi:10.1088/1757-899X/296/1/012027

Indonesian teacher engagement index: *a rasch model analysis*

Sasmoko[1,2], B S Abbas[3], Y Indrianti[2] and S A Widhoyoko[4]

[1]Faculty of Humanities, Bina Nusantara University, Indonesia
[2]Research Interest Group in Education Technology, Bina Nusantara University, Indonesia
[3]Faculty of Engineering, Bina Nusantara University, Indonesia
[4]Podomoro University, Indonesia

e@sasmoko.com

Abstract. The research aimed to calibrate Indonesian Teacher Engagement Index (ITEI) using instrument with RASCH MODEL. The respondents were 672 teachers of elementary, junior high, high school and vocational school. The number of items planned was 165 items with the initial reliability of 0.98. The ITEI scale uses Likert Scale (1 to 4) which was converted from ordinal scale to Equal Interval Scale. RASCH MODEL analysis was done by selecting based on Outfit Mean Square (MNSQ) between 0.5-1.5 as a good item, and measuring Point Measure Correlation (Pt Mean Corr) with the criterion of 0.4-0.85. Moderate Outfit Z-Standard (ZSTD) was ignored because the sample was> 500. Conclusions: ITEI is valid with 30 items and reliability of 0.97, and less engage teachers significantly at $\alpha < 0.05$.

1. Introduction

The research on teacher engagement has evolved since the 1970s, presenting various results, but conical at one conclusion that teacher engagement is an ideal condition of teachers who have a significant impact on students' life [1,2]. Engaged teachers are able to positively influence both behaviors and thought so as to achieve results that can improve teaching and learning [3]. Teacher engagement refers to the combination of teacher characteristics, knowledge and motivation and its participation in the profession will support teachers in shaping teacher engagement [4,5].

The Indonesian Teacher Engagement Index (ITEI) instrument is structured with reference to the concept of characteristics that teachers must have in Indonesia. ITEI is developed into 5 dimensions, that is how the teachers are able to show positive psychology, have an active role in building positive education, able to show good performance, have a supportive competence, posses national character as characteristic of Indonesia and able to show nationalism leadership Engagement [6–10].

Positive psychology shows the level of mental health and well-being of teachers who are able to be understood on the meta-psychological level because it relates to the potential, motive and capacity of a teacher [6,11,12]. Positive psychology includes wisdom and knowledge, courage, humanity, justice, temperance and transcendence [13,14]. Positive education is formed when teachers are able to develop and have positive emotions, positive engagement, positive accomplishment, positive purpose, positive relationship and positive health [10]. Teacher performance measures can be seen when teachers are able to meet task demands and responsibilities as a teacher (task performance), able to demonstrate effective behaviours that support the main tasks and responsibilities as a teacher (contextual performance), minimize unproductive behaviour in work (Counterproductive behaviour) and adapt skilfully with the changes (adaptive performance) [7,8,15].

Teacher competence in Indonesia consists of pedagogic competence, personality competence, social competence, and professional competence [16,17]. The national character that the teachers must possess refer to the ideology and the foundation of the Republic of Indonesia named Pancasila consisting of the character of Godhead, the character of the Justification and Civilized Humanity, the character of the Unity of Indonesia, the character of Democracy Led by Wisdom of Wisdom and the character of Social Justice. While nationalism leadership is formed from the ability of teachers in identifying themselves as educators who have a leadership spirit that represents the interest of Indonesia as a whole, build a positive perception not only towards their regions but also Indonesia show themselves as reliable educators, demonstrate competence in managing multi ethnic learners, become role models in developing the paradigm of students in viewing their areas as an integral part of Indonesia, motivate students to promote national interest and creative and innovative in handling problems with the principle of national stability [18–20]

The formulations of this research problem are 1) finding the validity and reliability of ITEI instrument profile for Indonesian context and 2) finding the tendency of the teacher engagement in Indonesia.

2. Research Method

The respondents of this research are 672 elementary, junior high, high school and vocational high school teachers who are spread in Java. This research is a preliminary research conducted in order to build valid and reliable Indonesian Teachers Engagement Index (ITEI) instrument. The instrument was constructed using one of the stages in Neuroresearch, namely, the exploratory research [21].

The Indonesian Teacher Engagement Index (ITEI) instrument is structured with reference to the concept of characteristics that teachers must possess in Indonesia [6–10]. ITEI is developed with 5 dimensions, those are:

2.1 How the teachers are able to show positive psychological condition (positive psychology)

This dimension is characterized by six indicators embodied in 6 virtues and 24 strength characters, such as 1) Wisdom and Knowledge, cognitive strength that entail the acquisition and use of knowledge consisting of creativity, curiosity, open mindedness, love of learning and perspective; 2) Courage, an emotional strength that involves authenticity, bravery, persistence, and zest; 3) Humanity, an interpersonal strength that involves (tending and befriending) others entailing kindness, love, and social intelligence; 4) Justice, a civic strength that underlies healthy community life entailing fairness, leadership, and teamwork; 5) Temperance, a strength that protects against excess entailing forgiveness, modesty, prudence, self-regulation; and 6) Transcendence, a strength that forges connections to the larger universe and provides meaning entailing appreciation of beauty and excellence, gratitude, hope, humour and religiousness[13,14].

2.2 How the teachers can play an active role in building positive education

This dimension consists of six indicators, such as 1) positive emotion, a positive emotional experience such as joy, gratitude, and hope; 2) positive engagement, e.g. interest, engagement, curiosity, and absorption; 3) positive accomplishment, an incentive to achieve meaningful results; 4) positive purpose, in which the teachers are able to contribute to others and to the community; 5) positive relationships related to social and emotional skills to encourage the establishment of positive relationships; And 6) positive health, a healthy and optimal physical and psychological condition[10]

2.3 How teachers can demonstrate good performance,

Performance is a multi-dimensional concept consists of four indicators: 1) task performance, in which teachers are able to perform their main duties and responsibilities as teachers; 2) contextual performance, in which teachers are able to exhibit positive behaviours that support the effectiveness of performance; 3) counterproductive work behaviour, in which teachers are able to minimize behaviours that reduce the effectiveness of performance and 4) adaptive performance, in which teachers are able to adapt and be proactive with changes [7,8,15,22,23]

2.4 How teachers possesses supportive competencies

Teachers have four main competencies that must be possessed: pedagogic competence, social competence, personal competence and professional competence [16]

2.5 How teachers possess the character of nationality as a characteristic of Indonesia

This dimension is based on the five precepts of Pancasila as the philosophy of the Indonesian nation, so this dimension consists of five indicators, such as 1) Character of Godhead; 2) Fair and Civilized Humanitarian Character; 3) Character of Unity of Indonesia; 4) Character of Democracy Led by Wisdom of Wisdom and 5) Character of Social Justice.

2.6 How teachers are able to demonstrate nationalism leadership engagement.

This dimension was developed by [15] consists of seven indicators, such as 1) Able to identify themselves as educators who possess a leadership spirit that represents the interests of Indonesia as a whole; 2) Have a positive perception of the region as well as Indonesia that Bhinneka Tunggal Ika as its own machine to fight for the integrity of Indonesia; 3) Demonstrate as trustworthy educators; 4) Possess competence in managing multiethnic students; 5) Able to be an example in developing the paradigm of students in seeing their areas as an integral part of Indonesia; 6) Able to motivate students to prioritize national interests; And 7) Creative and innovative in dealing with issues with the principle of national stability.

The ITEI instrument is constructed using the Likert Scale rating with a score rating of 1) indicating that the statement is very unsuitable to the condition that the teachers experience daily and never do at all, 2) showing that the statement is less in line with the condition that the teachers experience daily and do it occasionally, 3) indicating that statement adequately describes the condition they experience daily and several times do it and 4) showing that the statement strongly describes the condition they experience daily and always do it.

Construct validity of ITEI through instrument item validity is done by Rasch Model analysis. Rasch Model is a model developed by Dr. Georg Rasch in 1960. Rasch model is a model of calculation analysis where item parameters can be estimated independently of the characteristics of sample calibration parameters(Masters, 1982; Curtis & Boman, 2007; Lewandowski, Co-investigator, & Lewandowski, 2015). The advantage that can be obtained by using the Rasch Model is that the interaction between the respondents and the items in the instrument uses the same logit scale intervals as well as able to analyze polytomous items[27]. The Rasch Model is able to demonstrate a response to statistical independence of a number of items[28].

Because the ITEI instrument uses the Likert Rating Scale, each item checked with four answer choices will be modeled into three thresholds. The threshold of each item has a predicted difficulty (F) with the following equation :

$$P_{ni1}(X = 1/\beta_n, \delta_i, F_1) = \frac{e^{(\beta-[\delta+F])}}{1+e^{(\beta-[\delta+F])}} \quad (1)$$

RASCH MODEL is trying to measure the instrument from the mean value outfit Mean Square (MNSQ). MNSQ shows the size of randomness or distortion in a measurement. The value of MNSQ has some implications on the measurement, in which the MNSQ value indicates that the item has good conditions for measurement in the range of 0.5 - 1.5. The MNSQ is able to determine how well each item contributes to defining a common construction [29].

This analysis also attempts to measure the instrument in terms of Standardized fit Statistics (ZSTD) which is an analysis to see the suitability of data with the model. The value of ZSTD illustrates that the data has a logical approximation which is ranged between -1.9 - 1.9 [30].

The third value seen by RASCH MODEL analysis is the value of Point Measure Correlation (Pt Mean Corr) with a range of values 0.4 - 0.85.

Initial reliability test results of instrument items consist of 165 items indicate that good item reliability is 0.90.

```
-------------------------------------------------------------------------------
| Person      672 INPUT       672 MEASURED               INFIT          OUTFIT   |
|           TOTAL      COUNT     MEASURE  REALSE    IMNSQ   ZSTD  OMNSQ    ZSTD|
| MEAN      608.5      165.0        2.27     .16     1.05     .2   1.00      .0|
| S.D.       24.9         .0         .53     .05      .21    1.2    .27     1.3|
| REAL RMSE     .17 TRUE SD      .50  SEPARATION  2.95  Person RELIABILITY  .90|
|------------------------------------------------------------------------------|
| Item      165 INPUT       165 MEASURED               INFIT          OUTFIT   |
|           TOTAL      COUNT     MEASURE  REALSE    IMNSQ   ZSTD  OMNSQ    ZSTD|
| MEAN     2478.2      672.0         .00     .08     1.01     .0   1.00     -.1|
| S.D.      131.5         .0         .57     .02      .24    3.2    .34     3.5|
| REAL RMSE     .08 TRUE SD      .57  SEPARATION  6.76  Item   RELIABILITY  .98|
-------------------------------------------------------------------------------
```

Figure 1. Reliability Test Results

The results of analysis by using RASCH MODEL regarding the MNSQ, ZSTD, and PT mean Corr are as follows:

From the results of the analysis, there are 30 valid and eligible items. ITEI instrument reliability test results with 30 valid items obtains :

```
 Calculating Fit Statistics
>=====================================<
Standardized Residuals N(0,1)  Mean: .01 S.D.: .99
Time for estimation: 0:0:0.250
 Processing Table 0
F:\KONFERENSI INTERNASIONAL\CAPEU UNESA 2017\Equal Interval Fit 672 ITEI.prn
-------------------------------------------------------------------------------
| Person      672 INPUT       672 MEASURED               INFIT          OUTFIT   |
|           TOTAL      COUNT     MEASURE  REALSE    IMNSQ   ZSTD  OMNSQ    ZSTD|
| MEAN      109.5       30.0        3.07     .63     1.00     .0    .98      .0|
| S.D.        8.9         .0        1.61     .54      .31    1.1    .37     1.1|
| REAL RMSE     .83 TRUE SD     1.38  SEPARATION  1.66  Person RELIABILITY  .73|
|------------------------------------------------------------------------------|
| Item       30 INPUT        30 MEASURED               INFIT          OUTFIT   |
|           TOTAL      COUNT     MEASURE  REALSE    IMNSQ   ZSTD  OMNSQ    ZSTD|
| MEAN     2451.9      672.0         .00     .08     1.00    -.1    .98     -.2|
| S.D.       77.4         .0         .46     .01      .22    3.2    .20     2.4|
| REAL RMSE     .08 TRUE SD      .45  SEPARATION  5.50  Item   RELIABILITY  .97|
-------------------------------------------------------------------------------
```

Figure 2. ITEI Reliability Test Results for 30 items

After obtaining the valid items, the actual condition of the population can be immediately found out by doing SPSS analysis which results as follows :

Descriptives

			Statistic	Std. Error
ITEI Equal Interval Scale (Y)	Mean		3.0668	.06211
	95% Confidence Interval for Mean	Lower Bound	2.9449	
		Upper Bound	3.1888	
	5% Trimmed Mean		3.0184	
	Median		2.7700	
	Variance		2.592	
	Std. Deviation		1.61005	
	Minimum		.37	
	Maximum		6.18	
	Range		5.81	
	Interquartile Range		1.68	
	Skewness		.861	.094
	Kurtosis		-.286	.188

Figure 3. SPSS Test Results

IOP Conf. Series: Materials Science and Engineering **296** (2018) 012027 doi:10.1088/1757-899X/296/1/012027

3. Result and Analysis

The Rasch Model analysis shows that of 165 items in the ITEI instrument, there are 30 items that have good conditions for measurement. While other items are less good and less productive (MNSQ values are between 1.5 - 2.0 or MNSQ <0.5) for measurement although no items degrade the quality of the measurement system (MNSQ> 2.0). Of these 30 items, the dimension of nationalism leadership engagement is not represented, so there is a change represented in the following picture:

Table 1. ITEI Dimensions, Indicators and Number of Items before and after analysis

Dimension	Indicator	Number of ITEI Items before analysis	Number of ITEI Items after analysis
Positive Psychology	Wisdom & knowledge	7	1
	Courage	5	2
	Humanity	3	1
	Justice	4	1
	Temperance	5	0
	Transcendence	5	3
Positive Education	Positive emotion	3	0
	Positive engagement	2	1
	Positive accomplishment	2	0
	Positive purpose	4	0
	Positive relationships	3	0
	Positive health	2	1
Teacher performance	Task performance	3	1
	Contextual performance	5	3
	Counterproductive behaviour	3	0
	Adaptive performance	3	0
Teacher competence	Pedagogic competences	48	9
	Personality competences	18	1
	Social competences	6	1
	Professional competences	9	0
Character nationality	God Character	3	0
	A Just and Civilized Humanitarian Character	2	1
	Character of Unity of Indonesia	2	1
	Character Democracy led by the wisdom of wisdom	3	2
	Character of Social Justice	3	1
Nationalism Leadership Engagement	Able to identify themselves as educators who possess a leadership spirit that represents the interests of Indonesia as a whole	3	0
	Have a positive perception of the region and also Indonesia that Bhineka Tunggal Ika as his own machine to fight for the integrity of Indonesia	1	0
	Show them as a trustworthy educator	3	0
	Posses the competence in managing multiethnic students	1	0
	Able to be a role model in building the paradigm of students in viewing their areas as an integral part of Indonesia	1	0
	Able to motivate students to promote national interests	2	0
	Creative and innovative in dealing with issues with the principle of national stability	1	0

The categorization for the ITEI instrument consists of four categories defined as the score 1 = Not Engage, score 2 = Less Engage, score 3 = Engage, and score 4 = very engaged. The SPSS test generates confidence interval with a significance level of 5% as follows:

The summary is as follows:

Table 2. Confidence Interval Instrument ITEI

Class Interval	Information
0.175 – 1.974	
1.975 – 3.774	
3.375 – 5. 574	
5.575 – 7.374	

The results of the analysis in the population show that the population is in the category of less engage.

4. Discussion

This research is a preliminary research conducted in order to construct an Indonesian Teacher Engagement Index instrument which will eventually be tested and implemented nationally to see the condition of teachers in Indonesia. Although the Rasch Model analysis eliminates one dimension, the five dimensions are able to describe an engagement condition in the teacher. The teacher relates closely to the students, the school, and the community, so that the character of the engaged teacher becomes the deciding factor that can affect the students' life [31–33]; The formation of values developed in schools brings the character of the nation manifested in the behaviour and competence the students possess[1,17].

Based on the results of the two types of research above, the ITEI instrument is worthy of being used as a measuring tool to rationally measure the state of teachers' engagement to complement the teacher profiling illustrated through the Teacher Performance Assessment and Teacher Competency Test.

References

[1] Rutter R A and Jacobson J D 1986 Facilitating Teacher Engagement *Educ. Res.*

[2] Burns A 2010 Teacher engagement in research: Published resources for teacher researchers Books from regional locations Gregory Hadley (ed.). Action research in action, Singapore: SEAMEO Regional Language Centre (2003). Pp. iv + 54. ISBN 9971-74-081-8. Gertrude Tinker *Lang. Teach.* **43** pp 527–36

[3] Stanton K C 2011 Engineering Faculty Motivation for and Engagement in Formative Assessment 171

[4] Gardee A and Brodie K 2015 A teacher's engagement with learner errors in her Grade 9 mathematics classroom *Pythagoras* **36** pp 1–9

[5] Kathleen Travis Knowles 1999 THE EFFECT OF TEACHER ENGAGEMENT ON STUDENT MOTIVATION AND ACHIEVEMENT *Dep. Hum. Dev.*

[6] Gable S L and Haidt J 2005 What (and Why) Is Positive Psychology? *Rev. Gen. Psychol.* **9** pp 103–10

[7] Koopmans L, Bernaards C M, Hildebrandt V H, Van Buuren S, Van der Beek A J and De Vet H C W 2013 Development of an individual work performance questionnaire *Int. J. Product. Perform. Manag.* **62** 6–28

[8] Koopmans L, Bernaards C M, Hildebrandt V H, Schaufeli W B, de Vet Henrica C W and van der Beek A J 2011 Conceptual Frameworks of Individual Work Performance *J. Occup. Environ. Med.* **53** pp 856–66

[9] Darling-Hammond L 2000 How teacher education matters *J. Teach. Educ.* **51** pp 166–73

[10] Norrish J 2013 An applied framework for Positive Education *Int. J. Wellbeing* **3** 147–61

[11] Alex Linley P, Joseph S, Harrington S and Wood A M 2006 Positive psychology: Past, present, and (possible) future *J. Posit. Psychol.* **1** 3–16

[12] King L A 2017 *Why Positive Psychology Is Necessary* pp 216–8

[13] Seligman M E P, Steen T A, Park N and Peterson C 2005 Positive psychology progress: Empirical validation of interventions. *Am. Psychol.* **60** pp 410–21

[14] Seligman M E P, Ernst R M, Gillham J, Reivich K and Linkins M 2009 Positive education: positive psychology and classroom interventions *Oxford Rev. Educ.* **35** pp 293–311

[15] Yusoff R Bin, Ali A and Khan A 2014 Assessing reliability and validity of job performance scale among university teachers. *J. Basic Appl. Sci. Res.* **4** pp 35–41

[16] Anon 2005 Undang Undang Republik Indonesia No 14 Tahun 2005 tentang Guru dan Dosen

[17] Kementerian Pendidikan Nasional D J P M P dan T K 2011 *Pedoman Pelaksanaan Penilaian Kinerja Guru*

[18] Schildkraut D J 2005 The rise and fall of political engagement among latinos: The role of identity and perceptions of discrimination *Polit. Behav.* **27** pp 285–312

[19] Bass B M, Avolio B J, Jung D I and Berson Y 2003 Predicting unit performance by assessing transformational and transactional leadership. *J. Appl. Psychol.* **88** pp 207–18

[20] Susanto H 2013 Understanding of Regional History and Perception of *Int. J. Hist. Educ.* **XIV**

[21] Sasmoko; Ying Yi 2015 Construct Validity in NeuroResearch *Adv. Sci. Lett.* **21** pp 2438—41

[22] Rotundo M and Sackett P R 2002 The Relative Importance of Task, Citizenship and Counterproductive Performance to Global Ratings of Job Performance: A Policy-Capturing Approach *J. Appl. Psychol.* **87** pp 66–80

[23] Campbell J P and Wiernik B M 2015 *The Modeling and Assessment of Work Performance* **2**

[24] Curtis D D and Boman P 2007 X-ray your data with Rasch *Int. Educ. J.* **8** 249–59

[25] Lewandowski C M, Co-investigator N and Lewandowski C M 2015 Self-directed Learning Oriented Assessments in the Asia-Pasific *Eff. Br. mindfulness Interv. acute pain Exp. An Exam. Individ. Differ.* **1** pp 1689—99

[26] Masters G N 1982 A Rasch model for partial credit scoring *Psychometrika* **47** pp 149–74

[27] Sousa L B, Prieto G, Vilar M, Firmino H and Simões M R 2014 The Adults and Older Adults Functional Assessment Inventory: A Rasch Model Analysis. *Res. Aging* **37** 7pp 87–814

[28] Goh H E, Marais I and and Ireland Michael James 2015 A Rasch Model Analysis of the Mindful Attention Awareness Scale *Assessment* pp 1–12

[29] Kim B S K and Hong S 2004 A Psychometric Revision of the Asian Values Scale Using the Rasch Model *Meas. Eval. Couns. Dev.* **37** pp 15–27

[30] Karabatsos G 2001 The Rasch model, additive conjoint measurement, and new models of probabilistic measurement theory. *J. Appl. Meas.* **2** pp 389–423

[31] Cardwell M E 2011 Patterns of Relationships Between Teacher Engagement and Student Engagement *Fish. Digit. Publ*

[32] Basikin 2007 Vigor, dedication and absorption: Work engagement among secondary school English teachers in Indonesia *AARE Int. Conf. Perth, Aust. 27-28 Novemb. 2007* pp 25–9

[33] Boyd A C 2011 Washington State High School Science Teacher Engagement in and Motivation for Formative Classroom Assessment 169

IOP Conf. Series: Materials Science and Engineering **296** (2018) 012028 doi:10.1088/1757-899X/296/1/012028

The role of students' self-confidence in relation with swimming routines, frequency, and tutor in swimming class

S Hartoto, F D Khory and B B Prakoso

Department of Sport Education, Universitas Negeri Surabaya

setiyohartoto@unesa.ac.id

Abstract. It is compulsory for prospective physical education teachers to have the ability to perform swimming. The average of students' passing in swimming class has reached 72%. Most students who failed to pass the class are those who have had aquaphobia, the condition in which one failed to perceive a situation in a positive and objective, some of which are hard to detect. This perception may come from past experience and it could diminish students' confidence. Furthermore, the lack of confidence in students may cause unsatisfactory learning results. Therefore it is critical for the teachers to have a comprehensive knowledge of their students' past experience in formulating a lesson. This research used descriptive qualitative approach. The aim of this article is to investigate the correlation between students' confidence level and swimming routines, frequency, and tutors in order to succeed swimming class. This article will attempt to describe the results of a research conducted to 139 students of Department of Sport Education Universitas Negeri Surabaya as prospective physical education teachers in Indonesia who took swimming class. Past experience and confidence level are measured by a questionnaire. The results of the research show that students who have a higher level of confidence are those who follow practice routines with adequate frequency and helped by a compatible tutor.

1. Introduction

Swimming learning at Physical Education Teacher Education is important for prospective PE teachers so they can master swimming skills and can teach swimming. These competencies are important to be possessed by prospective PE teachers as a requirement for a PE teacher candidate to teach their students a swimming skill in the future. However, swimming teaching for prospective PE teachers is not an easy task. The proof of PE teachers' graduation rates in swimming lessons from 2006 until 2014 the prospective PE teachers in the Department of Sport Education at Universitas Negeri Surabaya ranged from 66% to 82% of the total prospective PE teachers studying swimming. If the average graduation rate of prospective PE teachers who studied swimming is still in the range of 72%. To be able to graduate in swimming learning, prospective PE teachers are required to master knowledge in swimming lessons and can practice at least two swimming styles. The results of self-evaluation of swimming learners, those who entered the group did not pass, dominated by those with a history of aquaphobia [1].

Their aquaphobic state tends to have fear when they are in water. The level of their fear of water is feared can reduce the ability of learners to be able to master themselves in the activity in the water. In fact, the first step to make learners can easily master swimming skills is to generate confidence to be able to master themselves in the activity in water [2]. Confidence is believed to be the psychological aspect closest to the learning process for students to achieve high achievement [3]. For that, mastery of a person with full confidence is needed in learning swimming. According to Melon Dash, one's of self-control in the activity in water can be divided into five categories, i.e.: calm, nervous, fear, terror, and panic [2]. Confidence contains about ones of self-confidence to be successful [3]. The high self-confidence of learners in the learning process is very important. High-confidence of learners will be able to overcome the fear of failure and maintain self-persistence in the trial and error process so that they will continuously try to do the exercises [4].

However, to build of high self-confidence of learners' in learning is not an easy task, especially for prospective PE teachers who have an excessive fear of water. Teachers are required to know the swimming experience of prospective PE teachers as the basis of the right dose of swimming learning. The level of the learning experience of student can be utilized to know the readiness of learners' in receiving learning materials. The learning experience of swimming can reflect the learning style of a learner. Learning styles in swimming can be visual, auditory, cognitive, and kinesthetic [5]. In addition, from the initial experience of learning a good pool is expected to provide opportunities to learners can master themselves in the activity in the water. Conversely, with bad experiences while learning swimming in the past will give a traumatic impression that appears aquaphobia. Rahman et al. [6] provide an alternative for measuring the initial experience of swimming learning through four things, i.e.: the learners' self-perception of swimming skills, the early age of swimming learning, the first tutor to learn about swimming, and the place for activities in the water [6].

The first tutor to learn swimming becomes important, given that swimming is a sport that demands a person's high skill level. The first tutor becomes the key to the successful learning of a learner's swimming. Tutors can provide motivational motivation for learners to achieve achievement in learning [7]. Strengthening motivation becomes important for swimming learning so that learners get a fun learning experience. For that, tutors are very important in swimming learning but as in learning skills in general, learners through three stages of learning, i.e.: verbal-cognitive stage, motor stage, and autonomous stage [8]. This theory suggests that the need for suitable tutors and exercises with sufficient duration and frequency for learners can master swimming skills. Tutors can be a source of knowledge by learners through modeling and translate skills to be learned [9]. In addition to tutors as skill translators who want to learn, tutors are also an effective feedback loop on learners' performance while doing learning activities [10].

This study aims to find out, the readiness of learners in the Department Sport Education Universitas Negeri Surabaya in swimming learning through confidence level search, tutor, frequency, and exercise routines. Furthermore, it would know the relationship between the four variables on the basis of the development of learning materials.

2. Methods

This research is included in a type of correlational research that seeks to explain the relationship between self-confidence, tutor, frequency, and swimming exercise routines owned by prospective PE teachers in the Department of Sport Education Universitas Negeri Surabaya. The number of prospective PE teachers involved in the study was 139 who were still within the 2nd year. Confidence level measurements utilize a self-confident questionnaire successfully developed by Hartoto at 2016. The questionnaire used is a questionnaire that is closed by using a Likert scale of 5, i.e.: Strongly Disagree to Strongly Agree. Components of the self-confidence questionnaire as many as five things that have been valid and reliable, i.e.: (1) confidence in physical appearance ($r = 0.830$, $p < 0.01$); (2) learning ability ($r = 0.800$, $p < 0.01$); (3) social acceptance ($r = 0.708$, $p < 0.01$); (4) behavior ($r = 0.835$, $p < 0.01$); and (5) physical ability ($r = 0.785$, $p < 0.01$). Cronbach's Alpha value of 0.913 indicates that the items that are arranged are reliable. Based on the above explanation, the instrument in the form of a confident questionnaire has been declared valid and reliable [11]. While tutors, frequencies, and routines are known through the accompanying questionnaire with an open answer. Data analysis will use descriptive statistics, Anova, and Ancova in the level of 0.05.

3. Results

Based on the results of research can be analyzed data and presented the results as in table 1 as follows:

Table 1. Number of students frequency based on types of tutors and routines in swimming learning

Variable	Sub-Variable	Frequency	Percent	Valid Percent	Cumulative Percent
The first tutor to learn swimming	Parent	47	33,8	33,8	33,8
	PE teachers	14	10,1	10,1	43,9
	Coach in Club	3	2,2	2,2	46,0
	Friends	57	41,0	41,0	87,1
	Others (brother, uncle, grandma, coach in extracurricular, and self-training)	18	12,9	12,9	100,0
	Total	139	100,0	100,0	
Swimming exercise routines	Irregular	103	74,1	74,1	74,1
	Regular	36	25,9	25,9	100,0
	Total	139	100,0	100,0	

Based on table 1 above it can be seen that from the number of prospective teachers as many as 139 students there are as many as 41% of prospective PE teachers have the first tutor to learn swimming is a friend, 33.8% are their parents, and the rest are trainers in clubs, PE teachers, and self-training . More than half or exactly as many as 74.1% of prospective teachers do exercises with tutors regularly while the rest or 25.9% is irregular. In adolescence, learners do tend to choose to learn with friends. Peer-tutoring becomes an alternative choice for them to learn to collaborate with the learning environment [12].

Furthermore, analyzed levels of confidence based on the group of prospective teachers according to tutors and routines in swimming practice. The results of the analysis can be seen in table 2 as follows.

Table 2. Self-confidence level differences by tutor type and routine in swimming practice

Variable	Sub-Variable	N	Mean	Std. Dev.	F/t	Sig.	fyi
The first tutor to learn swimming	Parent	47	166,32	15,587	2,023	0,095	Same
	PE teachers	14	157,43	21,353			
	Coach in Club	3	188,33	16,442			
	Friends	57	164,84	16,951			
	Others (brother, uncle, grandma, coach in extracurricular, and self-training)	18	164,11	21,978			
	Total	139	165,01	17,953			
Swimming exercise routines	Irregular (1)	103	162,61	15,618	-2,304	0,026*	2>1
	Regular (2)	36	171,86	22,244			

Note: * significant at the 0.05 level

Based on table 2 above can be explained that the different types of tutors do not affect the confidence level of prospective PE teachers in swimming learning. However, those who studied swimming regularly with tutors had higher levels of confidence than those who studied swimming on random. That is, anyone who becomes the first tutor is not a problem in shaping confidence. However, the competence of tutors to be variables that need to be considered in order to be able to provide learning in accordance with the learning needs of learners. High tutor competence can provide students with comprehensive explanations and knowledge to learn about swimming. Tutors with high competencies have an impact on learners' learning motivation [7]. High motivation to learn will give a strong impetus to learn swimming so as to beat the fear of learners in the activity in the water. So the selection of appropriate tutors is very important in order to increase confidence.

The next data analysis is to know the relationship between four variables namely, confidence level, tutor, frequency, and routine. Pool learning frequency for one week on average is 1.85 days with at most 7 days and at least 1 day. The results of the analysis can be seen in table 3 as follows.

Table 3. Relationship of self-confidence with routines, frequency, and swimming tutor

Source	Type III Sum of Squares	df	Mean Square	F	Sig.
Corrected Model	4820,433[a]	6	803,405	2,674	,018
Intercept	160367,954	1	160367,954	533,744	,000
Frequency of Swim Learning	140,409	1	140,409	,467	,495
Routines of Swim Learning	481,483	1	481,483	1,602	,208
The First Tutor	1754,577	4	438,644	1,460	,218
Error	39660,560	132	300,459		
Total	3829086,000	139			
Corrected Total	44480,993	138			

a. R Squared = .108 (Adjusted R Squared = .068)

Based on table 3 above it can be explained that simultaneously differences in tutors, routines, and frequency of student learning within a week significantly influence the confidence level of prospective PE teachers in swimming learning. Tutor relationships with learners can provide an advantage in swimming learning. The proximity of tutors and learners is expected to provide ease of tutors in choosing learning styles that fit the needs of learners, whether it be visual, auditory, cognitive, and kinesthetic [5]. However, from the results of the analysis test, the difference in confidence level, students who used to learn swimming regularly have higher levels of confidence than those who do not learn regularly. That is, the routine becomes very important to give the positive concept to the students themselves to have good swimming skills so that confidence can be formed. Self-Confidence is closely related to self-esteem. Self-confidence in the self-image obtained from the results of self-evaluation. The evaluative dimension of one's self is called self-esteem. People with low self-esteem will affect their low beliefs as well. People with low self-esteem are unable to assess themselves thoroughly [13].

4. Conclusion

Tutor differences are not related to the level of confidence, but those who study regularly with tutors have a higher level of confidence compared with those who learn randomly. The appropriate tutor will provide comfort in learning so it is expected to make the learner has a learning frequency so high the confidence level can be high. In fact, only the difference of routine that gives significant impact to the level of confidence but simultaneously the difference of tutors, routines, and frequency of student learning within a week significantly influence the confidence level of PE teachers in swimming learning. Finally, the results of the research show that students who have a higher level of confidence are those who follow practice routines with adequate frequency and helped by a compatible tutor.

References

[1] Khory and Dwi F 2009 Model Pembelajaran Renang pada Individu yang Mengalami Aqua-Phobia (Studi pada Mahasiswa S1 Angkatan 2008 Jurusan Pendidikan Olahraga FIK Unesa yang Memprogram Mata Kuliah Renang I di Semester Genap 2009) *J. Ordik* **7**

[2] Stillwell B 2007 Swimming Instruction for Those Fearfull of Water *J. Palaestra, ProQuest Fam. Heal.* **23** pp 36–42

[3] Komarudin 2015 *Psikologi Olahraga: Latihan Keterampilan Mental dalam Olahraga Kompetitif* (Bandung)

[4] Lawrence D 2006 *Enhancing Self-Esteem in the Classroom* (California: SAGE Publications Inc)

[5] Singh M M 2005 *A Generation and Classification of the Possible Learning Styles Used in the Acquisition of Basic Swimming Skills* (Canada: Library and Archives Canada)

[6] Rahman A and et al 2014 The Prevalence of Naturally Acquired Swimming Ability Among Children in Bangladesh: a Cross-Sectional Survey *J. BMC Public Heal. 2014* **14** p 404

[7] d'Arripe-Longueville F, Gernigon C, Huet M L, Cadopi M and Winnykamen F 2002 Peer tutoring in a physical education setting: Influence of tutor skill level on novice learners' motivation and performance *J. Teach. Phys. Educ.* **22** pp 105–23

[8] Schmidt R A and Wrisberg C A 2000 *Motor Learning and Performance: A Problem-Based Learning Approach* (USA: Human Kinetics)

[9] Jambor E A 1996 Beyond Language Barriers Teaching Techniques for Swimming *J. Phys. Educ. Recreat. Danc.* **67** pp 34–6

[10] Best C, Lieberman L and Amdt K 2002 Effective use of interpreters in general physical education *J. Phys. Educ. Recreat. Danc.* **73** pp 45–50

[11] Hartoto S 2016 Validasi Kuesioner Percaya Diri dalam Belajar Renang *Seminar Nasional dan Workshop Keolahragaan Tahun 2016, dengan tema "Membangun Prestasi Olahraga Indonesia Melalui Dunia Pendidikan dan Kebugaran Jasmani Bangsa"* (Surabaya)

[12] Wood D 2003 The why? what? when? and how? of tutoring: The development of helping and tutoring skills in children *Literacy, Teaching, Learn.* **7 (1)**

[13] McKay M and Fanning P 2000 *Self-Esteem* (Canada: New Harbinger Publications Inc)

IOP Conf. Series: Materials Science and Engineering **296** (2018) 012029 doi:10.1088/1757-899X/296/1/012029

Structuring students' analogical reasoning in solving algebra problem

S Lailiyah[1], T Nusantara[2], C Sa'dijah[2], E B Irawan[2], Kusaeri[1], and A H Asyhar[3]

[1] Faculty of Education and Teacher Training, Islamic State University of Sunan Ampel, Jl. Ahmad Yani 117, Surabaya, East Java, Indonesia

[2] Faculty of Mathematics and Science, State University of Malang, Jl. Semarang 5, Malang, East Java, Indonesia

[3] Faculty of Science and Technology, Islamic State University of Sunan Ampel, Jl. Ahmad Yani 117, Surabaya, East Java, Indonesia

lailiyah@uinsby.ac.id

Abstract. The average achievement of Indonesian students' mathematics skills according to Benchmark International Trends in Mathematics and Science Study (TIMSS) is ranked at the 38th out of 42 countries and according to the survey result in Program for International Student Assessment (PISA) is ranked at the 64th out of 65 countries. The low mathematics skill of Indonesian student has become an important reason to research more deeply about reasoning and algebra in mathematics. Analogical reasoning is a very important component in mathematics because it is the key to creativity and it can make the learning process in the classroom become effective. The major part of the analogical reasoning is about structuring including the processes of inferencing and decision-making happens. Those processes involve base domain and target domain. Methodologically, the subjects of this research were 42 students from class XII. The sources of data were derived from the results of thinks aloud, the transcribed interviews, and the videos taken while the subject working on the instruments and interviews. The collected data were analyzed using qualitative techniques. The result of this study described the structuring characteristics of students' analogical reasoning in solving algebra problems from all the research subjects.

1. Introduction

The report on assessment framework TIMSS 2011 stated that the average achievement of participants was 386 which is meant Indonesia is in the low level [1]. Average achievement of Indonesian participants in TIMSS 2011 was decreased from the average achievement in TIMSS 2007 that is 397, the TIMSS 2011 framework was like the TIMSS 2007 framework. This finding showed that the mathematics skills of Indonesian students ranked 38th of 42 countries. The research finding of The Programme for International Student Assessment (PISA) 2009 said that the mathematics skills of Indonesian students are at the 61st level from 65 countries with average score 371. Students' skill was seemed low especially in finding algorithms, interpreting the data, and using the steps in solving the problems. The research finding of TIMSS 2011 and PISA 2009 indicates that Indonesian students already have their mathematics skill, but that skill is not sufficient to solve regular problems about manipulating mathematical form and choose the accomplishment strategies and to solve non-regular problem such as mathematical problems as well.

The low skill of Indonesian students based on the TIMSS 2011 and PISA 2009 report become an important reason to do a deeper research about reasoning in mathematics. The low achievement of Indonesian students on the TIMSS 2011 needs specific study related to the content domain and cognitive domain. Mullis stated that the average percentage of the correct answer on the content domain and cognitive domain is 22% of Indonesian participants who are able to solve algebra problems and 17% of Indonesian participants who are able to solve the reasoning problem. Based on these results, it is needed to do an in-depth study of reasoning and algebra [2].

Learning mathematical concepts need to have special skills. Reasoning is a skill that is used both for teaching and learning [3]. This reasoning must have harmony to real life. Reasoning and proving become major goals in mathematics and learning process as one of the skills that must be mastered by students in learning mathematics [4]. The curriculum of NCTM that was adapted to Indonesian curriculum, especially for the curriculum 2006, states that teaching and learning mathematics at secondary schools aims to make the students are able to master the concepts, do reasoning, solve the problems, communicate their ideas and have the manner to respect the usefulness of mathematics in life [5]. Based on that curriculum, it can be said that reasoning is a very important skill in learning mathematics.

In mathematics, reasoning is divided into three types. They are deductive reasoning, analogical reasoning and inductive reasoning [3]. Analogical reasoning is reasoning skill that made to better learning in mathematical context. The advantages of analogical reasoning, if it is applied to mathematics teaching, are able to: increase students' creativities, develop students' reasoning abilities and motivation, increase their problem-solving abilities, make students remember the mathematics concepts in a long-term memory, associate the abstract concepts of mathematics to the students' real world, and give other examples from mathematics analogy examples. Analogical reasoning is a very important component in mathematics learning because it is the key to creativity and it can make the teaching-learning process in the classroom become effective [6]. Many cognitive processes involve analogy-making in one way or another: perceiving a stone as a human face, solving a problem in a way similar to another problem previously solved, arguing in court for a case based on its common structure with another case, understanding metaphors, communicating emotions, learning, translating poetry from one language to another [7].

Analogical reasoning plays a significant role in problem-solving [8]. The ability to use base problem (source domain or base domain) that has an identical structure in solving the target problem (target domain) can increase the performance of problem-solving. The definition of analogical reasoning in this study is a thinking process of inferring by using the same framework between the base problem and the target problem. Ruppert said that the analogical reasoning components include 4 things, they are the structure of encoding and inferring, mapping, applying, and verifying [9]. The component of analogical reasoning that will be applied in this study is stated by Ruppert.

Structuring is one of the important components in analogical reasoning because the rule of problem-solving in analogical reasoning from base domain to target domain is based on the structure-mapping. Gentner said that structure-mapping was first introduced to clarify the importance of structural similarity between the source domain and target domain or the relation between objects for each domain [10]. The purpose of structuring work-mapping is to catch the psychology process while doing the analogical mapping and tell the way of someone in using the analogy to make a conclusion.

Algebra is one of four cognitive domains from TIMSS that has the highest percentage that is 30%. The question number 6 is an algebra question in TIMSS 2011 about the quantity that full fills two differences which are presented using the balance of weights. The analogical problem consisted of the sixth question which is about the balance problem between first weights and second weights. The average percentage of the correct answer for the question number 6 for TIMSS is 18%. The percentage of students who answered the question correctly is low because the students could not identify the relation happened on the base domain (on those two weights balance) so that students were not able to

IOP Conf. Series: Materials Science and Engineering **296** (2018) 012029 doi:10.1088/1757-899X/296/1/012029

do the structuring in analogical reasoning. Therefore, it needs further discussion about structuring algebra problems.

2. Method

This study is a qualitative research and aims to discover the structuring of students' analogical reasoning in solving an algebra problem. This research was held at 3 Senior High Schools in Surabaya. The students who had been chosen were decided based on the following consideration: the students have good command in Mathematics, they are recommended by teachers, and they are communicative. In analyzing the result, this study used constant ratio between two subjects chosen from each group.

The data were collected from the result of algebra measuring instruments. The source data in this study were taken from the result of thinks aloud, transcribed interview, and video that was recorded while the subjects doing the worksheet and during the interview. The algebra measuring instrument in this study was presented in Figure 1 below.

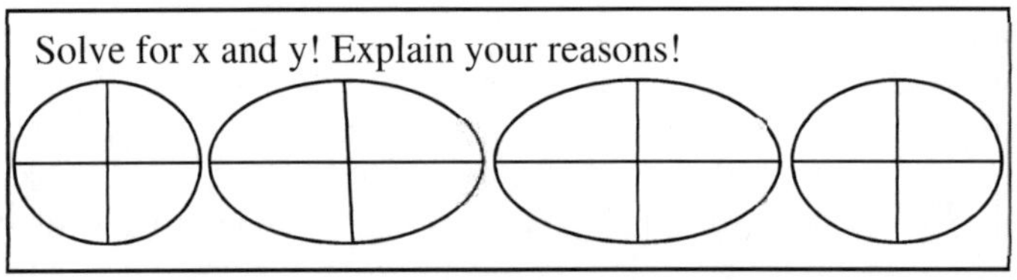

Figure 1. Algebra Measuring Instrument in This Study

The data had been collected by giving the questions to a student to be completed or solved. In the problem-solving process, students were asked to say loudly what they are thinking. The researcher then recorded the verbal expression from them and took a note of their expression including the unique things done by students when they solve the problem. When a student finished doing that question, the data collection techniques mentioned above were also applied for the other students till the researcher got the expected subject. This kind of data collection technique is called think out loud or think aloud.

The collected data were analyzed using qualitative data analysis. One of the models applied is interactive analysis technique. In this study, the arrangement of units is based on the problem that would be investigated; structuring, mapping, applying and verifying. And then, the students' spoken or written answers were categorized and coded.

3. Discussion

Structuring in analogical reasoning divided into two types; internal structuring and connective external structuring [11]. As for the results of this study are described as follow:

3.1. Internal Structuring

Internal structuring is an analogical reasoning that is applied by identifying every mathematics object in every base domain. The subjects included in internal structuring group are subject 1 (S1) and subject 2 (S2). The analogical reasoning of S1 and S2 have the same characteristics, that is subjects identifying relation item in each of base domains and subjects mapping the characteristic code form base domain to target domain.

In the process of reasoning, S1 and S2 can identify the base problem by linking parts only to the base domain (first circle and fourth circle) so that characteristic codes are formed. S1 and S2 can map the characteristic code to the target domain. S1 and S2 can apply the characteristic code to the target domain so that completion is obtained. Although S1 and S2 can make the connection of the source characteristic source code, map, and apply the characteristic code to the target problem, these two subjects do not check the correctness of the answer. The truth of both subjects' answers was clarified at the time of the interview. As for the occurrence of internal structuring on analogy reasoning is presented in Figure 2.

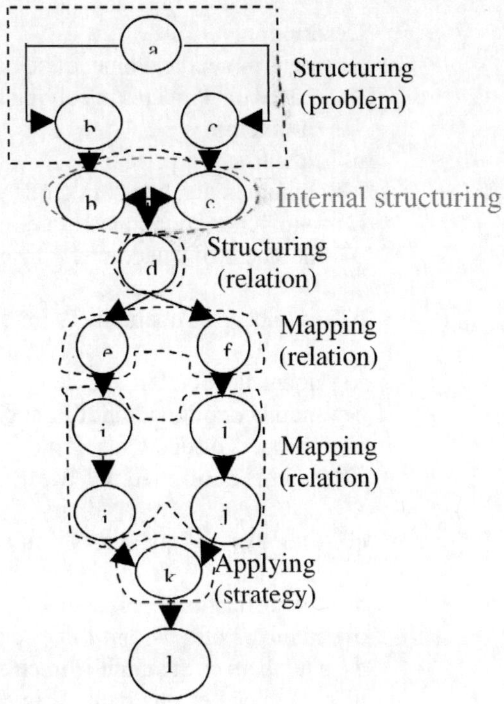

Description:
a = determine the value of x and y based on 4 images in solving math problems.
b = identification circle 1.
b'' = create characteristic code starting from quadrant I, II, III and IV in circle 1.
c = Identification circle 4.
c'' = create characteristic code starting from quadrant I, II, III and IV in circle 4.
d = draw a conclusion from the characteristic code on L I and L IV.
f = identification circle 3.
i = summation or subtraction from quadrant I, II, III, and IV in circle 2.
i' = the first two linear equations are formed.
k = summation or subtraction from quadrant I, II, III, and IV in circle 3.
l = the second two linear equations are formed.

Figure 2. Internal Structuring in Analogy Reasoning

3.2. Connective External Structuring

Connective external structuring is an analogical reasoning done by identifying every mathematics object in corresponding parts of the source domain and target domain. The subjects included in internal structuring group are Subject 3 (S3) and Subject 4 (S4). The analogical reasoning of S3 and S4 have the same characteristics, that is the subject identify the parts corresponding to the base domain and target domain on connective (connecting the part corresponding among all base domain and target domain with finding identical relationship) and the subject maps the relationship/characteristic code partially to the other corresponding parts.

In the process of reasoning, S3 and S4 have recognized the problem of solving the mathematical problem; S3 and S4 identify the corresponding sides between base domain and target domain. The relationships that are (identical) in the corresponding parts of all circles are then mapped to the corresponding parts of the other on all circles. Based on this S3 and S4 have known the relationship between problems. S3 and S4 have also known the appropriate strategy that will be used in solving the mathematical problem that is mapping the relationship to another part between the base domain and target problem and solved with the appropriate strategy. After getting the values of x and y, S3 and S4 verify the answers. Based on the processes occurring in S3 and S4, there are some common characteristics of the occurrence of connective external. As for the occurrence of connective external structuring on analogy reasoning is presented in Figure 3.

IOP Conf. Series: Materials Science and Engineering **296** (2018) 012029 doi:10.1088/1757-899X/296/1/012029

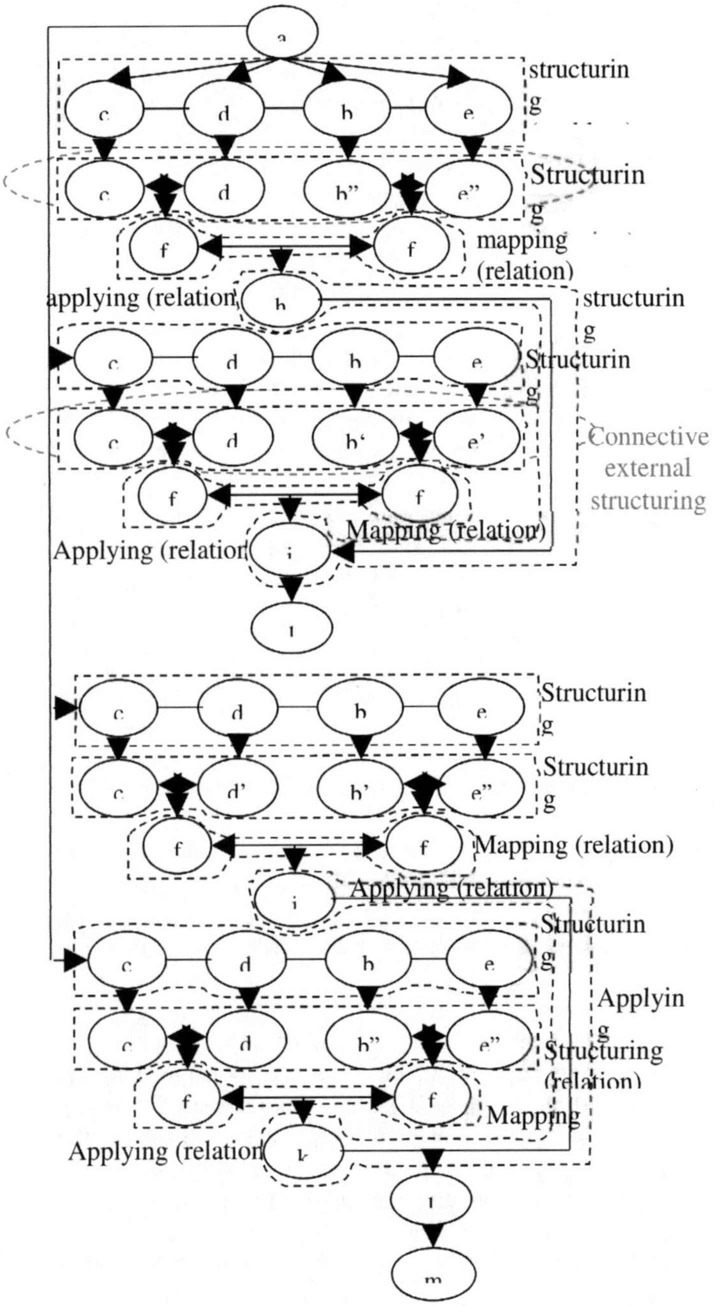

Description:

a = determine the value of x and y based on 4 images in solving math problems.
b = identification circle 1.
b'= numbers quadrant I in circle 1.
b" =numbers of quadrant II in circle 1.
b"'=numbers of quadrant III in circle 1.
b""=numbers of quadrant IV in circle 1.
c = identification circle 2.
c'= numbers of quadrant I in circle 2.
c" =numbers of quadrant II in circle 2.
c"'= numbers of quadrant III in circle 2.
c""= numbers of quadrant IV in circle 2.
d = identification circle 3.
d'= numbers of quadrant I in circle 3.
d" =numbers of quadrant II in circle 3.
d"'= numbers of quadrant III in circle 3.
d""= numbers of quadrant IV in circle 3.
e = identification circle 4.
e'= numbers of quadrant I in circle 4.
e" =numbers of quadrant II in circle 4.
e"'= numbers of quadrant III in circle 4.
e""= numbers of quadrant IV in circle 4.
f = summation of numbers in circle 1 and circle 4 equal summation of numbers in circle 2 and circle 3.
h = the first two linear equations.
i = the second two linear equations.
j = the third two linear equations.
k = the fourth two linear equations.
l = solving two linear equations.
m = finish.

Figure 3. Connective External Structuring in Analogy Reasoning

4. Conclusion

Internal structuring is a structuring that occurred only in the base problem. The characteristics code from mathematics problem is built internally in the base problem. The first step of doing internal structuring is by identifying the mathematics object consisted only in the base problem to get the characteristics codes. Therefore, the second step of this internal structuring is finding the identical characteristics code from each problem and inferring. The third step of this structuring is mapping the relation of the conclusion from base problem to target problem completely. The mapping result is applied to the target problem and then the subject chose the right strategy to solve that problem. This internal structuring is in line with the model of representative analogical reasoning stated by Holyoak, complete structure-mapping by Gabora and Saab, representative chart of analogical reasoning structure-mapping from Gentner, and general relation structure of analogical reasoning [10, 12-14].

In this connective external structuring, the characteristics codes are created among parts that are related to the base problem and target problem. The first step to do this connective external structuring is identifying the mathematical objects in the related parts between base problem and target problem so that the characteristics code is created. The second step in this connective external structuring is finding the identical characteristics codes in the corresponding parts between base problem and target problem and making the conclusion. The third step in this connective external structuring is mapping the conclusion of those correlations into other parts that are corresponding to the base domain and partial target domain. The result of that mapping was applied in others parts that are correlated with the base problem and target problem and choosing the right strategy for solving the problem.

Based on the result of this study, there are some suggestions that are: (1) in this study, there are some subjects and when they do reasoning there was also reflection. Therefore, the relation between analogical reasoning and reflection need to be analyzed. (2) In this study, students' creativities in creating characteristics code are needed for extensive external structuring. Therefore, the relation between external structuring and creativity need to be discussed. (3) this study only analyzes the structuring in students' analogical reasoning to solve the mathematics problem, but structuring and mapping are two analogical reasoning that has a correlation to each other. Because of that, mapping and analogical reasoning in solving mathematic problem needs to be analyzed.

References

[1] OECD 2010 *Student Performance in Reading, Mathematics, and Science* **1** http://dx.doi.org/10.1787/9789264091450-en[
[2] Mullis I V S, Martin M O, Foy P and Arora A 2012 *TIMSS 2011 International Result in Mathematics* (Boston: TIMSS & PIRLS International Study Center)
[3] Mofidi S A and Parvaneh A 2012 *Indian Journal of Science and Technology* **5 (6)** pp 2916—22
[4] NCTM 2000 *Principles and Standards for Schools Mathematics*. (United States of America: NCTM)
[5] BSNP 2006 *Standar Isi Untuk Satuan Dasar dan Menengah* (Jakarta: BSNP)
[6] Gust H and Ka4i-Uwe K 2006 *Proc. Cognitive Science Society in cooperation with the 5th International Conference of Cognitive Science in the Asian-Pacific region (CogSci/ICCS)* (Germany: Lawrence Erlbaum) pp 417–1422
[7] Kokinov B and R M French 2003 *Encyclopedia of Cognitive Science* Vol. 1 pp.113—8
[8] English L D 2004 *Mathematical and Analogical Reasoning of Young Learners* (New Jersey: Lawrence Erlbaum)
[9] Ruppert M 2013 *Proc. Eighth Congress of European Research in Mathematics Education (CERME 8)* (Turkey: Manavgat-Side, Antalya)
[10] Gentner D 1983 *Journal of Cognitive Science* **7** pp 155-70
[11] Lailiyah S 2017 *Jurnal Pendidikan Sains* **5 (2)** pp 38-45
[12] Holyoak K J 2012 *The Oxford Handbook of Thinking and Reasoning* (New York: Oxford University Press) pp 234-59
[13] Gentner D and L Smith 2012 *Encyclopedia of Human Behavior* (2nd Ed) pp 130-36
[14] Gabora L and Adam S 2011 *Proc. of the Annual Meeting of The Cognitive Science Society* (Houston TX: Cognitive Science Society) pp 3506-11

IOP Conf. Series: Materials Science and Engineering **296** (2018) 012030 doi:10.1088/1757-899X/296/1/012030

Process skills approach to develop primary students' scientific literacy: *A case study with low achieving students on water cycle*

Suryanti[1], M Ibrahim[2] and N S Lede[3]

[1]Teacher Education Elementary School of Universitas Negeri Surabaya
[2]Biology Education, Universitas Negeri Surabaya
[3]Post Graduate of Universitas Negeri Surabaya

suryanti@unesa.ac.id

Abstract: The results of the Program for International Student Assessment (PISA) study on the scientific literacy of Indonesian students since the year 2000 have been still far below the international average score of 500. This could also be seen from the results of the science literacy test of 5th-grade students of primary school in Indonesia which showed that 60% of students are still at level ≤ 3 (value < 500). The students' science literacy skills need to be improved by applying learning with a process skills approach. This study aims to describe the findings of classroom action research using a process skills approach to the science literacy level of primary students ($n = 23$). This research was conducted in 2 cycles with stages of planning, implementation, observation, and reflection. Students' ability in scientific literacy was measured by using description and subjective tests of context domains, knowledge, competencies, and attitudes. In this study, researchers found an improvement in students' science literacy skills when learning using a process skills approach. In addition, students' scientific attitude is also more positive. In activities for learning science, students should be challenged as often as possible so that they have more practice using their scientific knowledge and skills to solve problems presented by teachers in the classroom.

1. Introduction

Mastering literacy is important for all people in order to be able to compete in the global era both as members of society, citizens, and citizens of the world [1-3]. Every citizen has a scientific literacy ability which allows them to survive in the wild as well as in the workplace, with the knowledge, understanding, skills, and values that make up that competence. Literacy activities are more than just reading and writing, and include thinking skills using knowledge sources in print, digital, visual, and auditory [4].

Scientific literacy can be trained through studying. According to King, the purpose of teaching science is producing students who have competence in scientific literacy. The skills of teachers in planning the learning process is the key to successful learning [5]. One alternative way to improve literacy abilities of science students in the learning process is by applying the process skills approach. Scientific literacy can be enhanced through research-based learning, including *discovery learning* [6]. According to Rauf, et al, [7], process skills in science are very important in learning how to gain knowledge. Aktamis and Ergin [8] stated that process skills include the skills of each individual that are used in daily life which can improve the quality and standard of life by understanding the nature of science. Ango [9] stated that science process skills are basic and essential components of the process of learning science under the guidance of a teacher. In line with this opinion, Ibrahim, et al., [10] stated that mastering process skills mean mastering skills to conduct research and solve problems. Problem-solving and research skills are life skills and are the highest learning outcomes for students.

This research aims to describe the level of scientific literacy of primary students in studying science with a process skills approach. The results of this research are expected to contribute to the improvement of science students' scores in scientific literacy in Indonesia conducted by PISA every 3 years. The results of the 2009 PISA study found that the average literacy score in science was 383, literacy in reading was 402 and literacy in math was 371 [11].

The results of research that have been conducted by Foulds and Rowe [12] show that when science teaching and increasing process skills are integrated into the learning activities of students, this can improve the students' understanding. Gormally, Cara., etc. [13] in their research found an increase in the scientific literacy of students using lab inquiry. The research of Sullivan [14] found that an understanding of environmental robotics with an open approach is able to encourage students to use scientific literacy and process skills as well as improving understanding. Remziye, et al [15] in their research found that learning by using inquiry approaches can improve process skills and attitudes about science.

Previous researchers, in their research on high school and college students, implemented inquiry-based activities which necessitate process skills to enhance understanding of what is being studied. In this study, researchers use a process skills approach to measure the scientific literacy intelligence of science students, in particular, primary school students. Scientific literacy includes context, knowledge, competence, and attitude, which can only be measured by testing high-level thinking ability. Scientific literacy is regarded as a key learning outcome in education for all students aged 15 years, regardless of whether they continue studying science or not, for participating in society [16]. Although not yet 15 years of age, primary students need to be introduced to scientific literacy in the context appropriate to their cognitive development, for example, their immediate environment and personal setting. The students who were the subjects of this research at a minimum had access to an on-site classroom and learning resources. In addition, their initial scientific literacy results were very low. Teaching that has occurred up until this point centered on the teacher and the students were not actively involved in the learning process. As a result, students are not used to using scientific knowledge through a series of scientific activities. In addition to this, measuring the students' knowledge was generally done focusing on memorized information, so only lower level knowledge was being measured.

The focus of this study is to describe the scientific literacy ability of science students learning about the water cycle. Basic competencies include describing the need for water-saving and describing the process of the water cycle and the way that human activities that may be affected. The scientific literacy indicators that were observed include identifying scientific issues, explaining phenomena scientifically and using scientific evidence.

2. Research method

This research is classroom action research with a descriptive, qualitative design, which aims to describe or reveal more deeply about the process and learning outcomes related to the scientific literacy level of primary school students. Before the action research began, students were given a pre-test and after the action research, they were given a post-test.

The subjects in this study were the 5th graders of Inpres Naikoten 1 Primary School, Kupang City, NTT Indonesia. There were a total of 23 students consisting of 12 boys and 11 girls studying water cycle. The subjects of the study were chosen after considering where the relevant needs were.

The research procedure included the planning phase, implementation, observation, and reflection. Instruments used in the research included scientific literacy tests, student activity observation sheets, and student science attitude questionnaires. Student activity observation sheets included formulating problems and hypotheses, using tools and materials as directed, observing,

IOP Conf. Series: Materials Science and Engineering **296** (2018) 012030 doi:10.1088/1757-899X/296/1/012030

analyzing data, and making conclusions. Student questionnaires were structured to determine students' opinions about vacant land empty of cultivated crops, excessive use of water, and so forth. The scientific literacy test was conducted twice; during the water cycle material and the material about the impact of human activities on the water cycle. The test was multiple choice. Here is an example of scientific literacy, which falls into the category of high-level thinking skills.

The multiple choice questions in each cycle were 10 questions; with 2 questions of level 1, 1 question of level 2, 2 questions of level 3, 1 question of level 4, 2 questions of level 5, and 2 questions of level 6. Before being used, the test was validated by 2 experts. The test was valid and needed only small revisions. The data about the scientific literacy level of students was analyzed descriptively. The indicators used to determine the success of this research was if there is an increase in the middle classification of scientific literacy level, 80% of students are able to answer questions at level 4 and the observation of scientific attitude reaches 80% (very good).

3. Results and Discussion

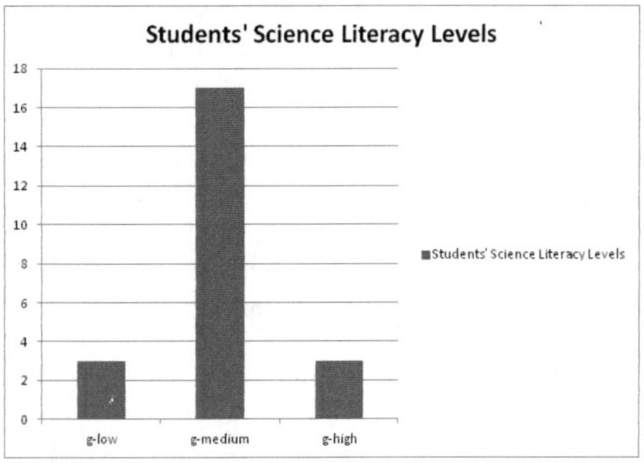

Data on the literacy test results in the areas of knowledge, context, and competency are presented as followed. It appears that students' science literacy ability in Cycle 1 learning is 52.17% ≤ level 3 and 47.83% are at level 4. None can reach level 6. After learning cycle 2, 8.70% students are ≤ 3 and as many as 91.30% are at level ≥ 4.

Figure 1 shows the increase in students' science literacy levels in the range of (g) > 0.7 (g-high) scale of 3 people, 0.7 ≥ (g) ≥ 0.3 (g-medium) of 17 people, G) < 3 (g-low) as many as 3 people. After applying the process skills approach, there is an increase in scientific literacy skills from cycle 1 to cycle 2. This result shows that by undertaking process skills activities that are hands-on, it increased student learning outcomes. In process skills activities, students were directly involved in activities such as watching videos about the water cycle, asking questions, conducting activities as directed, discussing observations and making conclusions. By watching the video and asking some questions students will construct their own thoughts and discuss this with their peers. This is in accordance with the opinion of Semiawan, et al., [17] who explained that process skills are associated with the physical and mental skills needed for the fundamental abilities possessed, controlled and applied in scientific activity when scientists find a new discovery. Skills-based learning in the science process is learning that integrates the skills of the scientific process into a presentation of the materials in an integrated way. This learning emphasizes the process of seeking knowledge rather than the transfer of knowledge, learners are viewed as the learning subjects that need to be actively involved in the learning process, and the teacher is only a facilitator who guides and coordinates learning activities.

The above explanation is supported by the statement of Jerome Bruner and his colleagues who provide important theoretical support known as discovery learning, which emphasizes the importance of helping students acquire knowledge in a way that can train students' intellectual abilities, stimulate their curiosity and motivate their competence. This is what is meant by acquiring knowledge through discovery learning. In his book *Toward a Theory of Instruction*, Bruner points out "if we teach science, we will not produce small life libraries about science, but we want to make children think mathematically for themselves and participate in the process of acquiring knowledge. Knowing is a process, not a product" [18].

The result of an increase in student scientific literacy through applying the process skills approach shows the effectiveness of the approach used in teaching. Information submitted by teachers to students about learning materials related to the concept of water cycle and saving water can be applied to daily life and students can draw conclusions from the problems provided by teachers. This is supported by Piaget's theory which explains that the main factor that encourages the cognitive development of a person is the motivation or power from within the individual himself to want to learn and interact with his environment [19]. Therefore, in order for students to be motivated to learn, teachers must responsible for conducting interesting lessons, one of which is by applying the process skills approach in learning science. The results of the science literacy of the students in the domain of knowledge are in line with the results of the attitude domain which shows a positive attitude. The results of the attitude towards science domain are presented in Table 1 below.

Table 1. Scientific literacy attitude domain

No.	Statement	%	Category
1	Create wastewater storage at home	97	Very Good
2	Get drinking water as much as possible and not drink it until it runs out	100	Very Good
3	Be indifferent when I see the tap running in the school toilet	98	Very Good
4	Do not bury plastic waste in the soil	97	Very Good
5	Always promote a water-saving attitude to friends	99	Very Good
6	There are no restrictions on groundwater drilling by the government	95	Very Good
7	Provide suggestions to parents to create a rain water catchment basin	97	Very Good
8	I chose to take a bath from the with a dipper rather than with a shower	100	Very Good
9	The government establishes a program of absorbing wells at several points of the city	96	Very Good
10	Turn off the water tap during brushing and washing hands	100	Very Good

Scientific literacy in the domain of science attitudes became very good in relation to their responses about wastewater capture, taking drinking water, not burying plastic rubbish, and so on. The results from this attitude domain can reawaken the active role of the students to acquire concepts from the learning activities, then the students can apply the concepts obtained to everyday life. This explanation is supported by Piaget's theory that the child builds his own knowledge from his own experience within his environment. In Piaget's view, knowledge comes from action, and cognitive development largely depends on how far children are actively manipulating and actively interacting with their environment. The tendency of primary school children is to move from the concrete things, and then look at needs in an integrated way. Based on the above trends then, learning is an active process, constructive, and goal-oriented, all of which depends on the mental activity of learners. The provision of direct learning experiences in science lessons is strongly emphasized through the use and development of process skills and scientific attitudes with a view to understanding concepts and solving problems [21]. This is in accordance with the student's activity observation results in Table 2.

Table 2. Student activity in classroom learning

No.	Student Activity	Percentage (%)
1.	Formulate questions	100
2.	Make a hypothesis	89
3.	Using tools and materials according to instructions	100
4.	Cooperate in the activities of observation	87
5.	Discuss and analyze the observation data	98
6.	Discussion to summarize	89

Based on Table 2 above, students were involved in scientific investigation activities $\geq$ 80%. The process skills approach integrated into science lessons can actively involve students in the learning process [17], as well as strengthening the revelation of Dimyati and Mudjiono [20] on the advantages of science process skills. Providing opportunities for students to work with science, rather than just telling or listening to stories about science, can lead to more active students, as well as science process skills that can help students learn scientific processes and knowledge at the same time.

High student activity in this study is supported by several theories, such as Piaget, who states that learning is an active process, constructive, goal-oriented, and learning all depends on the mental activity of learners [21]. Bruner also revealed that discovery learning is the foundation of the discipline of science, the need for the active involvement of students in the learning process, and he believes that this actually takes place through personal discovery. The goal of education not only trains the knowledge base but also creates opportunities for students' invention and creativity [22]. Bruner suggests that students should learn through active participation with concepts and principles in order for them to gain experience, and conduct experiments that allow them to discover the principles themselves [23].

Jerome Bruner is a Harvard psychologist who provides important theoretical support for discovery learning, an instructional model that emphasizes the importance of helping students to understand the structure or key ideas of a discipline, the active involvement of students in the learning process and the belief that true learning takes place through personal discovery. Discovery learning emphasizes the inductive reasoning and investigation processes that characterize the scientific method [24].

The most visible and influential student activity is the discussion between peers and teachers. This is in line with what described by Vygotsky's theory which states that if children work or handle tasks that cannot be solved on their own, then they can be solved with the help of peers or adults.

4. Summary

There is an increase in students' scientific literacy levels after they engage in lessons using the process skills approach. This happens because in the learning process the students are actively involved in learning both physically and mentally (hands-on activity) through observing activities, making inquiries, conducting activities as directed, analyzing data, and drawing conclusions. The students' scientific attitude also experiences positive development. This shows that the ability of scientific literacy can be increased through learning activities.

In order to increase scientific literacy, it is best if classroom learning is designed to be problem-solving and challenging to students. To present problems in the classroom, the teacher can present phenomenon from the surrounding the environment in the form of video or live activities, demonstration, and storytelling related to the topic being studied.

References

[1] Holbrook J and Rannikme M 2009 The Meaning of Scientific Literacy Estonia: University of Tartu *Int. J. Environ. Sci. Educ.*
[2] Genci M 2015 The Effect of Scientific Studies on Students' Scientific Literacy and Attitude *Ondokuz Mayis Univ. J. Fac. Educ.* 34 141–52
[3] Zuriyani E 2016 Literasi Sains dan Pendidikan
[4] Mulyana A 2016 Tahapan Pelaksanaan Gerakan Literasi Sekolah
[5] Ainurrahman 2011 *Belajar dan Pembelajaran* (Bandung: Alfabeta)
[6] Widodo W 2016 Literasi Sains Mahasiswa Calon Guru IPA Berdasarkan Permintaan Kognitif (Cognitive Demand) PISA: Bagaimana Langkah Selanjutnya *Prosiding Seminar Nasional 2016* (Surabaya: FMIPA UNESA)
[7] Rauf A A R 2013 Inculcation of Science Process Skills in a Science Classroom *Can. Cent. Sci. Educ. Asian Soc. Sci.* 9 1911–2025
[8] Aktamis H and Ergin O 2008 The Effect of Scientific Process Skill Education on Student' Scientific Creativity, Scientific Attitudes and Academic Achievements *Asia-Pacific Forum Sci. Learn. Teach.* 9 1–21
[9] Ango M L 2002 Mastery of Science Process Skills and Their Effective Use in the Teaching of Science: An Educology of Science Education in the Nigerian Context *Int. J. Educol.* 16
[10] Ibrahim M 2010 *Dasar-dasar Proses Belajar Mengajar: Hakikat SAINS, Kurikulum Tingkat Satuan Pendidikan, Media Pembelajaran, Taksonomi Tujuan Pembelajaran, Perancangan Pembelajaran Sampai Kepada Ciri Guru Profesional* (Surabaya: Unesa University Press)
[11] OECD 2010 PISA 2009 Results: What Students Know and Can Do – Student Performance in Reading, Mathematics and Science 1
[12] Foulds W and Rowe J 1996 The Enhancement of Science Process Skill in Primary Teacher Education Students *Aust. J. Teach. Educ.* 21
[13] Gormally C, Brickman P, Hallar B and Amstrong N 2009 Effects of Inquiry-based Learning on Students' Science Literacy Skills and confidence *Int. J. Scholarsh. Teach. Learn.* 3
[14] Sullivan F R 2008 Robotics and science literacy: Thinking skills, science process skills, and systems understanding *J. Res. Sci. Teach.* 45 373–394
[15] Ergul R, Simsekli Y, Calis S, Ozdilek Z and Sanli M 2011 The Effects of Inquiry-Based Science Teaching on Elementary School Students' Science Process Skills and Science Attitudes *Bulg. J. Sci. Educ. Policy* 5
[16] Inzanah 2012 *Pengembangan Perangkat Pembelajaran IPA Berbasis Kurikulum 2013 untuk Melatih Literasi Sains Siswa SMP* (Universitas Negeri Surabaya)
[17] Semiawan C 1992 *Pendekatan Keterampilan Proses* (Jakarta: Grasindo)
[18] Dahar R W 2006 *Teori-Teori Belajar dan Pembelajaran* (Jakarta: Erlangga)
[19] Ratumanan T G and Lourens T 2011 *Penilaian Hasil Belajar Pada Tingkat Satuan Pendidikan* (Surabaya: Unesa University Press)
[20] Dimyati and Mudjiono 2009 *Belajar dan Pembelajaran* (Jakarta: Rineka Cipta)
[21] Hidayat T 2014 Makalah Teori Piaget dan Penerapannya dalam Pembelajaran di SD
[22] Nur M 2014 *Model Pembelajaran Berdasarkan Masalah* (Surabaya: Pusat Sains dan Matematika Sekolah Unesa)
[23] Slavin R E 1994 *Educational Psychology: Theory and Practice 4th* (Massachusetts: Allyn and Bachon Publishers)
[24] Arends R 2008 *Learning to Teach Buku 2* (Yogyakarta: Pustaka)

IOP Conf. Series: Materials Science and Engineering **296** (2018) 012031 doi:10.1088/1757-899X/296/1/012031

Gender, academic achievement, and ownership of ATM as predictors of accounting students' financial literacy

Susanti[1] and H T Hardini[2]

[1]Universitas Negeri Surabaya - Faculty of Economics
[2]Universitas Negeri Surabaya - Faculty of Economics

susanti@unesa.ac.id

Abstract. This study examined the relationships between GPA, gender, and ownership of ATM on accounting students' financial literacy (n = 184). Financial literacy was assessed using a paper-and-pencil objective (multiple choice) test measuring general knowledge of finance, income, money management savings, loans, and investment. Gender and GPA data were obtained from the university records. Regression analysis found that GPA and ownership of ATM were associated with financial literacy, but gender was not. Female students with an ownership of ATM and those with a high GPA were found to be superior to males. The implication of this research is that students are expected to increase their GPA and utilize financial facilities in the form of ownership ATM and other financial instruments so as to increase financial literacy. In addition, the need for financial literacy training from related parties to improve financial literacy for students who have low financial literacy.

1. Introduction

1.1. Background

The presence of diverse financial services has a positive side that this will provide improved prosperity for middle-class people who are unfamiliar with access to financial services. An example is banking services. Banking parties are increasingly competitive in innovating the products they offer to customers. It also makes access and regional and geographical conditions become one target of banking to be closer to the community. It also suggests that financial services do not belong to a group of people in urban areas alone.

In order for making the finances to be processed carefully, efficiently, and efficiently by utilizing financial services, it is important for individuals to understand about financial literacy. Financial literacy is a set of skills and knowledge that allows individuals to make effective decisions on their investments in order to improve their finances. Financial literacy is an effort to increase the public's sensitivity to the financial services sector, beginning with knowing, then believing, to being skilled at becoming actively involved, in other words achieving a well-literate society in the financial services sector.

The National Institute for Literacy [1] states that literacy is defined as an individual's ability to read, write and speak, calculate and solve problems at the required level of proficiency, in individuals, families, and communities. Financial literacy is financial knowledge and ability to Applying it in everyday life with the aim of achieving prosperity [2]. [3] Financial literacy puts more emphasis on the ability to understand the basic concepts of economics and finance, to how to apply them appropriately. [4,5] and Chen and Volpe state that financial Literacy as knowledge to manage finances. The financial literacy indicators according to Chen and Volpe [3] are general knowledge of finance, savings, and loans, insurance, and investment. Meanwhile, according to Jumpstart Coalition states, financial literacy indicators are income, financial management, savings and investment, and loans or credit.

The low level of financial literacy in Indonesia makes the Financial Services Authority (OJK) issued several policies and programs to increase the financial literacy of the community one of them by launching the Indonesian National Strategy of Financial Literacy (SNLKI) consisting of three pillars with one of the pillars is the education and national campaign of literacy Finance. The policies and programs are not only for business actors but also for various groups, from housewives to students and students. Students are one of the components of society which is big enough to contribute to the economy because in the future the students will enter the world of work and begin to be independent in their financial management [6].

Nidar and Bestari [7] mentioned demographic factors affecting financial literacy including parental education, pocket money, education level, faculty, parent income and insurance. Research conducted by Mahdzan and Tabiani [8] found that influential demographic factors were age, gender, education level, number of children, marital status, and work experience. Cude at al [9] also resulted in the conclusion that the GPA factor affects the financial literacy, the higher the GPA score, the financial literacy will also be healthier or better. Nevertheless, the results of research conducted by Rita and Pesudo [10] state that the GPA has no significant effect on student financial literacy.

2. Research methods

This research is included in this type of quantitative research. Independent / independent variables in this study consisted of three variables namely, GPA (X1), Gender (X2), and ownership of ATM (X3). Sedangan dependent variable / bound that is financial literacy (Y). In the gender variables, researchers used the dummy method to see the effect of the dependent variable. Dummy variables here consist of two categories: male and female, therefore, used a dummy variable. Dummy variables are used because men and women belong to the nominal scale. The population in this study is all students of Department of Economic Education Prodi S1 Accounting Education and accounting year of the year 2013/2014 and 2014/2015 amounted to 342 students. While the sampling is determined by the proportional random sampling method where the target of the sample being studied is taken proportionally based on the total population of each class. So the number of samples taken on each class is not the same. Determination of the number of samples using the Slovin formula is

$$n = \frac{N}{(1+Ne^2)} \quad (1)$$

Description = n = Number of Sample
N = Population
e = significance level (0.05)

$$n = \frac{N}{(1+Ne^2)} \quad (2)$$

$$n = \frac{342}{(1+342x0{,}05^2)} \quad (3)$$

$$n = \frac{342}{(1+0{,}855)} \quad (4)$$

$$n = \mathbf{184} \quad (5)$$

In this research, data collection technique used research instrument in the form of questionnaire/questionnaire. Before given to the questionnaire respondents first been tested the validity and reliability in order to be able to measure students' financial literacy well, as table 1.

Table 1 Total Population and Sample Research

Department / Study Program	Force	Population	Sample Calculation	Sample
	2013 A	50	50/342x184 = 26,90	27
Accounting	2013 B	39	39/342x184 = 20,98	21
	2014 A	49	49/342x184 = 26,36	26
	2014 B	50	50/342x184 = 26,90	27
Accounting	2013 A	36	36/342x184 = 19,36	19
Education	2013 B	35	35/342x184 = 18,83	19
	2014 A	42	42/342x184 = 22,59	23
	2014 B	41	41/342x184 = 21,78	22
Total		342	184	184

While the data analysis techniques use descriptive statistics to provide an overview of student financial literacy in the low or high category. Here are the criteria for students' financial literacy Interval [10] as follows:

$$\textbf{Interval} = \frac{\textbf{High Value} - \textbf{Value Values}}{\textbf{Many Class Intervals}} \tag{6}$$

$$\textbf{Interval} = \frac{\textbf{100} - \textbf{0}}{\textbf{2}} = \textbf{50} \tag{7}$$

So obtained interval data variable financial literacy as table 2 follows:

Table 2 Category of Financial Literacy Students

Financial Literacy	Category
0 < 50	low
51<100	High

While to test the influence between variables, we used multiple regression analysis. The form of multiple regression statistics in this study are as follows:

Y = α + b1D1+ b2X1 + b3X2

Where :

Y = Financial Literacy
α = Constanta
b1 = Coefficient of gender
b2 = Coefficient of IPK
b3 = Coefficient of ownership ATM
D1 = Gender (dummy)
X1 = IPK
X2 = Ownership of ATM

3. Result of research

3.1. Literacy level Financial Accounting and Accounting Education students Faculty of Economics UNESA

The results of the descriptive analysis indicate that the students of accounting and accounting education force of 2013/2014 and 2014/2015 have financial literacy with an average of 51.00 with high category (100> 51). As many as 94 students (51%) studied were in high financial literacy classification, while 90 students (49%) were low (0 <50)

3.2. Classic assumption test

3.2.1 *Normality Test. To determine normally distributed research data, it can be shown in p-plot chart as follows:*

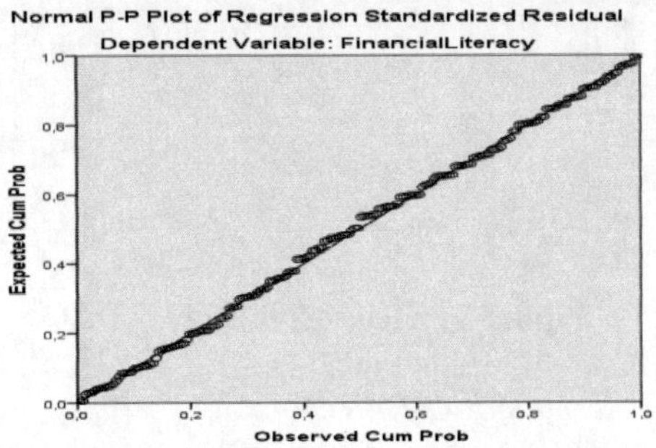

Graph 1. P-Plot Test Data Normality

Based on the results of spss test version 22, obtained p-plot table approaching slashes so it can be said normally distributed data.

3.2.2 *Multicollinearity test.* The multicollinearity test is used to determine whether there are independent variables that have similarities between the independent variables in a model. The output of SPSS version 22 can be shown in table coefficient as follows:

Table 3. Coefficients

Model		Unstandardized Coefficients		Standardized Coefficients			Collinearity Statistics	
		B	Std. Error	Beta	T	Sig.	Tolerance	VIF
1	(Constant)	-246,252	19,375		-12,710	,000		
	Gender	-,071	1,673	-,002	-,043	,966	,983	1,018
	GPA	84,260	5,939	,684	14,189	,000	,805	1,242
	Ownership of ATM	12,073	2,506	,234	4,817	,000	,796	1,257

a. Dependent Variable: Financial Literacy

Based on the result of SPSS test version 22 above, obtained multicollinearity free data shown at VIF value less than 10.

Heteroscedasticity test. Heteroscedasticity test is a test that serves to know whether or not the deviation of classical assumption Heteroscedasticity that is the variant inequality of the residual for all observations on the regression model. Based on the result of SPSS test version 22, obtained free data of Heteroscedasticity indicated on spreading data spreads do not form a pattern.

Here are the test results through SPSS version 22 as figure 1:

IOP Conf. Series: Materials Science and Engineering **296** (2018) 012031 doi:10.1088/1757-899X/296/1/012031

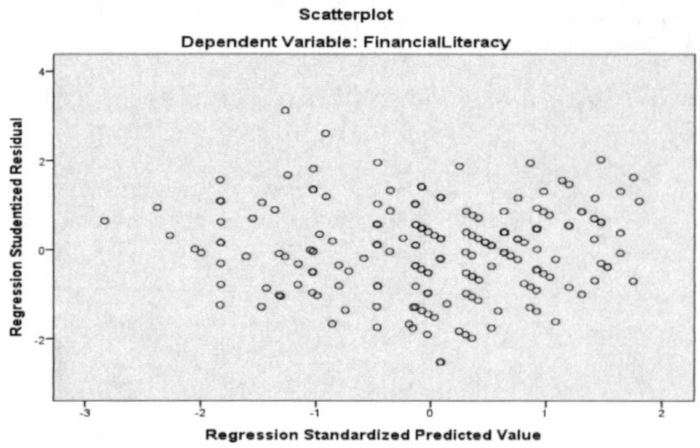

Figure 1. Heteroscedasticity Test

3.3. Test Results coefficient of determination (R), Test t, and Test F

3.3.1 Test Result R. The R test is used to find out how big the simultaneous contribution of the independent variable to the dependent variable. Here is a summary model table that shows the value of R (Coefficient of determination) of a model.

Table 4. Model Summary

Model	R	R Square	Adjusted Square	RStd. Error of the Estimate	Durbin-Watson
1	,814a	,663	,657	10,893	1,218

a. Predictors: (Constant), Owner Savings Account, Gender, GPA
b. Dependent Variable: Financial Literacy

Based on SPSS version 22 results obtained data that Adjusted R Square value of 0.657. This means that the factors of financial literacy are influenced jointly by gender variables, ATM ownership and GPA of 65.7%. The rest is 34.3% influenced by variables outside the study.

3.3.2 Test Result t. To test the hypotheses 1, 2 and 3 in this research, it can be shown t-count results in table 3 that the above coefficient table should be greater than t table.

Ha 1: There is a GPA influence on accounting financial literacy and accounting education faculty of economics Universitas Negeri Surabaya.

Based on the above coefficient table, the t-hit value for the GPA of 14.189> 1.973, so that Ha1 is accepted. So it can be concluded that there is influence between variables GPA to financial literacy accounting students and accounting education faculty of economics Universitas Negeri Surabaya.

Ha 2: There are gender influences on the financial literacy of accounting students and accounting education faculty of Economics of Universitas Negeri Surabaya. Based on the above coefficient table, the t-count value for the gender variables is 0,043 <1,973, so Ha 2 is rejected. So it can be concluded that there is no influence between gender variables on accounting student financial literacy and accounting education faculty of economics Universitas Negeri Surabaya.

Ha 3: There is an influence of ATM ownership on the financial literacy of accounting students and accounting education of the faculty of economics of Universitas Negeri Surabaya. Based on the above coefficient table, the t-count value for the ATM ownership variable is 4.817> 1.973, so that Ha 3 is accepted. So it can be concluded that there is an influence between variable ownership of ATM to financial literacy accounting student and education accounting faculty of economics of Universitas Negeri Surabaya.

3.3.3 *Test Result F.* F test is used to test hypothesis 4 (Ha 4) that is influenced simultaneously between gender variables, IPK and ownership of ATM to financial literacy of Accounting and Accounting education student of Economic Faculty of UNESA. The results of the F test can be shown in the following ANOVA table:

Table 5. – ANOVA

Model		Sum of Squares	Df	Mean Square	F	Sig.
1	Regression	41980,191	3	13993,397	117,928	,000b
	Residual	21358,803	180	118,660		
	Total	63338,995	183			

a. Dependent Variable: Financial Literacy

b. Predictors: (Constant), Owner Savings Account, Gender, GPA

The acceptance criteria of Ha is if F arithmetic> F table.

Ha 4: There is a simultaneous influence between cumulative achievement index, gender and savings account ownership on financial literacy of accounting students and accounting education faculty of economics Universitas Negeri Surabaya

In the table above the value of F table: F arithmetic is 117,928> 2.65 then Ha4 accepted. So it can be concluded there is a simultaneous influence between the cumulative achievement index, gender and savings account ownership of financial literacy accounting students and accounting education faculty of economics State Universitas Negeri Surabaya.

4. Discussion

There is a GPA influence on the financial literacy of accounting and accounting education students Faculty of Economics, Universitas Negeri Surabaya. The result of the statistical analysis indicates that the GPA (Grade Point Average) has a positive and significant influence on the financial literacy of accounting and accounting students FE UNESA. The influence here means that the higher the student's GPA, the financial literacy will also increase. Conversely, the lower the student's GPA, the lower the financial literacy. Thereby causing their financial knowledge to differ from each other. Differences in the ability to understand the concept of finance at the time of lectures make students with high GPA tends to have a better financial concept than the students with lower GPA. In the end, the financial literacy of students with high GPA is better than those with lower GPA.

The statistics show that female students have higher financial literacy compared to male students. According to the researchers' observations, the difference in financial literacy between male and female students is because female students are more diligent to understand matters relating to financial understanding. Female students are also more diligent to make financial planning both in the short and long term. This is because women are more anxious and afraid of financial problems, causing them to be more careful. The results of this study support the theory of Cude et al. [9] which revealed that the higher the GPA score, the financial literacy will also be healthier or better.

There is no influence of gender on accounting financial literacy and accounting education Faculty of Economics, Universitas Negeri Surabaya. The results showed that gender does not affect student financial literacy. This shows that both male students as well as in the faculty of economics, especially the accounting and accounting education professions have no difference in terms of understanding financial information. The results of this study do not support the theory that the students' financial literacy is influenced by gender. [3]. The results of this study are in accordance with research by Rita and Pesudo [10] that gender variables do not significantly affect the financial literacy of the students because they are suspected to be familiar with the forms of financial management and financial products.

There is the influence of ATM ownership to the financial literacy of accounting and accounting education students Faculty of Economics, Universitas Negeri Surabaya. The results of this study indicate that the ownership of ATM significantly affects the financial literacy of Accounting and Accounting education students. Students who use ATMs have relatively high financial literacy compared to students who do not use ATMs. Mandell [11] explains that students who use ATMs can manage their finances practically as they can easily conduct financial transactions considering the ATM function as a debit card that can be used to pay a certain expense without giving cash. He also found that the use of ATMs can affect a person's financial literacy level. There is a simultaneous influence between genders, GPA and ownership of ATM affect the financial literacy of accounting and accounting education students Faculty of Economics, Universitas Negeri Surabaya.

5. Conclusion and recommendations

Based on the results of research and discussion, it can be concluded that there is a GPA influence on the financial literacy of accounting and accounting education students Faculty of Economics, Universitas Negeri Surabaya. There is no influence of gender on the financial literacy of accounting and accounting education students Faculty of Economics, Universitas Negeri Surabaya. There is an influence of ATM ownership to financial literacy of accounting and accounting education student Faculty of Economics, Universitas Negeri Surabaya. There is simultaneous influence between gender, GPA and ownership of ATM influences to financial literacy of accounting and accounting education student Faculty of Economics Universitas Negeri Surabaya

References

[1] Remund D 2010 Financial Literacy Expliced, The Case for a Clearer Definition in a Increasingly complex Economy *J. Consum. Aff. Summer* **44** 276–95
[2] Lusardi A and Mitchell O S 2007 Baby Boomer Retirement Security: The Roles Of Planning, Financial Literacy, And Housing Wealth *J. Monet. Econ.* **54** 205–24
[3] Chen H and Volpe R P 1998 An Analysis of Personal Financial Literacy Among College Students *Financ. Serv. Rev.* **7** 107–28
[4] Danes S M and Hira T K 1987 Money Management Knowledge Of College Students *J. Student Financ. Aid* **17** 4–16
[5] Keown and Anne L 2011 The Financial Knowledge of Canadians *Canadian Social Trends*
[6] Nababan, Darman, Sadalia and Isfenti 2012 Analisis Personal Financial Literacy dan Financial Behavior Mahasiswa Strata I Fakultas Ekonomi Universitas Sumatera Utara *J. Media Inf. Manaj.* **1**
[7] Nidar, Rahman S, Bestari and Sandi 2012 Personal Financial Literacy Among University Students (Case Study at Padjadjaran University Students, Bandung, Indonesia) *World J. Soc. Sci.* **2** 162–71
[8] Mahdzan, Nurul S and Saleh T 2013 The Impact of Financial Literacy on Individual Saving: An Exploratory Study in The Malaysian Context *Transform. Bus. Econ.* **12**
[9] Cude B J, Lawrence F C, Lyons A C, Metzger K, LeJeune E s, Marks L and Machtmes K 2006 College Students and Financial Literacy: What They Know What We Need to Learn *Conference Eastern Family Economic and Resource Management association*
[10] Rita, Maria R, Pesudo and Beaya C A 2014 Apakah Mahasiswa Sudah Melek Keuangan? *J. Din. Akuntansi, Keuang. dan Perbank.* **3** 58–65
[11] Mandell and Lewis 2008 *The Financial Literacy of Young American Adults: Results of The 2008 National Jump$Tart Coalition Survey of High School Seniors and College Students* (Washington DC: The Jump$tart Coalition for Personal Financial Literacy)

Geography literacy can develop Geography skills for high school students: *is it true*?

W S Utami[1], I M Zain[1] and Sumarmi[2]
[1]Geography Education Faculty of Social Science and Law, Universitas Negeri Surabaya
[2]Geography Education Universitas Negeri Malang

wiwikutami@unesa.ac.id

Abstract. The most important issue related to education in Indonesia is the low quality of student learning and competence. The basic thing that is important to be studied is the demands of 21st-century skills that are difficult to fulfil with the low competence of student learning. Low competence of student learning demonstrated by low capacity of scientific literacy includes geography literacy. Geography skills of Indonesian students are also low. It is shown from the students' ability to use maps to describe and to analyze is low. The purpose of this study is to determine the correlation between the literacy skills of geography to develop geography skills of high school students in Surabaya. Written and performance tests were given to the sample of 29 high school students. The results of the tests we analyzed based on Geography literacy and its correlation to Geography skills in terms of the ability to use the media, map, and analyze the phenomenon of the geosphere. The results showed that the students who have low literacy geography have difficulty in using map.

1. Introduction

Understanding and responding to the challenges and opportunities of the world in the twenty-first century will require many skills; the capacities to think and communicate mathematically and scientifically will remain at a premium. Geographic literacy will also be necessary for some reasons of enhancing the economic competitiveness, preserving the quality of life, sustaining the environment, and ensuring national security. As individuals and as members of society, human face decisions on where to live, what to build, where to travel, how to conserve energy, how to wisely manage scarce resources, and how to cooperate or compete with others [8, 11]

Making all of these decisions, personal and collective, requires a geographically informed person or someone who can see the meaning in the arrangement of things on Earth's surface, who can see relations between people, places, and environments, who can use geographic skills, and who can apply spatial and ecological perspectives on life situations. Geographic skills make a person can understand the connections between patterns of rivers and the physical processes that create them, between patterns of cities and the human processes that create them, and between what happens in the places in which we live and what happens in the places throughout the world near and far.

The goal of the National Geography Standards is to enable students to become geographically informed through knowledge and mastery of three things: (1) factual knowledge; (2) mental maps and tools; (3) and ways of thinking [8]. UNESCO writes literacy is a fundamental human right and the foundation for lifelong learning. It is fully essential to social and human development in its ability to transform lives. For individuals, families, and societies alike, it is an instrument of empowerment to improve one's health, one's income, and one's relationship with the world. The uses of literacy for the exchange of knowledge are constantly evolving, along with advances in technology. From the internet to text messaging, the ever-wider availability of communication makes for greater social and political participation [16].

Geographic literacy could be defined as the competence of turning understanding and comprehension of geographical knowledge into a skill because literacy consists of solving problems, reasoning, critical and creative thinking processes. Geographic literacy skills, bring this competence to a conscious level and contribute to students' geographic literacy skills and processes in their professional lives [3]. In this context, geographic literacy does not only mean geographical knowledge. Literacy is a systematic approach towards events, founding, situations, and places which requires understanding, comprehension, analytical and synthesis skills. On Earth where we live as a part of the geographical environment, human beings and nature are affected by each other directly or indirectly.

A geographically literate person has the skill to look for solutions which cause the least damage to nature and reduce negative side effects. Additionally, he/she has the skill to understand the relationships between different cultures and people. To achieve this, a person should first learn the features of where he/she lives, why he/she lives there, what the surrounding events and discoveries are, and how and when the relationships between these could affect him/her. In order to do this, one should have a good geographical vantage point.

Geographic literacy was described in the literature as one of 34 new literacy fields, along with environmental literacy, media literacy, economic literacy, visual and technology literacy [2]. According to the United Nation (UN), literacy is described as "individuals acquiring the basic living skills, being able to solve the problems they may face in life, understanding the societies and the economies of the 21st century, and actively participating in them".

In a study carried out in the USA in 1988, the National Council for Geographic Education [NCGE] attempted to determine the geographic literacy level of Americans. The results showed that14% of Americans could not show their country's location on a map, and 25% did not know the location of the Pacific Ocean [15]. In Indonesia, the literacy skills of learners are still low. Results of Indonesian National Assessment Program (INAP) explained that 73.61% of learners have the less knowledge about science literacy skills and 46.83% students have less ability to read. The purpose of this study is to analyze the literacy skills of geography and to develop geography skills of the students.

2. Literature review

Literacy is described as "individuals acquiring the basic living skills, being able to solve the problems they may face in life, understanding the societies and the economies of the 21st century, and actively participating in them. Geographic literacy could be defined as the competence of turning understanding and comprehension of geographical knowledge into a skill because literacy consists of solving problems, reasoning, critical and creative thinking processes" [3].

In this context, geographic literacy does not only mean geographical knowledge. Literacy is a systematic approach towards events, founding, situations, and places which requires understanding, comprehension, analytical and synthesis skills. On Earth where we live as a part of the geographical environment, human beings and nature are affected by each other directly or indirectly. A geographically literate person has the skill to look for solutions which cause the least damage to nature and reduce negative side effects. (IGU, 2008)

Many researchers have described the concept of geographic literacy as the knowledge of a location or the skill of positioning a location on a map [13,15]. However, some researchers state that knowledge of a location is not enough to describe geographic literacy, instead proposing that having the knowledge of the location of places is the basic level of geographic literacy (Keeling, 2003; Bein, 1990; Donovan, 1993). Backler and Stoltman stated that having location knowledge is only the first step to geographic literacy and the real literacy is the ability to use geographic knowledge and skills [16]. In another description, Eve, Price and Counts (1994) described geographic literacy as the skill of reading a map, having location knowledge and the skill of understanding interactions between cultures and people in other areas. The concept is not limited to reading graphics, reading maps or skills related to field observations; it also requires having a geographical view [2].

Geographic literacy is divided into three groups: low, middle, and high level. Low-level geographic literacy consists of the capability of knowing a location. Middle level requires one to understand human interactions with the environment in terms of cause-effect relationships. This level consists of the skills of questioning, verifying, evaluating and analyzing. High-level literacy is a critical geography approach [10]. At this level, people understand the difference in the powers and natural resources of countries related to geography. For this reason, geographic literacy, instead of being plain knowledge, it is a capability which helps in coping with daily problems [13].

In a different study, Oigara measured the low-middle-high level geographic literacy skills of university students. In the research, it was found that the geographical knowledge of students was generally weak [13]. Torrens measured high school students' location knowledge of European countries and major cities. As a result of the research, it was found that the basic location knowledge of students was insufficient [15].

3. Methods
This research was conducted on a class of XI student of social sciences program at a high school in Surabaya, Indonesia. The total is 28 students of 11 girls and 17 boys. Geography literacy in the study includes media literacy, literacy maps, and knowledge of geography related to world issues and current events (population, natural resources, and weather, health, urbanization, deforestation, climate change). Geography literacy skills are assumed to affect the geography skills. High literacy skills showed high geography skills. Data of geography literacy skills is gotten through a written test and a performance test. Data were analyzed descriptively.

4. Results and Discussion
It is explained like any social or physical scientist, that the geographer's goal is to ask and answer questions in order to gain knowledge. By asking a question, the geographer will develop a hypothesis and measure variables in order to confirm or deny the hypothesis, just as any scientist would do [9]. The geographer is not limited, however, to any one discipline, type of data, or set of tools. He or she is free to use any available information or tools in order to understand how and why things appear as they do in Earth space. That is being said that geographers do use a method that requires certain skills. These are the skills your students must learn in order to "do geography."

The National Geography Standards identify five skills that will guide a geographic inquiry. These five skills enable geographers to (1) ask geographic questions about spatial distributions and processes, (2) acquire geographic information about distribution, (3) organize geographic information, (4) analyze geographic data, and (5) answer geographic questions [8]. Literacy ability of geography in the study includes media literacy, literacy maps, knowledge of geography related to world issues and current events (population, natural resources, and weather, health, urbanization, deforestacy, climate change). Literacy geography student of SMAN 18 Surabaya can be seen in the following figure.

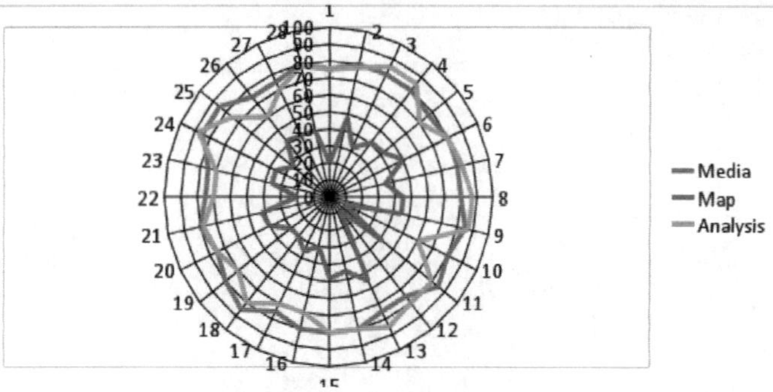

Figure 1. Literacy ability geography

Based on the result of written test and performance tests conducted by 28 students, the result is below average. This proves the average literacy 64.11 geographies. Low literacy skills, especially with the ability to use a map to show the location, describe the characteristics of an existing location on the map that the average 36.35. Students' ability to use the media to explain the phenomena, the symptoms and the process is already well (79.18).

Table 1. Students' Geography Literacy

Geography Literacy Component	Average Value
Media Literacy	79.18
Literacy map	36.35
Knowledge of geography	76.79
Average Geography Literacy	64.11

Students' skills in analyzing and studying the phenomena of the geosphere are also good (76.79). Students with literacy skills in analyzing the phenomena of good geosphere turned out to have the literacy skills and media literacy skills using a good map scoring the average of 70.39. When linked together between geography and literacy skills to read a map and to show the location and assess the characteristics of the region on the map is visible in Figure 2 below.

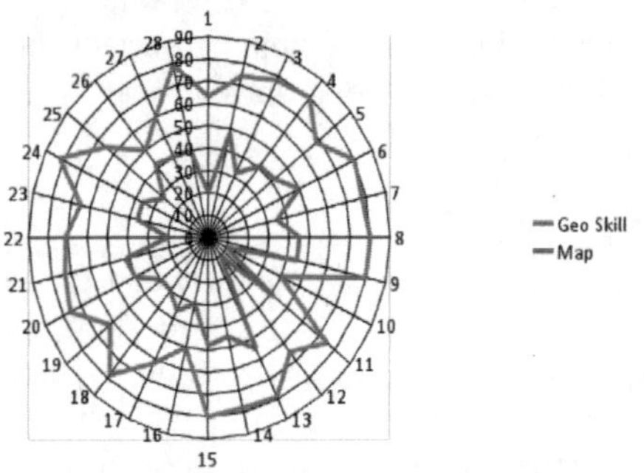

Figure 2. Geography skill and literacy map

Geography skills of students grades on average 70.39. Geography skills are not supported by the literacy skills using a map as a tool for navigation skills and the identification of the place. The score of literacy map of the students is 36.35. Geography skills are well supported by media literacy skills and knowledge of geography. And location is the first of four concepts which Gershmel identifies as the core constituents of geography: 1) location, i.e., knowing where things are; 2) place, i.e., understanding the unique character and differences of places; 3) links, i.e., knowing connections between different locations; 4) regions, i.e., comprehending spatial patterns, both formal and functional, at a larger scale [4].

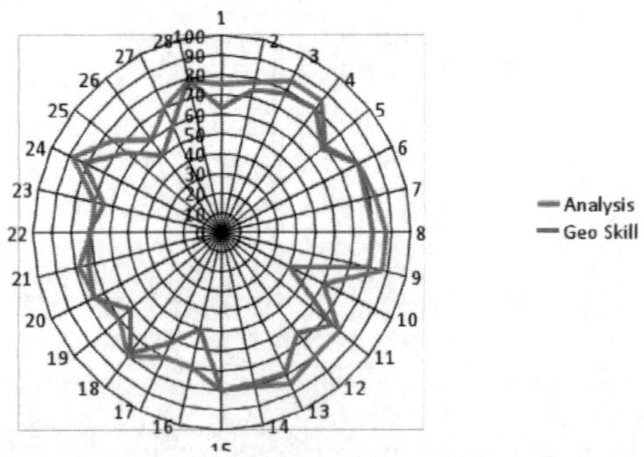

Figure 3. Geography skill and knowledge of geography

Knowledge of geography supports students' geography skills. There is an average score of 76.79 on geography knowledge. Students who have a good knowledge of geography also has good geography skills. Geographic literacy could be defined as the competence of turning understanding and comprehension of geographical knowledge into a skill because literacy consists of solving problems, reasoning, critical and creative thinking processes. Today, in order to impart literacy skills to individuals, teachers should first have these skills. It is important that teacher candidates should attain these skills before graduation. When these conditions are established, it will be easier for teacher candidates to be aware of their geographic literacy skills, bring this competence to a conscious level and contribute to students' geographic literacy skills and processes in their professional lives [3].

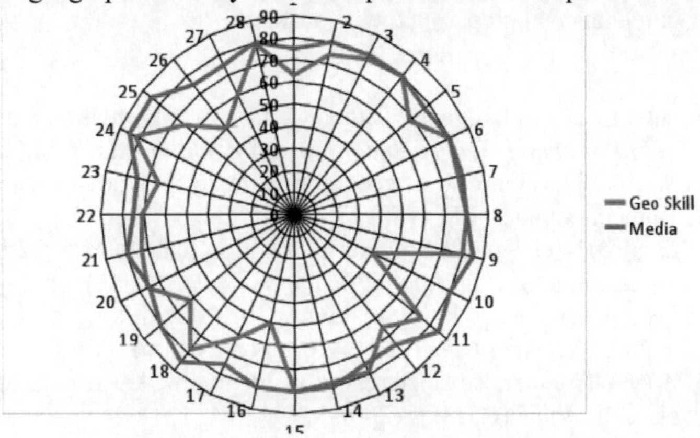

Figure 4. Geography skill and literacy media

Pellegrino described the specific skills seemed to be "21st-century skills" may be defined, categorized, and determined differently from person to person, a place to place, or school to school, the term does reflect a general-if somewhat loose and shifting-consensus. An illustrative overview of the knowledge, skills, work habits, and character traits commonly associated with 21st-century skills. One of them is Information and communication technology (ICT) literacy, media and internet literacy, data interpretation and analysis, computer programming [11].

The convergence of media and technology in a global culture is changing the way we learn about the world and challenging the foundations of education. It is no longer enough to read the printed word; children, youth, and adults too. They need the ability to both critically interpret the powerful images of a multimedia culture and express themselves in multiple media forms. Media literacy education provides a framework and a pedagogy for the new literacy needed for living, working, and citizenship in the 21st-century. Moreover, it paves the way to mastering the skills required for lifelong learning in a constantly changing world [14].

It is important for students to master the media and ICT today. Students' skills in using media and ICT to analyze the problems. In a global media culture, people need two skills in order to be critical thinking and self-expression. Media literacy instils both of these core skills, enabling the future student to sort and understand geosphere phenomena. The results of performance tests indicate that the ability of the average student in good media literacy (79.18). It supports students' geography skills. In lessons, media help students to analyze geographic knowledge related to various issues that occur in Indonesia such as deforestation, forest fires, urbanization, an outbreak of dengue fever.

5. Conclusion

One of the character traits commonly associated with 21st-century skills including critical thinking, problem-solving, reasoning, analysis, interpretation, and synthesizing information, Information and communication technology (ICT) literacy, media and internet literacy, data interpretation, and analysis, computer programming. 21st-century skills require that students have the literacy skills of science is no exception geography literacy. In general science literacy of students is still low. Low literacy is still low competence related science student learning. Geography literacy knowledge and skills related to geography. Literacy geography visits media literacy, literacy map and geographic knowledge. The results showed geography literacy skills supported by good knowledge of geography. What needs to be studied further is the low literacy map high school students. This affects the low literacy geography.

Acknowledgement
Muhammad Arsyad (orcid.org/0000-0001-5619-4543) Department of Agricultural Socio-economics, Faculty of Agriculture, Hasanuddin University Agricultural Economics and Rural Development. Thank you for your guidance in preparing the manuscript.

Reference

[1] Bascom Johnathan 2011 Geography Literacy and Moral Formation among University Students, *Review of International Geographical Education Online @RIGEO* **1 (2)**
[2] Bennett M W 1997 Development of geographic literacy in students with learning disabilities (Eric Document Number: ED 418034)
[3] Dimenly Yurdal 2014 Geographic literacy perception scale (GLPS) validity and reliability study *Mevlana International Journal of Education (MIJE)* **4 (1)** pp 1—15 Available online at http://mije.mevlana.edu.tr/ http://dx.doi.org/10.13054/mije.13.43.4.1
[4] Gershmehl P 2005 *Teaching Geography* (New York: Guilford Press)
[5] Kitchin R M 1996 Are there sex differences in geographic knowledge and understanding? *The Geographical Journal* **162 (3)** pp 273—86
[6] National Geographic-Roper Public Affairs 2008 *Geographic literacy study* Prepared for: The National Geographic Education Foundation National Geographic Society NW, Washington, D.C.
[7] National Geographic-Roper Public Affairs 2006 *Geographic literacy study* Prepared for: The National Geographic Education Foundation National Geographic Society NW Washington, D.C.
[8] National Geographic Society (NGS) 2008 *Geography for life: The national geography standards* (Washington, DC: Author)
[9] Netemeyer R G, Bearden W O, and S Sharma 2003 *Scaling procedures: issues and applications*, *sage* (Thousand Oaks)
[10] Oigara J 2006 *A multi-method study of background experiences influencing levels of geographic literacy* Unpublished Ph.D Thesis (Binghamton University, USA)
[11] Pellegrino James W 2014 Assesment as Positive Influence on 21st Century Teaching and Learning: A System Approach to Progress, *Psicologia Educativa 20 (2014) 65-77*, www.enseiver.es/psed
[12] Saarinen T F and C L Mac Cabe 1994 World patterns of geographic literacy based on sketch map quality. *Professional Geographers* **47 (2)** pp 196—204
[13] Stoltman, J P 1991 Teaching geography at school and home *Eric Clearing House for Social Studies: Social Since Education* (Bloomington: ED 335284)
[14] Thoman Elizabet, Tessa Jolls 2005 *Literacy for 21st Century, An Overview and Oriented Guide to Media Literacy Education* (CML, Centre for Media Literacy)
[15] Torrens P M 2001 Where in the world? Exploring the factors driving place location knowledge among secondary level students in Dublin, Ireland *Journal of Geography* **100 (2)** pp 49—60
[16] United Nations-UN 1986 *International literacy year* The General Assembly, Recalling its resolution 41/118 of 4 December

IOP Conf. Series: Materials Science and Engineering **296** (2018) 012033 doi:10.1088/1757-899X/296/1/012033

The implementation of flipped classroom model in CIE in the environment of non-target language

Xiao Renfei[1, 2], Ali Mustofa[2], Zhang Fang[2, 3] and Su Xiaoxue[1, 2]

[1]College of International Cultural Exchange, Central China Normal University, Hubei Wuhan, China
[2]Pusat Bahasa Mandarin, Universitas Negeri Surabaya, East Jawa, Indonesia
[3]College of Humanities, Jianghan University, Hubei Wuhan, China

alimustofa@unesa.ac.id

Abstract. This paper sets a theoretical framework that it's both feasible and indispensable of flipping classroom in *Chinese International Education (CIE)* in the non-target language environments. There are mainly three sections included: 1) what is flipped classroom and why it becomes inevitable existence; 2) why should we flip the classroom in *CIE* environments, especially in non-target language environments; 3) take *Pusat Bahasa Mandarin Universitas Negeri Surabaya* as an instance to discuss the application of flipped classroom in non-target language environments.

1. Introduction

The current society is economical so that we permanently pursue how to reach the maximum benefits even in learning and teaching. While it's generally acknowledged the learning is influenced by internal and external environments, with the updating of education technologies, it becomes a somehow new issue when we should know how to integrate the new blending learning environments. The teacher cannot only be the initiator in class but also to integrate the intrinsic and extrinsic factors, such as inspiring students' motivation, managing their cognitive load, utilizing the physical social contexts, etc. Therefore, when the technology becomes an integral part of educational settings, the teachers will be in search for ways to incorporate technology into their classrooms to create better learning opportunities for their students [1]. However, in Indonesia, learning system is generally *Teaching Centered Learning (TCL)* [2] and rarely use modern educational technology, which is defined by Hamdan, McKnight & Arfstrom [2] as a traditional teacher-centered model. The learning system really needs reforming into *Student Centered Learning (SCL)* where the teacher is not only the learning source materials but also the learning partner for the students who help the students to facilitate their studying. And how to use the modern technology to stimulate students' initiative is an indispensable question in the current informationalization age.

It's less than a decade that most national universities established *Chinese major*, *Chinese Department* or *Pusat Bahasa Mandarin/Confucius Institutes* in *Indonesia*. In other words, *Chinese International Education (CIE)* don't have much time which means, on one hand, we maybe won't influenced by the traditional mode, but on the other hand, we should cross the river by feeling the stones. The following is the design and its experiments of flipped classroom in Confucius Institute at *Universitas Negeri Surabaya (UNESA)*.

2. What's the flipped classroom?

In traditional didactic teaching model, the teachers are the main resourses who transferring their knowledge and information to the students, it is a one-way track so that making the students become passive and without creativity. How to maximize the students' autonomous learning ability has been pedagogically thinking so that produced the method of communication, the method of the task, humanistic method and so on in the course of second language teaching. No matter how these methods student-centered, the teaching mode is still in the range of that class is class and homework is homework until the emergence of the flipped classroom which turned the class and homework model up-side-down.

It is said that the sparks of flipped classroom have been generated since last century when Alison King published "From Sage on the Stage to Guide on the Side" which focuses on the importance of the use of class time for the construction of meaning rather than information transmission. While not directly illustrating the concept of "flipping", King's work is often cited as an impetus for an inversion to allow for the educational space for active learning [3, 4]. Then Eric Mazur played a significant role in the development of concepts influencing flipped teaching through the development of an instructional strategy he called peer instruction, which moved information transfer out of the classroom and information assimilation into the classroom, allowed him to coach students in their learning instead of lecture.

The real and somehow similar to modern flipped classroom started from Lage, Platt and Treglia (2000)'s study, whose paper "Inverting the Classroom: A Gateway to Creating an Inclusive Learning Environment" asserts that one can leverage the class time that becomes available from the inversion of the classroom (moving information presentation via lecture out of the classroom to media such as computers or VCRs) to meet the needs of students with a wide variety of learning styles. And in the same year, J. Wesley Baker presented a paper discussing what he termed the "classroom flip" at an education conference in what may be the first published mention of the word "flip" associated with this model of teaching and learning [5].

The flipped classroom really getting attraction was in 2007, when two chemistry teachers—Jonathan Bergmann and Aaron Sams from Woodland Park High School carried out their classes by recording their lectures and posted them online in order to accommodate students who couldn't attend the class due to many factors (it happens often because the school located in a rural area which is difficult for the students attending the class, and they also must spend a lot of time to go to the school by using the buses). They uploaded the synchronous videos of interpretation and PowerPoint presentation demo on the internet to let the students download and study before the class, leaving the class time for the students to finish their homework, or help the students who with difficulties in the experiments.

However, perhaps the most recognizable contributor to the flipped classroom is Salman Khan who began recording videos at the request of a younger cousin he was tutoring because she felt that recorded lessons would let her skip segments she had mastered and replay parts that were troubling her in 2004. Later on, he founded Khan Academy whose slogan is "a free world-class education for anyone anywhere" based on this model in 2006, which becomes so influential that for some it turns into synonymous with the flipped classroom and it can now provide more than 5600 videos from various studies. Up to now, the flipped classroom has become a widely accepted model and it has been applied in many educational institutions. Apart from Khan Academy, MEF University Model is also very influential internationally, for it is the first university in the world that adopted the "flipped classroom" educational model university-wide [6].

So flipped classroom was usually described as a setting where that "which is traditionally done in class is now done at home, and that which is traditionally done as homework is now completed in class" (Bergmann and Sams 2012, p.13), i.e. "events that have traditionally taken place inside the classroom now take place outside the classroom and vice versa" [7]. However, this definition is only the tip of the proverbial iceberg. A successfully flipped classroom functionally should increase the interaction between students and teachers (Roehl et al., 2013), contribute to student engagement, team based skills, personalized student guidance, classroom discussion, and creative freedom of faculty

(Millard, 2012), and be "effective in helping students learn the content, and increased self-efficacy in their ability to learn independently" (Enfield 2013, p. 14).

3. Why flipped classroom in CIE?

Since Teaching Chinese as a Foreign Language (TCFL) or Teaching Chinese as a Second Language (TCSL) becoming a subject, the language skills, such as listening, speaking, reading and writing training are recognized as the center aspects, whose fundamental task is to teach language, different from the teaching of linguistics, i.e. the teaching model should be language skills training rather than knowledge learning. The orientation of communication determined that it demonstrates a kind of flipping from the very beginning. For example, there is a kind of reflection of the flipped idea that the teacher asks students to preview the new words before the class, and then ask them to dictate in the class, which is so extensive that traditional demonstration classes are all teaching in that way. As TCFL/TCSL turned to Chinese International Education (CIE), there are ever some debates, such as what is the core task of CIE, what role should cultural teaching take and how about their relationship, etc. However, most experts confirmed that the essence of CIE is language teaching.

The technical nature of language teaching determines that TCSL/CIE must be pedagogically student-centered, to which the Grammar Translation Method who is prevalent in 18-19 centuries cannot compare, and the task-based interactive construction learning mode becomes the inevitable choice. With the constant development technologies, the deep integration of technologies and curriculums provides more possibilities for interactive learning. Among those strategies, the flipped classroom is one of the most effective models in constructing abilities.

Nevertheless, the truly modern Chinese flipped classroom in TCSL environments has been around for less than a decade. With the worldwide spreading of Khan Academy mode, Chinese flipped classroom is popping up all the time. So far, according to our survey, there are at least 100 universities, such as Beijing Language and Culture University (BLCU), Peking University (PU), engaging in the practice of Chinese flipped classroom. Usually speaking, as long as the university is in process of constructing of informatization courses, there exists a Cloud-platform (or their own, or cooperate with other specialized organization), where provides a comprehensive management platform for the teachers' curriculum resources construction, class management, achievement analysis of students, etc. The Department of Teaching Affairs also encourages the teachers applying for the project to construct their courses under the supervising of the university.

Differ from the systematic and organizational constructing, it seems casual and lack of attention in overseas environment where the flipped classroom actually should be more needed. The reasons for necessity are as following:

- Unlike TCFL class where there are more than 20 or 30 hours per week, CIE class usually has less than 12 hours per week. Let's take Universitas Negeri Surabaya as an instance, students in Chinese Department have only 12 hours Chinese courses each week. And some other universities are even less.
- Unlike TCFL environment where it's easy to find language partners, it's difficult to find a language partner in overseas training, then it becomes a practical way through following videos the teacher recorded.
- In addition, Unlike TCFL and Chinese Department students whose Chinese courses are compulsory and with an affirmative schedule, many overseas students who studying Chinese are just for interest or some other motivation so that their time is limited and their schedule is flexible.
- Unlike TCFL class where we almost have the same demands in the same class, CIE overseas always need to accommodate to individual requirements.
- There maybe exists another problem that the university curriculum system may take Chinese as a foreign language like English in CIE overseas environments, but the difference is Chinese learners are mostly beginners.

And along with the network technology developing, the improvement of people's living standards and increasing purchasing power of computer/smartphone, it becomes more feasible of flipped classroom in the overseas CIE environments.

- Firstly, flipped classroom is a product of the era of multi-media, therefore, the relatively complete multimedia technology platform is a prerequisite. And students are generally with their personal computer or smartphone (according to our survey, more than 70% students have computer or smartphone; almost all students have access to a computer and master the basic operation method), it laid a material foundation for the flipped classroom implementation.
- Secondly, it's mutually adapted to each other between basic principles of CIE and flipped classroom: a) flipped classroom which demands sufficient and efficient preview can fully carry out the principle of "essence interpret and practice as much as possible" which is fundamental in CIE and improve the efficiency of students' learning; b) flipped classroom is a kind of method which absorbs the advantages of cooperative learning, where with the substitution of "student-to-student" interaction, "teacher-to-student" is no longer the only mode of communication; c) flipped classroom can fully arouse the students' enthusiasm and initiative.

Confucius Institute Headquarters/Hanban may have been gotten the needs of online Chinese courses for overseas students, so they began to set up Confucius Institute Online (CIO) from several years ago. Up to now, there are already more than 307.000 live classes and 7.537 lessons. It can be stated that Confucius Institute Online has the greatest influence in the overseas CIE's environments.

4. How do we flip in UNESA?

However, when CIO can't apply the courses which demanded in an overseas environment, or the courses provided by CIO are not suitable for overseas students, then it becomes necessary to integrate the resources and establish own resources. While it's difficult to find a platform to construct and manage the courses recourses in CIE overseas environment, there are some practical Apps which can help us integrating resources, among those, "Micro Tas (微助教)" is the easiest one to operate and convenient, entertaining the process of evaluation and teaching. However, on the transmission and management of resources, the function of "Micro Tas" is limited, and only non-audio and non-video resources can be transmitted, so under the auxiliary of other social networking platforms, for instance, QQ group, Wechat group, and so on, it can play a more important role in classroom flipping.

Pusat Bahasa Mandarin (Confucius Institute) at Universitas Negeri Surabaya undertakes three sites of Chinese teachings: Chinese major, Chinese Department language teaching, Chinese elective courses, and social students' Chinese language teaching. Depending on the subject, there are some differences in the form and procedure of flipped classroom.

- Chinese major students:
  1) Set up a class on "Micro Tas" where existing four basic tasks of registration, question answering, discussion and courseware managing and ask students to join in.
  2) Set up a QQ group for video and audio management and ask students join in.
  3) Uploading teaching materials to "Micro Tas" and QQ group.
  4) Give clear instructions to students for the tasks they need to complete at home.
  5) Check the students' learning by answering and discussing on "Micro Tas".
  6) Further discussing and solve the problems in class.
- Chinese elective courses and social students:
  1) Set up a class on "Micro Tas" for them.
  2) Set up a QQ group and ask students join in.
  3) Make full use of the resources of CIO and supplement our own resources.
  4) Give clear instructions to students for the tasks they need to complete at home.
  5) Check the students' learning by answering and discussing on "Micro Tas".
  6) Further discussing and solve the problems in class.

All in all, the situation of Chinese department of each university is not the same. CIO is hard to accommodate the different needs of multi-level, diversified teaching. Meanwhile, CIO has already supplied a dimensioned of courses for the elective students and social short-term classes, we should get full use of it. "Micro Tas" with the help of social networking platforms is potential to combine these aspects together.

5. Conclusion

The above supplies a theoretical framework for the inversion of classroom instruction in a non-target environment. It's assumed that it serves the principles of personalized-differentiated learning, student centered instruction, and constructivism. In other words, the students have a great variety of hands-on activities, promoting more meaningful learning. However, it doesn't mean that the traditional methods don't have their advantages of their existence, as a matter of fact, for example, Grammar Translation Methods, they are also commonly used and sometimes effective in non-target language teaching environment. After all, with the advent of the new technology revolution, the utilizing of new education technologies is an unreturnable trend in the history of education.

Nevertheless, the effect of flipped classroom in the non-target environment is still subject to further investigation. We just got some responses from participants that they benefited from attending flipped classroom, for example, the flipped classroom can contribute to personalizing of learning at their own pace, advancing their preparation and improving understanding of the content of lessons, overcoming the limitations of class time, increasing the participation, and so on.

References

[1] Koehler M J, Mishra P, Hershey K and Peruski L 2004 With a Little Help From Your Students: A New Model for Faculty Development and Online Course Design *J. Technol. Teach. Educ.* **12** 25–55

[2] Hamdan N, McKnight K and Arfstrom K M 2013 A Review of Flipped Learning (New York: Pearson)

[3] King A 1993 From sage on the stage to guide on the side College Teaching **41** 30–5

[4] Halili S H and Zainuddin Z 2015 The Classroom: What We Know And What We Don't The Online Journal of Distance Education and e-Learning **3** 28–35

[5] Baker J W 2000 *The Classroom Flip: Using Web course management tools to become the Guide by the Side Selected papers from the 11th International Conference on College Teaching and Learning, ed J. A. Chambers* (Jacksonville, FL: Florida Community College at Jacksonville)

[6] Sahin M, Kurban F and Caroline 2016 *The Flipped Approach to Higher Education: Designing Universities for Today's Knowledge Economies and Societies* (UK: Emerald)

[7] Lage M, Platt G and Treglia M 2000 Inverting the Classroom: A gateway to Creating an Inclusive Learning Environment *J. Econ. Educ.*

Acknowledgments

Financially supported by teaching research project- *Research on the Teaching of Chinese Wisdom Classroom in the Informationalized Technological Environment* from Central China Normal University, the project number: 201648.

IOP Conf. Series: Materials Science and Engineering **296** (2018) 012034 doi:10.1088/1757-899X/296/1/012034

Designing learning management system interoperability in semantic web

Y Anistyasari, R Sarno, N Rochmawati

Department of Informatics, Universitas Negeri Surabaya, Indonesia

yenian@unesa.ac.id

Abstract. The extensive adoption of learning management system (LMS) has set the focus on the interoperability requirement. Interoperability is the ability of different computer systems, applications or services to communicate, share and exchange data, information, and knowledge in a precise, effective and consistent way. Semantic web technology and the use of ontologies are able to provide the required computational semantics and interoperability for the automation of tasks in LMS. The purpose of this study is to design learning management system interoperability in the semantic web which currently has not been investigated deeply. Moodle is utilized to design the interoperability. Several database tables of Moodle are enhanced and some features are added. The semantic web interoperability is provided by exploited ontology in content materials. The ontology is further utilized as a searching tool to match user's queries and available courses. It is concluded that LMS interoperability in Semantic Web is possible to be performed.

1. Introduction

In recent years, e-learning systems have become very popular in all fields of higher education. The internet and web technology offer great solutions for presenting, publishing and sharing learning content and information. The extensive adoption of e-learning across the World Wide Web (WWW) has set the focus on the interoperability requirement, especially in terms of learning resources across different Learning Management Systems [1]. LMS is a high-level strategic solution for planning, sharing, and managing all learning processes within an organization, including online, virtual classroom, and distance learning [2]. In addition, interoperability is the ability of different computer systems, applications or services to communicate, share and exchange data, information, and knowledge in a precise, effective and consistent way. Interoperability is required to build the knowledge-intensive, open and accessible learning services [3,4].

Recent developments of LMS specifications such as Learning Object Metadata (LOM), Shareable Content Object Reference Model (SCORM), Learning Design and other pedagogy research in semantic e-learning have presented an inclination of Semantic Web technologies to encourage existing content-focused learning services to semantic-aware and personalized learning services [5–7]. Ontology-based technologies and intelligent agents are expected to support semantic information processing on Semantic Web. Most LMSs currently work in a closed system. Some systems still operate their own framework for learning content description rather than adopting the standard LOM. The concept of course, student, educational resource, summary or grade must be formally described in order to be shared among all the systems in an educational institution [5–8]. These interoperability issues affect the flexibility of the teaching-learning process and head to a decrease in end-user satisfaction and learning success.

Some studies have been conducted to satisfy the completeness of LMS in university [1,3–11]. Previous research used the concept of synchronization between learning materials in LMS. However, students are unable to follow courses at another university. Only administrator can perform the synchronization. On the other hand, lifelong learning networks must be able to connect some LMSs without any access restriction for users. Semantic web technology and the use of ontologies are able to provide the required computational semantics and interoperability for the automation of tasks in LMS. Semantic Web is a collection of standards, data structures, and software that improve the online experience more detailed and intelligent. The Semantic Web offers students the possibility of having a rich of related content delivered to their desktop without explicitly identifying or requesting it. However, studies about LMS interoperability in the semantic web has not been exploited deeply. Therefore, this work proposes a design of LMS interoperability in the semantic web.

2. Design of LMS interoperability in semantic web

2.1 LMS design

This work assumes that there are four types of the operator for proposed LMS interoperability. They are portal and local administrators, lecturers, and students. Portal and local administrators are responsible for managing a central server that regulates LMS metadata and managing LMS for each university, respectively. Students can find and enroll in their university's LMS as well as they find and join the course from other universities' after paying tuition fees. Lecturers will be part of the LMS after they have been enrolled in university's LMS as a lecturer. They can teach students from other universities. In addition, lecturers can create new course materials. The design of proposed LMS is depicted in Figure 1. This design refers to Moodle context design which consists of three main components: users, front page, and course category.

2.2 Database design

Database design of proposed LMS interoperability is adjusted from Moodle table with some additional tables. Table relationships are depicted in Figure 2. Tables are grouped according to their functions. Group A is an association of tables to store user information (administrator, lecturers, and students). Group B and C respectively are an association of tables to store payment data and serves as a course table group. User log data is stored in group D. In addition, Group E is responsible for managing students' grade data while group F is to classify users and courses. Finally, group G is used for managing the student assignments and table H is to store information of university local-LMS.

2.3 LMS interoperability design

As aforementioned, portal and local LMS layers are proposed in this work. Local LMS only contains web services, while data services are on the portal. The user must log in with a username and password to access portal LMS through local LMS. The system then searches which university has that username and password and verifies them. If the login is successful, the user is registered and listed in `mdl_log`. If a user searches for the desired course, the system will find the course storage and display the search results. If users want to access the course, it will check the user's role as a lecturer or student. Another condition is if the user is a student then it will check the completeness of a course requirement including tuition fee payment as show figure 1 below:

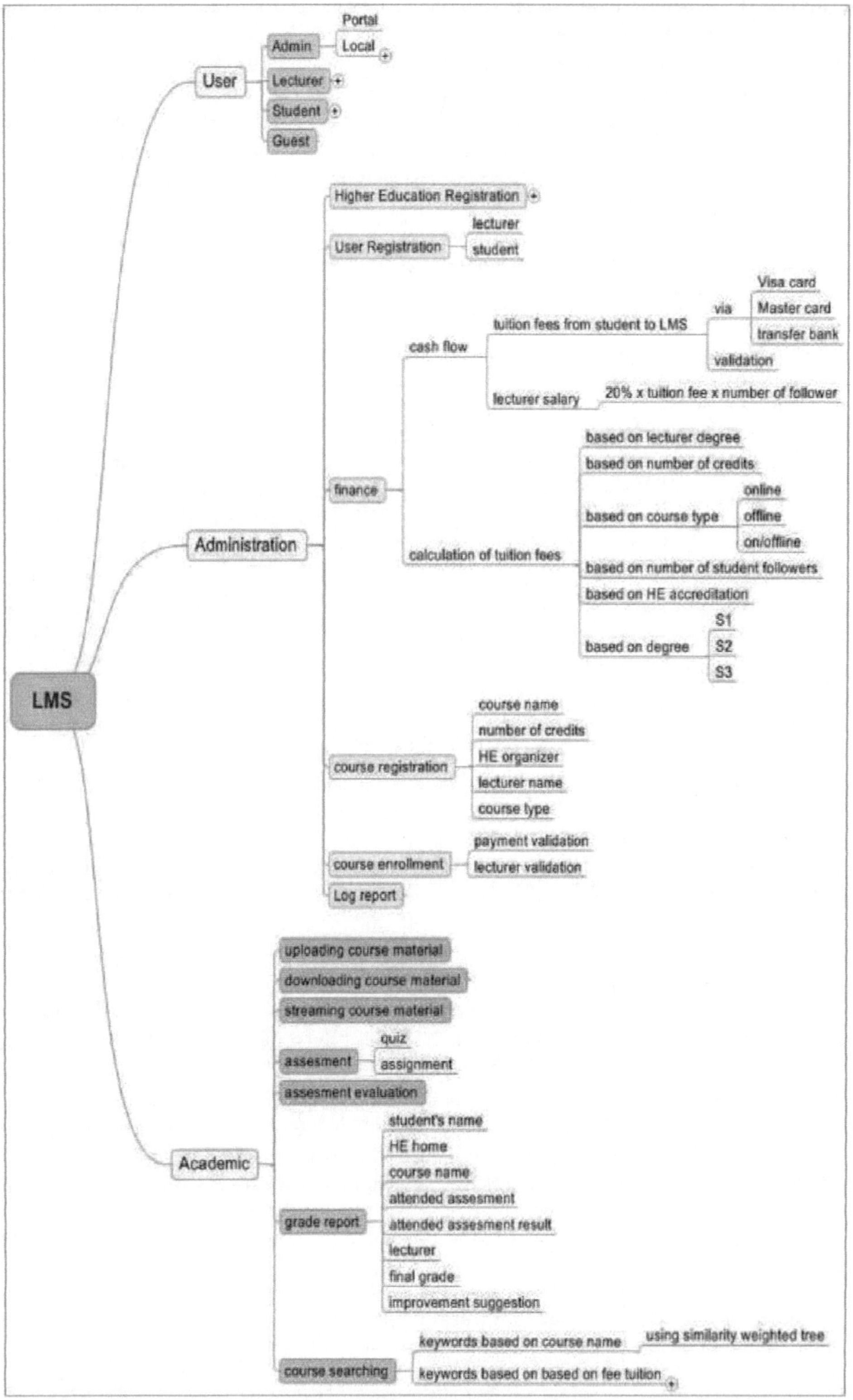

Figure 1. LMS design

Content is assumed as web service in local-LMS since it is an API Web (Application Programming Interface Web) that is accessed by HTTP. The first step, a user must be authenticated by the LMS through a login process. After has been successfully authenticated, the user can log in to LMS and connected to the portal. Users can search for desired courses. The searching process is performed through weighted ontology or other artificial intelligence methods. LMS then finds related courses displays the results. Finally, students can perform various activities such as courses enrolment, discussions, downloading materials, uploading assignments, doing assessments, and following streaming courses as show figure 2 and 3 below:

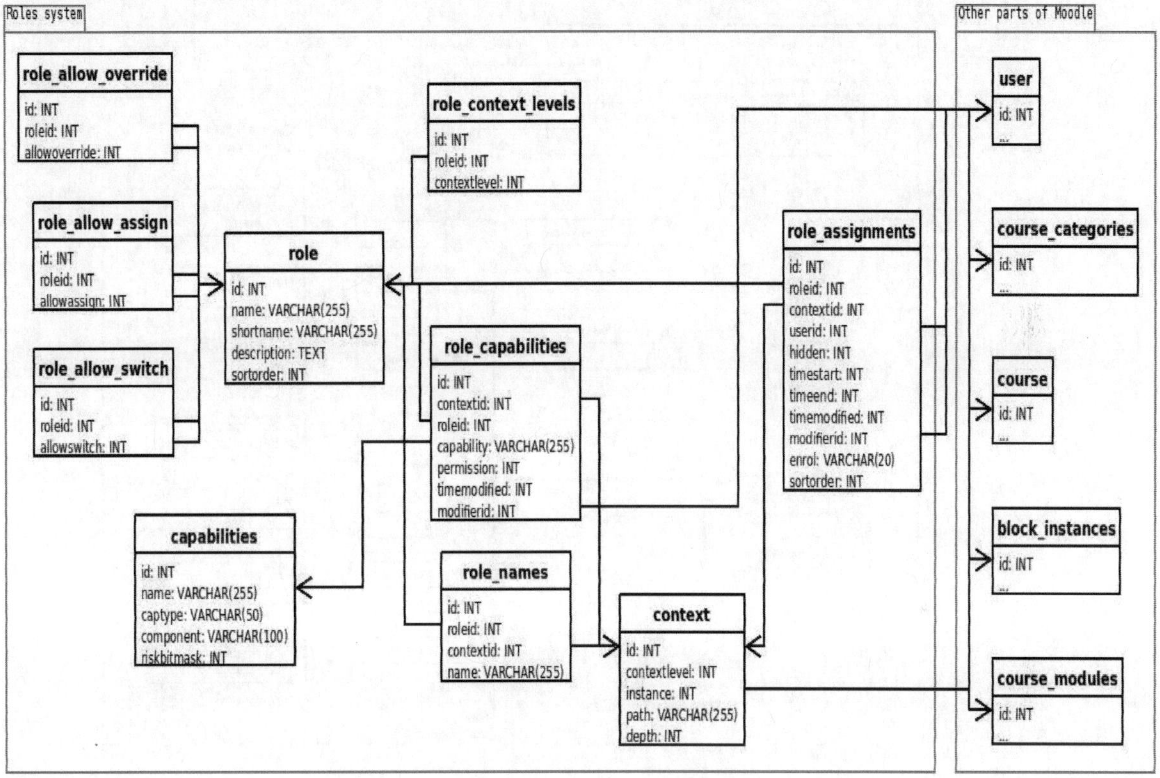

Figure 2. Moodle tables relationships

In this work, semantic search and matching for course subject developed using weighted ontology. Searching and matching tools can provide contextualized queries and searches across repositories of teaching material or repositories in different local-LMS. Furthermore, searching and matching tools are suitable for revealing university capability to the outside world to attract student enrolment. Weighted ontology [11] is built on Learning Content Management System.

Each stored document is extracted and is weighted based on term density which is formulated in (1). The more relations that a term has, the greater weight it has. Let's assume *in(c)* and *in (o)* are in-degree of the concept (c) and in-degree of all layered structure, respectively. Furthermore, let's denote *out(p)* and *out(o)* as outdegree of the concept (p) and out-degree of the hierarchical structure.

$$w_{(c)} = \frac{in(c) + out(p)}{in(o) + out(o)} \tag{1}$$

IOP Conf. Series: Materials Science and Engineering **296** (2018) 012034 doi:10.1088/1757-899X/296/1/012034

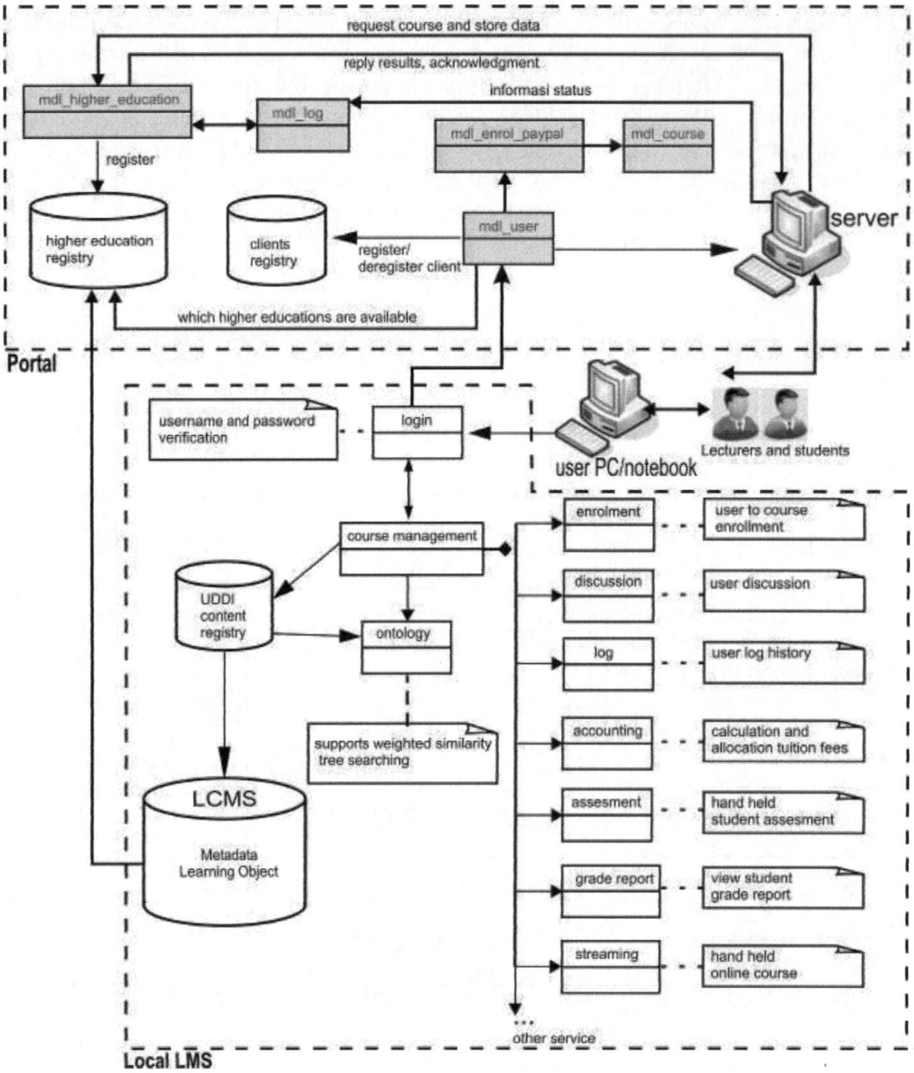

Figure 3. LMS interoperability design

Matching process between two ontology or more is referred to multiple matching. Multiple matching is assumed as *f* function that matches ontology $\{o_1, o_2, \ldots o_n\}$. An input alignment *A*, a set of parameter *p*, and a resource *r* produce an alignment *A'* among those ontologies. Therefore, $A' = f\{ o_1, \ldots, o_n, A, p, r\}$.

3. Conclusion

This paper proposed a design of learning management system (LMS) interoperability in the semantic web. The proposed interoperability is applied in Moodle as an LMS. Supplementary tables added to provide the capability of LMS interoperability in the semantic web. The semantic web interoperability in this work is provided by using an ontology for content materials. The ontology is further utilized as a searching tool. If the previous tuition fees are determined subjectively (no reference to determine tuition fees), in this study tuition fees are calculated based on lecturer education, course taught by teams or individuals, the number of credits, types of classes (online, offline, or on/ offline), accreditation of universities, and the number of enthusiasts. This study also calculates the financial results of the distribution of tuition payment. Tuition payments can be paid via credit card or bank transfer. Obviously, LMS interoperability in Semantic Web is possible to be performed. The next step of this work is to apply this proposed design.

References

[1] Ford C, Mcnally D and Ford K 2017 Understanding Faculty Use of the Learning Management System *Online Learn Consort J* pp 50–67

[2] Hua K L, Anistyasari K, Hsu C H, Chin T L, Yang C L and Wang C Y 2017 Multicast scheduling for stereoscopic video in wireless networks *Multimed. Tools Appl* **76** pp 1–28

[3] Garcia R and Pariente T 2009 Interoperability of learning objects copyright in the Luisa semantic learning management system *Inf. Syst. Manag* **26** pp 252–61

[4] Alier M F, Guerrero M J C and Gonzalez M A C 2010 Interoperability for LMS: the missing piece to become commonplace for e-learning innovation *Int. J. Knowl. Learn* **6** pp 130–41

[5] Tiropanis T and Davis H 2009 Semantic technologies for learning and teaching in the Web 2.0 era *Intell. Syst* **December** pp 49–53

[6] Garcia-Gonzales H, Gayo J E L and Paule-Ruiz M 2016 Enhancing e-Learning content by using Semantic Web technologies *IEEE Trans. Learn. Technol* **PP** 1

[7] Alsultanny Y 2006 e-Learning System Overview based on Semantic Webq *Electron. J. e- Learn.* **4** pp 111–8

[8] Manickasankari N, Arivazhagan D and Vennila G 2014 Ontology-based Semantic Web Technologies in E-learning Environment using Protégé *Indian J. Sci. Technol* **7** pp 64–7

[9] Anistyasari Y and Sarno R 2011 Weighted ontology for subject search in Learning Content Management System *Proceedings of the 2011 International Conference on Electrical Engineering and Informatics* (ICEEI)

[10] Kadir A Z A and Aziz N S 2016 Learning Management System of higher education institution *Indian J. Sci. Technol2* **9**

[11] Huang W, Webster D, Wood D and Ishaya T 2006 An intelligent semantic e-learning framework using context-aware Semantic Web technologies *Br. J. Educ. Technol* **37** pp 351–73

IOP Conf. Series: Materials Science and Engineering **296** (2018) 012035 doi:10.1088/1757-899X/296/1/012035

Implementation of science process skills using ICT-based approach to facilitate student life skills

Y S Rahayu, Yuliani and B R Wijaya

Biology Department, Faculty of Mathematics and Natural Sciences, Universitas Negeri Surabaya

yunirahayu@unesa.ac.id

Abstract. The purpose of this study is to describe the results of the implementation of a teaching-learning package in Plant Physiology courses to improve the student's life skills using the science process skills-based approach ICT. This research used 15 students of Biology Education of Undergraduate International Class who are in the Plant Physiology course. This study consists of two phases items, namely the development phase and implementation phase by using a one-shot case study design. Research parameters were the feasibility of lesson plans, student achievement, Including academic skills, thinking skills, and social skills. Data were descriptively Analyzed According to the characteristics of the existing data. The result shows that the feasibility of a lesson plan is very satisfied and can be improvements in student's life skills, especially with regards to student's thinking skills and scientific thinking skills. The results indicate that the science process skills using ICT-based approach can be effective methods to improve student's life skills.

1. Introduction

In 21st Century Information and communication technologies (ICT) have become very common today in many aspects of life. ICT is also becoming very important in the social and economic life of society. The field of education is also now taking advantage of ICT in learning, but the effect is not covering other aspects. More education leads to traditional educational activities, where professors become the center of information and attention of students. The demands of 21st-century education forced the shifting of paradigms. Students are required to master ICT effectively as one's ability to compete in the 21st century [1, 2].

Student results based on observations conducted by researchers showed that academic achievement by using traditional methods alone had unsatisfactory results. These obstacles arise because the materials studied abstract so that the student is difficult to establish the concept of knowledge. Learning by using ICT has many advantages, one of which is learning can be done independently. Students can use computer animation to understand the concepts learned. Attewell *et al.* believe that technology can make learning becomes easier [3], but it takes too much training, preparation and production of appropriate materials for such learning to be more effective [4].

As the education world continues to grow, various phenomena in the development of real-life demands graduates have a variety of skills to face the competition. National Council for Curriculum and Assessment (NCCA) conduct online surveys to evaluate the response of academics know the priority that should be practised in the world of education. One result of the survey is important to develop life skills [5]. MONE explaining the concept of life skills *(life skills)* as follows: the generic life skills consist of personal skills *(personal skills)* and social skills *(social skills).* Personal skills also include academic skills (academic skills). While the specific life skills *(specific life* skills) consisting of academic skills *(academic skills)* and skills vocational *(vocational skills)*[6].

Tim Broad-Based Education Education Ministry argued that the purpose of life skill education is to: (1) to actualize the potential of learners that can be used to solve problems encountered, (2) provide an opportunity for the school to develop flexible learning, in accordance with the principle of broad-based education and (3) optimizing the use of resources in the school environment, by allowing the use of existing resources in the community, in accordance with the principle of school-based management [7]. Those goals are the main purpose of life skill education is to prepare students to be concerned is able, capable and skilled maintain the viability and growth in the future [8].

2. Methods

This study implemented a variety of learning models are packed with the application process skills-based approach to ICT for Plant Physiology courses with related concepts which include enzyme plant metabolism, photosynthesis, respiration, and nitrogen metabolism. This study consisted of two phases, the first phase of the development of learning tools, and the second stage is the implementation of learning tools Process Skills ICT to teach *Life Skills* students.

Assessment skills (life skills) includes several components that are tailored to the skills of the selected process as an approach to this learning process. life skills (life skills) that are Life generic for Skills consist of personal skills*(personal skills)* and social skills *(social skills).* Personal skills include academic skills (academic skills), while the specific life skills*(specific life* skills) consisting of academic skills *(academic* skills). Learning software development using 4-D models which consists of four stages namely *Define* (definition), *Design* (Design), *Develop* (Development) and *Disseminate* (deployment). The study design when implementing the learning device is a *One-Group Pretest-Posttest Design* developed by Campbell and Stanley [8]. Subjects were students of International Class is being reprogrammed subjects Plant Physiology, the student of class 2012, which amounted to 15 people. The research was conducted in the Department of Biological Science Education UNESA, the even semester 2013/2014 year.

3. Results and Discussion

3.1. Results

Summary of results score of implementation learning SAP at the meeting of 1, 2, 3, and 4 are presented in Table 3.1 show that for every aspect observed for each category include: (1) introduction, (2) the core activities and (3) cover respectively show a score of 4; 3.6; and 3.5 which are in both categories. For implementation learning instrument reliability value of 97%. These results indicate that the enforceability of the skills-oriented learning processes used to improve *life skills* students can be done well.

Student mastery learning data obtained by administration of *post-test* the first and second to students at the end of the lesson, Achievement test product is a test mastery of the material on the topic of water transport, translocation through the phloem, enzymes, photosynthesis and respiration, while the achievement test used to measure aspects of the process of *thinking skills* and aspects of *academic skills.* The results of the analysis of students' test results are shown in Table 3.2.

From the results of student learning completeness in Table 3.2 can be revealed that the results *test* of students is reaching the classical completeness with an average value of 76.42 which is the range of values with the letter B +. But there are three students who have not completed to obtain the value of B such provisions considered sufficient thoroughness to this course is B, such as students with no 1, 3 and 5. In addition to learning the results of the product, in this study also uses the test results to the process of learning know the *life skills* (thinking *skills* and *academic* skills)students.

IOP Conf. Series: Materials Science and Engineering **296** (2018) 012035 doi:10.1088/1757-899X/296/1/012035

Table 3.1 Implementation of Oriented Learning Skills Process

No.	Implementation stages of learning	Meeting 1, 2, 3, 4 Score O1	O2	X	$\bar{X}$	Criteria
1.	Introduction				4	Good
a.	Motivating students	4	4	4		
b.	Delivering Destination	4	4	4		
2.	Core				3.6	Good
a.	Convey information using ICT media	4	4	4		
b.	Dividing the students into groups of	4	4	4		
c.	Guiding / demonstrate the skill-oriented activities	4	4	4		
d.	Students perform observation and measurement activities	4	3	3.5		
e.	Students take notes and organize data of activities, observation or measurement	3	3	3		
f.	Students formulate conclusions	3	4	3.5		
g.	Students report and present the results of observations	3	3	3		
3.	Closing				3.5	Good
a.	Reflecting the learning outcomes	4	3	3.5		
	Total	37	36	36	3.6	
	Reliability	97%				

Description:
O1: Observer 1 O2: Observer 2 X: Average

Table 3.2 Mastery Learning Results Products (Content Mastery)

Students	Value of Learning Outcomes Products			
	Post Test 1		Post Test 2	
	Score	**Completeness**	**Score**	**Completeness**
1	66	C	60	NC
2	79	C	74	C
3	59	CC	52	NC
4	72	C	78	C
5	69	C	65	C
6	80	C	76	C
7	92	C	91	C
8	76	C	81	C
9	93	C	89	C
10	84	C	78	C
11	72	C	68	C
12	89	C	83	C
13	81	C	75	C
14	75	C	71	C
15	to 80	C	74	C
$\bar{X}$	78.5	C	74.3	C
average value overall: 76.42				

Description: C: Completed NC: Not Completed Range of values 1-100

Scores of the learning outcomes process are presented in Table 3.3.

Table 3.3 Mastery of Learning Outcomes Indicators for Measuring Life skills

Student	Learning Outcomes			
	Thinking Skill Score		Academic Skill Score	
	Score	Completeness	Score	Completeness
1	70	C	73	C
2	98	C	74	C
3	66	C	70	C
4	91	C	77	C
5	69	C	71	C
6	82	C	76	C
7	68	C	89	C
8	89	C	82	C
9	85	C	86	C
10	81	C	73	C
11	81	C	75	C
12	91	C	85	C
13	85	C	82	C
14	87	C	79	C
15	76	C	82	C
average	81.3	C	72.3	C
average Rating Overall: 76, 8				

Description: C: Completed Range of values 1-100

Based on Table 3.3 it can be seen that the average results learning process to test the 1st and 2nd measure thinking skills*(thinking skills)*and Academic Skills*(academic skills)*are respectively 81.3 and 72.3. These results suggest that mastery learning process for all students achieved the following study. Figures obtained are relatively varied but generally, show good results with the average value of B +.

Thinking skills*(thinking skills)*and Academic Skills*(academic skills)*students, it can be seen from the values obtained after the student answer the test results that reflect aspects of the process of learning *thinking skills* and *academic* skills.Details of *thinking skills* and *academic skills* of students are shown in Table 3.4.

Table 3.4 Summary of *Thinking Skills* and *Academic Skills* Score

No. LO	No. Item	Domains Cognitive	Test 1		Test 2		Mean
			Score		Score		
Thinking Skills							
1	1	C5	85	C	80	C	83
1	2	C6	80	C	85	C	83
1	3	C6	80	C	80	C	80
2	4	C5	86	C	87	C	87
3	5	C4	86	C	80	C	83
4	6	C4	90	C	90	C	90
5	7	C4	90	C	80	C	85
6	8	C5	85	C	85	C	85
Academic Skills							
1	9	C6	86	C	80	C	83
overall average: 84.33							

Description: C: Completed LO: Learning Objectives Range of values 1-100

Based on Table 3.4 can be disclosed that in test 1 and 2 percentage thoroughness of *thinking skills* and *of academic* achieve mastery skills, with an average reach 83 (rated A- on to the system font). Observation of social skills*(Social skills)* is recorded using the observation sheet social skills*(social skills)*when students perform lab activities and presentation of observations. The percentage breakdown aspects of social skills emerge, are presented in Table 3.5.

Table 3.5 Percentage of Social Skills

Aspects	Indicator	Aspects observed	Meeting								Average
			1		2		3		4		
			P1	P2	P1	P2	P1	P2	P1	P2	
social skills*(Social* skills)	1. The ability to work together in groups of	1.1 Understanding the role of members of the group	13	12	12	12	11	13	14	13	12.5 = 83.3%
	2. the ability to communicate with the environment	2.1 ability to communicate with other of members the group	13	14	13	13	13	14	13	13	13.3 = 89,3%
		2.2 ability to communicate with other groups	9	9	7	8	10	11	12	11	9,5 = 63,3%
		2.3 Being able to use language communicative orally	10	9	7	8	10	10	8	8	8,9 = 59%
		2.4 Can using communicative language in writing	11	12	10	10	12	11	10	9	10.5 = 70%

Description: P1 =observer 1 P2 =observer 2

According to Table 3.5 show that in general, assessment of social skills ranging from grades 59% to 88.9%. This means that if used a scoring system that applies in Unesa then the average value of the social skills of students still at a fairly wide range of values of between grades C to A. Social skills are still low since the ability to use verbal communicative language is followed by the ability to use language communicatively in writing.

This study begins to develop learning tools courses *Plant Physiology*-based process oriented ICT skills to improve *life skills* student incorporating ICT-based animation media. Further validation was done by a team of experts in the field of plant physiology to get feedback and suggestions regarding the feasibility of learning tools developed for the learning process using English as the language of instruction. SAP (Lesson Plan) prepared in accordance with good process skills approach basic skills processes and integrated process skills.

Based on the results of the analysis indicate that the validation and revisions have been made to the SAP (Lesson Plan) and the completeness of their supporters could feasibly be used to study skills-oriented processes and ICT-based and feasible to improve *life skills* (thinking *skills, academic skills*, and *social* skills)to students being programmed course *Plant* Physiology. MFIs are arranged oriented learning process skills approach to teaching *life skills* that include *thinking* skills, *academic* skills, and *social* skills.The analysis showed that students had no difficulty in following the process of learning the skills approach. The results of the validation of THB. THB products and processes from the material aspect also received good ratings and feasible to be implemented (data not shown).

Implementation observed learning involves three aspects: (1) The preliminary activities, (2) the core activities, and (3) Event cover. Worth mentioning that when designing learning device, prepared in accordance with the theory of *scaffolding* and cognitive apprenticeship expressed by Vygotsky. Theory *Scaffolding* is defined as the process of assistance from other people who have more knowledge (professors or fellow students) to people who are a little more knowledge to address the problems that go beyond the current level of development [8]. It is very necessary to improve the ability of students, especially in improving the ability of *life skills* that are indispensable as the provision of public life in order to be a successful man and capable of being in the environment with appropriately in accordance with its function.

Cognitive apprenticeship theory defined on the process whereby a person who is learning step by step acquire expertise in interaction with an expert, who serves as an expert, in this case, is an adult or an older person. In this case, can be interpreted faculty adviser subjects or peers who know more about the problem [9]. This can be proven at the time of the learning activity, students conduct discussions with peers and with faculty or co-assistant to solve the problems encountered at the time of observation or process skills activities. After completing this activity, students discuss with members of the group and then present the results of their discussion to the other groups.

Implementation of the learning device oriented ICT skills-based process, also in accordance with Annex Permendiknas No 22 of 2006 on the Content Standards Subjects Biology, which forms positive attitudes towards biology to realize the regularity and beauty of nature and exalt the greatness of God Almighty. Fosters scientific attitude, honest, objective, open, critical resilient and able to cooperate with others. This becomes very necessary and important for the students of Biology Education S1 as a form of imitation how they can translate the implementation in accordance with the demands of the Content Standards Curriculum. The importance of the use of ICT in learning is also supported by research done previously, that the use of ICT can enhance students' science learning outcomes [10]. Another study also describes the use of ICT in learning can increase student engagement in learning [11]. Activities that support it can be seen from the data processing that is adjusted on the basis of observations made, not made-up and be honest.

Besides the achievement of basic competency courses *in Plant Physiology* can be facilitated by using process skills approach to teaching *life* skills. Aspects *thinking skills* and *academic skills* can be seen in the learning activities with process skills approach that includes: making the formulation of the problem, identify variables, make observations, collect data and analyze the results of observations, and make inferences observations. *Social skills* can be seen clearly at the stage students discuss analyze and report the results of observation in the group. SAP enforceability of the instrument reliability of 97% indicates that the instrument used has a value that is reliable [12]. In this case, it can be assumed that the instruments used are reliable.

After the learning process-oriented skills, students who have achieved mastery to value B as many as 12 students from 15 students of International Class S1 is being reprogrammed subjects *Plant* Physiology. However, there is still a value of C according to Unesa pass mark is still within the range of passing. Mastery learning students after learning of ICT-based process-oriented skills is due to the learning process practised skills in an organized process. While it is known that the learning process skills will give a good impact for students, such as [13]: 1) Students undergoing the process to get the concept, formula or description of something so that students can understand it; 2) Students will actively participate in learning activities; 3) Allows students to develop a scientific attitude and stimulate curiosity; 4) Students will gain a sense of who actually lived because of their own to find a concept or generalization of the results of its work; 5) Understanding students about a concept or principle steadier allowing students to be able to apply them to other more relevant issues; 6) Students are satisfied with the results of observations and findings as one of the factors to foster intrinsic motivation in students; 7) Through this approach, the development of science and concept changes that might occur readily accepted; 8) Students are trained in activities required by science as practiced by scientists; 9) the student Skills acquired will be useful in everyday life; 10) The possibility of maximum utilization of the environment as a learning resource; 11). Familiarize students to express their opinions in a systematic way and respect the opinions of others.

Please also note that training process skills, students will be more easily controlled and appreciate the subject matter, because students are directly involved in the learning process. Provision of direct experience is emphasized through the use and development of process skills and scientific attitude with the goal of understanding the concepts and be able to solve the problem so that the results of student learning are maximized and students achieve mastery learning.

In addition, skills in the learning process of this subject also packaged based on ICT, given that many or most of the concepts contained in the subject of *Plant Physiology* are abstract. Therefore we need the help of the media that is able to bridge to translate what the abstract becomes easier to visually follow. Easy to follow visually abstract process is expected to be used as a bridge improve the cognitive abilities of students of both products and processes.

Achieving mastery learning outcomes-oriented process *thinking skills* and *academic* skills-oriented learning skills indicate that ICT-based process can improve the ability to *life skills* in students. Learning the process skills approach, giving students the opportunity to interact directly using the tools of the experiment so that *life skills* can be trained and the students can use the trial properly and in accordance with the objectives expected when students have activities in the Lab.

As it is known that the principle of learning-oriented life skills*(life skills)*more to the contextual learning, namely the relationship between real-life environment and experiences of students, as well as the process of learning the skills approach. Students learn the material in class, should not only be prepared only as researchers alone, but they must be possessed multidimensional thinking like that done by a researcher [14]. Skills are a very important process to achieve this and generate mindset to act as a science student, discovered by researchers [15]. Process skills can be described as the ability to use a researcher during their work duties and competence demonstrated during the resolve scientific problems [16]. There are a variety of skills in the process of skills, these skills consist of basic skills*(basic skills)*and integrated skills*(integrated skills).*Basic skills consist of six skills, namely: observing, classifying, predicting, measuring, conclude, and communicate. For skills integrated comprising: identifying variables, tabulating the data, present data in graphical form, describe relationships between variables, collect and process data, analyze research, develop hypotheses, defining variables operationally, designed the study, and carry out experiments [17]. These skills can simultaneously facilitate students on life skills.

Thinking skills*(thinking skills)*is the proficiency in using ratios or mind, which includes among other things: prowess identify and find information, process and make decisions, and solve problems creatively. Academic skills are often called intellectual prowess or scientific thinking skills which basically is the development of thinking skills in general, but it leads to activities that are scientific. These skills include skills to identify variables, explaining the relationship a particular phenomenon, formulating hypotheses, designing and conducting research. These skills are skills are integrated into process skills, this is in accordance with the capabilities and concepts underlying fundamental science as process skills. It is known that the ability and fundamental concepts that underlie science as process skills to teach *life skills* such as 1) Identify the questions that can be answered through scientific investigation; 2) design or conduct scientific investigations; 3) Develop descriptions, explanations, predictions and models using evidence; 4) Think logically and critically; 5) Recognize and analyze alternative explanations and predictions.

Based on Table 3.5 aspects of social skills are generally categorized either observed except for the aspects of communicative language use orally (59%) and aspects of communicating with other groups (63%). In accordance with Vygotsky's theory that emphasizes the social aspects of learning in the belief that social interaction with others spur the development of new ideas and enrich the intellectual development of students.

Based on the data, they found aspects of social skills are less well categorized. This is because, in reality, oral communication was not easy to do, let alone English. Often the student is difficult to accept the opinion of his interlocutor, not because of the content or ideas but because of the manner of delivery that is less pleasing. Therefore we need the ability how to choose your words and how to convey that is easily understood by his interlocutors, besides speaking in English is also not everyone can easily do. Students should develop the ability to be a good listener and appreciate the explanation submitted by other students. Students also should remain open and appreciate the ideas and different explanations, and consider alternative explanations, and students should strive for more learning to speak English.

Aspects of some social skills students gain both categories. This means learning skills-oriented ICT-based packaged process can teach social skills to students. The general condition which is a condition onset of activity on the process skills that students in the class and the social aspects of the open atmosphere that invites students to discuss. Social skills should mastery by students as prospective teachers in junior high or high in accordance with the purpose of the subjects of Biology at the junior / senior high school. Compliance with the demands of the subjects Biology / IPA in SMA / SMP can be used as modelling for prospective teachers to be able to package learning materials that will be facilitated by the students.

Obstacles or problems were found during the learning process most predominantly found was the lack of English language skills of students orally, however, the understanding or the ability of students in written English is better than his verbal ability. Therefore, the evidence described above indicate that ICT-based media can help improve students understanding of abstract concepts into a visual form that is easily understood.

4. Conclusions and suggestions

Based on the analysis, the conclusions that can be drawn from this study are as follows:

- The learning implementation process in the course *of Plant Physiology* to improve *life skills* through ICT-based process skills good category with a value of 97% reliability of the instrument.
- Cognitive learning outcomes of students average product are 76.4 (B + on the number of letters). For the ability of thinking skills*(thinking skills)*and scientific thinking skills*(academic skills)*students after applying the learning device *of Plant Physiology* through ICT-based process skills are respectively 85.7 and 83. That achievement is at A- value for the system font.
- The ability of social skills*(social skills)*students were in grades B to A- except for the ability to use communicative language orally (59% = value C) and the ability to communicate with other groups (63% = C).
- Obstacles encountered in the implementation of learning *Plant Physiology* English language skills of students orally is still inadequate as a provision in communicating effectively in the classroom.

Referring to the findings of the study, then things can be suggested that researchers are as follows:

- Improve English language skills of students of Biology Education International Class S1 to be able to communicate well in the classroom so that learning can take place effectively.
- Learning tools used in the S1 class Biology Education International will become more meaningful in achieving competency achievement indicator of subjects when using ICT-based media in order to bridge the abstract concept becomes easier to follow visually.

5. References

[1] Oliver R. 2002. The role of ICT in higher education for the 21st century: ICT as a change agent for education. HE 21 conference proceedings.

[2] Ananiadou, K. and M. Claro. 2009. 21st Century Skills and Competences for New Millennium Learners in OECD Countries. OECD Education Working papers, No. 41. OECD Publishing, Paris.
[3] Attewell J, et al. In 2009 the impact of mobile learning: examining what it means for teaching and learning. London, *Learning, and skills* network.
[4] Livingstone S 2014 critical reflections on the benefits of ICT in education. *Oxford Review of Education*
[5] Fitzpatrick S. Et al 2014 Priorities for primary education? From subjects to life-skills and children's social and emotional development. *Irish Educational* Studies.
[6] MONE 2006 guide to the preparation of educational unit level curriculum of primary and secondary education. Jakarta: Ministry of Education.
[7] MONE 2002 oriented education life skills (life skills) through a broad-based education approach (draft). Jakarta: Ministry of Education.
[8] HA Rachman 2009 dimensions of life skills (life skills) in teaching physical education. *Journal of Physical Education Indonesia.*
[9] Nur M 2005 study strategies. Surabaya: Center for Science and Mathematics School of the State University, Surabaya.
[10] Skryabin M, Zhang J, Liu L, Zhang D 2015 how the ICT development level and usage influence student achievement in reading, mathematics, and science *Computers & Education 49-58*
[11] Ezekoka and K. Gertrude 2014 maximizing the effects of collaborative learning through ICT *Social and Behavioral Sciences* 176 (2015) 1005 – 11
[12] Borich GD, 1994 *observation skills for effective teaching*. Englewood Cliffs: Merill Publisher.
[13] Hidayat, 2003 pengembangan perangkat pembelajaran biologi berorientasi keterampilan proses untuk menigkatkan hasil belajar siswa pada materi ekosistem. Tesis yang tidak dipublikasikan. Surabaya: PPS UNESA.
[14] Duran M and Ozdemir O 2009 a qualitative study on the effect of learners' scientific processing skills on their attitudes towards science *1st International Congress of Educational Research, Canakkale 18 March University, Canakkale*
[15] Erkol S and Ugulu I 2013 Examining Biology Teachers Candidates' Scientific Process Skill Levels and Comparing These Levels in Terms of Various Variables *Social and Behavioral Sciences 116 (2014) 4742 – 7*
[16] Monhardt L and Monhardt R 2006 creating a context forth a learning of science process skills through picture books *Early Childhood Education Journal*, 34 67-71.
[17] Dimyati M, Mujiono, 2002 *Belajar dan Pembelajaran*. Jakarta: Rineka Reserved.

IOP Conf. Series: Materials Science and Engineering **296** (2018) 012036 doi:10.1088/1757-899X/296/1/012036

21st centuries skill implication on educational system

T Wrahatnolo and Munoto

Electrical Engineering Department, Universitas Negeri Surabaya, Surabaya 60231, Indonesia

triwrahatnolo@unesa.ac.id

Abstract. The purpose of this article is to identify skill needed in 21st centuries and its implication on Indonesia's educational system. This research found that the 21st centuries skill application has more measurable benefits in some sections of life, such as critical thinking and problem solving, initiative, creativity, and entrepreneurship, communication, teamwork, metacognition (change of mindset), digital literature. This study applied qualitative data analysis. The data were taken from different sources and literature. The analysis showed that The 21st centuries education concept's implementation can be applied in the curriculum of the required subject that is addressed to achieve learning and innovation skills competence and also technology and information media skills competence. While supporting subject group directed to achieve life and career skills competence. All subjects are the derivation from core subject 3R, which are reading, writing, and arithmetic. Based on the description above, it can be concluded that 21st centuries skill needs; (1) a life planning; (2) flexibility and adaptability; (3) initiative and self-management (4) entrepreneurship; (5) social and cultural interaction; (6) productivity and accountability; (7) leadership; (8) critical thinking, (9) problem solving; (10) communication; (11) collaboration and teamwork; (12) lifelong learning; and (13) digital literation.

1. Introduction

Technology in 21st centuries is growing rapidly in the development of business and industry. The technology's growth also leads the type and skills structure experienced differently years to years so that we cannot predict a skill for two decades ahead. In fulfilling the 21st century's skill needs, we cannot know whether the obstacles that the alumnus will be faced and what kind of job that the alumnus needed when they entered the workforce. Skill is very dependent on highly contextdependent and multifaceted [1]. Individual's creative mind all depends on working environment, knowledge and problem-solving skill. The questions arise about how will the teacher design his/her learning activities? What is the effective strategy that the teacher needs to apply to the class? How does the teacher prepare a skilled millennium generation?

Acceleration of educational improvement is supported by media and digital technology usage which is known as information superhighway [2]. Since the internet was introduced in the commercial world in early 1970, information rapidly distributed to the entire world. In this century, education becomes more important to ensure students to get a skill, innovation ability, creativity, teamwork, and learn how to use life skill in their life. Problems arise from the variety of importance and excellence skills for their future. 21st centuries skill concept is one that has gained increasing currency as a reflection both of changing workplace needs and the evolving role of education [1]. Because of the need to gain access to fast information and fulfillment in 21st centuries, skill demands which are different from the past, making the knowledge-based education becomes important and needs a reformation in the education section. A good education is one that encouragingly makes a change by providing students with adequate knowledge and skills to face the future challenges.

2. 21st Century Skills in Education

The life of current students is very different with the pattern of life developed in the existing educational system. It shows lack of concerns towards knowledge and skills needed to understand existing perspective. Teachers, students, and authorities put forward a number of important skills for the student's future. The authorities prioritized employees' potential on the top list and they expected the importance of improving teacher's role in teaching based on the individual needs, especially subjects that emphasized on problem-solving skill. The implication of problem-solving skill in the learning process can be initiated earlier and become effective by practicing students with basic skills on practical application subjects. Emiliana Vegas [1] stated that a successful education system in school is one that attached problem-solving and combined the problem context with information, as for example, the use of mathematics and science to solve a practical problem.

This skill had to be known and owned by all employees in order to be skilled in the workforce. Harvey [3] viewed skill needs from individual and institutional perspective. Individual skills are skills or attribute of graduates to get a job. While institutional skill is related to the level of performance of educational institution's graduates. One-self attribute is a part of working characteristics, includes flexibility, adaptation ability, and problem-solving. Students demanded to: (1) have capacity and skill in analyzing and solving problems, to communicate idea and information, to do planning and organizing activities and collaboration with others; (2) have quality of confidence, hard works, optimism, a high self-esteem, and commitment using personal excellence; (3) have some related skills within their job and understanding of the work environment, choose a career and training; and (4) become confident, creative, and productive in using the new technology, primarily information and communication technologies, and to understand the effect from technology to society. Baruch [4] stated that individual has the responsibility for their own skill, while authorities provide a chance to develop them. It means every individual manage their careers based on the opportunity to their surroundings. Overtoom [5] stated that employee needs skills about self-management, teamwork, interpersonal ability, problem-based, and critical thinking in order to increase the productivity of their company.

Knowledge is an important asset for job seeker when they initiate their work, yet what is really sought is teamwork. Teamwork can be developed from an early age by enabling students to work in a team. Besides, communication is also an important aspect and is different for each person. An effective verbal communication is a fundamental tool of work. Currently, a verbal communication is not the only important aspect, but also writing skill, such as individual's ability in terms of writing sentences. Survey outcomes stated some of difference in college student's perspective on how this skill where almost half of the USA and Britain with range age from 18-25 years old show that their problem-solving skill is superior to Chinese college student [1] it may be a reflection on how this skill has or hasn't been a priority inside the education system in both countries.

Digital literacy is an important skill to help on codify information, problem-solving, and find meaning in word or data. Based on the study from The Economist Intelligence Unit (EIU), it stated that 21st centuries skills that fulfill the needs from authorities and wider society include leadership, digital literacy, problem-solving and communication complement the traditional skill, such as reading, writing, and counting [1]. Digital literacy is an ability to find, evaluate, utilize, share and make content using information technology. Every graduate assumed that they have already supplied with digital literacy because generally, the teacher had developed digital literacy to their students. Entrepreneurship is an important skill for students who will prepare them self in the workforce inside future's network. Entrepreneurship and creativity skills have been provided by the teacher in school, but still in theoretical context and less of implementation. A low level of digital literacy, entrepreneurship, and creativity in employee indicate that those skills still rated as personal responsibility so that they are out of the curriculum.

Currently, educational systems unable to provide job opportunity and proper skill needs for students. The authorities stated that they are satisfied with young employees when they started working for the company. But, there is skills gap that hampers their organizational performance. The educational systems should provide a skill as their provision to enter the workforce. The teacher recognized that the business world and industrial world are unsatisfied with the existing educational standards. The educational system is considered incapable of wielding wider skills to the students and even there has been a decrease in emphasis on skill in the teacher's learning process. The lack of time allocation in the curriculum and the tightness of educational authorities that only implies reading and counting are also suspected to be the biggest obstacle for teachers to teach 21st-century skills.

Information and Communication Technology (ICT) innovations play a role in changing learning processes, but the educational system controls the progressive transformation. The change of ICT technology is one of the important drivers of evolving skills. But, what is the effect of technology changes that they have to teach a skill? Data states that many teachers claim that the advancement of ICT has changed the way they teach. The teacher can apply technology in creative ways. Teachers and other stakeholders acknowledge that there is still a gap in digital literacy between regions. Some students feel that the education system in their country has effectively used ICT, some students think that their school is very good at using technology for the learning process. The majority of teachers argue that their students have a better understanding of ICT's in the classroom which brings the consequences of change, although teachers don't need to add more value to the use of ICT. Business executives agree that expanding access to technology for schools and universities is one way that the educational system in their country can benefit business. The criticism of 21st-century skill fulfillment is education emphasizes too much on content. One of the challenges for education world is that skill and knowledge can be obtained easily through the internet.

3. Method of the Study

This study applied qualitative data analysis. The data were taken from different sources and literature. The data then were discussed and analyzed using objective point of view related to the needed skills in education.

4. Results and Discussion

How are future skills taught? According to Soulé, "*Education systems need to provide students with hands-on learning that mirrors real-world problems and work opportunities in an interdisciplinary way.*" [6] The education system should provide a hands-on learning experience of the real world and employment opportunities in a variety of interdisciplinary ways to students. The 21st-century skills must be integrated into each subject, so the development of skills becomes integrated. Development of oral communication skills can be done by applying the use of dual-language; local language and international language, in the learning processes of all disciplines. It encourages students to gain public speaking experience, network developing, writing, and critical thinking. The key to skill development is the cross-curricular approach. According to Soulé, students with communication skill, collaboration, critical thinking, and problem-solving, and creativity and innovation related to their knowledge content will tend to be more successful, both at college, at work and as part of society [7]. The education system should provide a hands-on learning experience that reflects real-world problems and the opportunities for interdisciplinary work. This skill cannot be taught in a separate model but should instead lie along the curriculum. For students, they should be able to analyze information, manage resources, assess individual contributions to teams, and be responsible for specific tasks on all subjects. The implications for teachers are: (1) Teachers can develop this skill themselves and change

IOP Conf. Series: Materials Science and Engineering **296** (2018) 012036 doi:10.1088/1757-899X/296/1/012036

their pedagogical aspects, (2) Teachers need to improve skills to enable them to effectively teach content and skills, and (3) Teachers should teach students how to work effectively. While the implications for schools can be provided a free right in selecting, developing and applying 21stcentury skills in their curriculum. The greatest obstacle to incorporate 21st-century skills more broadly in education precisely lies in the rigor of the existing curriculum. The teacher found that the curriculum was too rigid on the development of wider skills for sufficient time allocation. The concept of 21st-century education has been adopted by the Ministry of Education and Culture of the Republic of Indonesia to develop new curriculum for elementary school, junior high school, high school and vocational high school since 2009, among others is 21st Century Skills [8].

According to Trilling and Fadel [8], 21st-century skills are grouped into three types, namely (1) life and career skills, (2) learning and innovation skills, and (3) media skills and information skills. These three skills are summarized in a scheme called the 21st-century skills-knowledge rainbow (a twist of 21st-century skills knowledge). The scheme was adopted by the non-profit organization P21 that developed examples of 21st-century education work worldwide through the http://www.p21.org/storage/images/stories/rainbow/Framework-copyrighted.png website. The concept of 21st-century inventiveness and the 3R subject is described as follows.

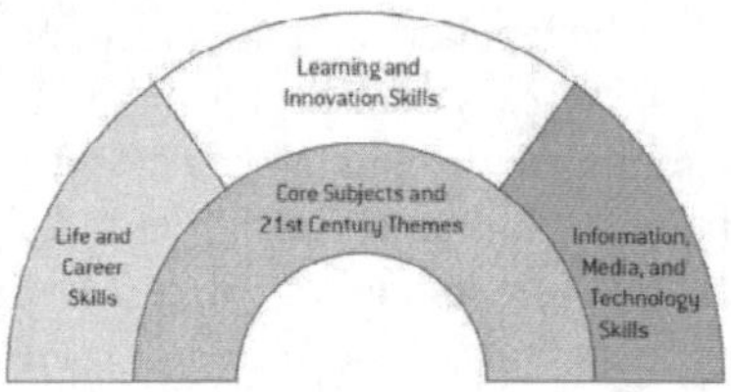

Figure 1. 21st-century knowledge-skills rainbow
(Source: Trilling dan Fadel (2009))

From arithmetic, the modern model of education develops with numbers. The term literacy and numeracy are considered capable of expressing the ability to produce something. The 3R concept is identical to the literacy, numeracy and ICT skills that develop in today's modern education system. In the context of 21st-century skills, 3R core subjects are defined as life and career skills, learning, and innovation skills and information media and technology skills.

Figure 2 shows a 21st-century life-skills sciences rainbow scheme adapted from P21. In the scheme developed by P21, it is confirmed by adding 3R core material in the educational context. 3R stands for reading, writing, and arithmetic. The idea of modern education comes from the syllables of reading and writing into literacy which is a model of learning in understanding ideas through the use of words.

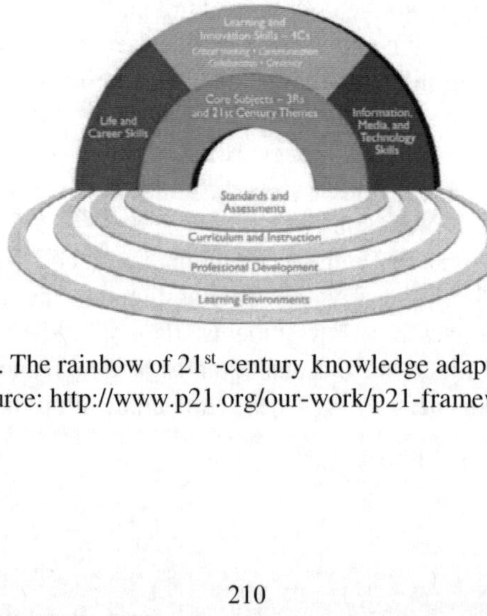

Figure 2. The rainbow of 21st-century knowledge adapted by P21
(Source: http://www.p21.org/our-work/p21-framework)

Life and career skills include: (a) having life planning, (b) flexibility and adaptability, (c) initiative and self-management, (d) entrepreneurship, (e) social and cultural interactions, f) productivity and accountability, and (g) leadership, and responsibility. Learning and innovation skills include (a) critical, creative, and innovative thinking, (b) problem solving, (c) communication, (d) collaborative and teamwork, and (e) lifelong learning. Technological and information media skills (L) include (a) information literacy, (b) media literacy and (c) information technology and communication literacy (Table 1).

Table 1. 21st Century Skill Needs

21st-century skill	**Description**
Career and life skill	1. Students have a plan: Students are able to set goals and targets to be achieved; defining specifications and quality standards; planning the use of resources including time, people, finances, and materials; understanding risk management; managing time and set priorities. 2. Flexibility and adaptability: Students can adapt to various changes and flexible in doing activities and learning process in the team. 3. Have initiative and self-management: Students can manage goals, target, time, media independently and manage themselves; have personal vision and goals; evaluate and monitor its own performance; have knowledge and beliefs of own ideas and vision; articulating their own ideas and vision; take responsibility; work ethically, work under pressure, target-oriented and test-resistant; taking initiative and making decisions, establishing evaluation criteria and participating in continuous improvement, reporting on progress and outcomes. 4. Entrepreneurship: Students are able to identify opportunities; assessing competitive advantage and ideas; identifying the customer or client needs; developing strategic objectives, dealing with stakeholders; using various methods of business communication as well as marketing and selling products or services. 5. Social and cultural interactions: Students can interact and work together effectively with different teams. 6. Productivity and accountability: Students can manage programs and activities oriented towards goals, targets, and outcomes. 7. Leadership and responsibility: Students can lead their friends and have responsibility for their duties, functions, and roles in their environment; managing a group of people, training, developing, motivating, providing feedback, monitoring, collecting, analyzing and organizing information, rational thinking, and managing and delegating tasks, coordination, and monitoring.

Innovation and Study Skill	1. Critical, creative and innovative thinking: Students are able to use inductive or deductive approaches to identify facts and phenomena; systemic thinking; think and work creatively and create new innovations; have creativity, embark on innovative ideas and solutions; determine the commercial viability of ideas; translate ideas into action; have sensitivity to strategic issues, politics, commercial, environment, culture, etc.; 2. Problem solving: Students are able to analyze facts and test assumptions; to define problem and factor contribution; to develop creative, innovative and / or practical solutions; to show initiatives in identifying and solving problems; to solve problems independently and in teams; to implement various strategies for solving the problem of designing contingencies; to develop and evaluate options; to make realistic decisions and action plans; to use math including budgeting and financial management to solve problems; to solve customer problems in relation to complex issues; to evaluate and monitor the process and results and make decisions and solve problems. 3. Communication: students are able to communicate clearly, straightforwardly and in collaboration with other team members; listening, understanding and speaking clearly; write right and true; negotiate effectively; have empathy, firmness, and wisdom, understand the wishes of others; building relationships and networks; sharing information and ideas recently; fluent in Indonesian and other languages; as well as having logic in summarizing information or data. 4. . Collaborate and teamwork: Students are able to work effectively with people of all ages, gender, race, religion or political persuasion; identify the strength of team members; acknowledge its own strengths and limitations; understand the role of team members and perform tasks; able to lead, coaching, mentoring and motivating others; providing and receiving constructive feedback and resolving disagreements. 5. Life-long learning: Students are able to understand the learning needs for change; manage your own learning; sharing knowledge and experience in the workplace; contribute to the learning community in the workplace; using various media for learning - mentoring, peer support, networking, communication information technology applications (ICTs), applying learning technical issues; has an ongoing learning spirit; being willing to learn in any arrangement on and from work; be open to new ideas & techniques; ready to invest time & effort in learning new skills.
Digital Literacy Skill	1. Literacy information: students are able to access information effectively (source information) and efficient (time); have a variety of basic ICT skills; using ICT to organize data; willing to learn new ICT skills; evaluate the information to be used critically and competently; use and manage information accurately and effectively to solve problems; 2. Media Literacy: students are able to select and develop media used to communicate; implement ICT as a management tool, and apply K3 in applying technology; 3. ICT literacy: students are able to analyze information media, and create appropriate media for communication.

5. Conclusion

Education in the 21st century is becoming increasingly important to ensure students have skills in learning and innovation, use of information technology and media, and can work and survive by utilizing life skills. As a first step, the implementation of 21st-century education concept can be applied to curriculum structure consisting of compulsory subjects aimed at achieving the competence of learning and innovation skills and technology and information media skills. While the supporting subject group is directed to achieve the competence of life and career skills. All subjects are a derivation of the 3R core subjects of reading, writing, and arithmetic. Based on the above description, it can be concluded that the needs of 21st century skills include (1) having life planning, (2) flexibility and adaptability, (3) initiative and self-management, (4) entrepreneurship, (5) social and cultural interaction, (6) productivity and accountability, (7) leadership, and responsibility, (8) critical thinking, creative and innovative, (9) problem solving, (10) communication, (11) collaborative and teamwork, (12) life and (13) digital literacy.

IOP Conf. Series: Materials Science and Engineering **296** (2018) 012036 doi:10.1088/1757-899X/296/1/012036

Reference

[1] The Economist Intelligence Unit Limited 2015 *Driving the skills agenda: Preparing students for the future, An Economist Intelligence Unit Report*

[2] Gates B, Myhrvold N and Rinearson P 1996 The Road Ahead *Penguin Books*

[3] Harvey L 2001 Defining and Measuring Employability *Qual. High. Educ.* **7** 97–109

[4] Baruch Y 1999 Response rate in academic studies-a comparative analysis *J. Hum. Relations* **52** pp 421–38

[5] Overtoom 2000 *Employability skills: An update. Eric Digest: Center on Education and Training for Employment* (Ohio: The Ohio State University)

[6] P21 (Partnership for 21st Century Skills) 2009 Learning environments white paper

[7] Anon 2016 P21's Framework for 21st Century Learning

[8] Trilling, Bernie and Fadel C 2009 *21st Century Skills: Learning for Life in Our Times*

IOP Conf. Series: Materials Science and Engineering **296** (2018) 012037 doi:10.1088/1757-899X/296/1/012037

Mathematics pre-service teachers' statistical reasoning about meaning

Y D Kristanto

Universitas Sanata Dharma, Yogyakarta

yosepdwikristanto@usd.ac.id

Abstract. This article offers a descriptive qualitative analysis of 3 second-year pre-service teachers' statistical reasoning about the mean. Twenty-six pre-service teachers were tested using an open-ended problem where they were expected to analyze a method in finding the mean of a data. Three of their test results are selected to be analyzed. The results suggest that the pre-service teachers did not use context to develop the interpretation of mean. Therefore, this article also offers strategies to promote statistical reasoning about mean that use various contexts.

1. Introduction

What do the pre-service teachers mean? It is worth to submit this question to find out the pre-service teachers' understanding about the concept of the mean. The answer to that question will also help the author to know about the context of the students, so we can provide appropriate treatment in the next statistical learning.

Reasoning the students towards the concept of the mean is very important [1]. This is because they will teach the concept at the junior or high school level. The following concept is the basic competency of junior high school which contains the mean concept. Analyze data based on data distribution, mean, median, mode, and data distribution to draw conclusions, make decisions, and make predictions [2]. The following concepts are the basic competencies of the senior high school that contain the mean concepts. Determine and analyze the centralized and distributed size of the data presented in the form of frequency distribution tables and histograms [2,3].

Based on two basic competencies in the field of mathematics, the competence which is able to analyze the mean that measures the center is a necessary competence for students. Therefore, pre-service teachers are required to be able to understand and interpret the mean concepts before they actually teach students in school.

However, some research indicates that there is still a lack of understanding of students about the concept of the mean [4–6]. They are still focused on the procedure for determining the mean, which is the ratio of the sum of all the data and the number of data. When they are asked to interpret the mean value, they are still unable to explain it clearly.

Interpretation of the mean can be done by connecting the mean with the context. There are four contexts that can be used to understand the mean concepts such as data reduction, fair share, a typical value, and signal in noise. The explanation of these four contexts will be further explained in the following discussion.

Formally, the mean is the central tendency of a data. However, the meaning of the mean can be understood from various contexts [7], at least for the purposes of teaching. There are four contexts that can be used to understand the mean, i.e. data reduction, fair share, a typical value, and signal in noise [6].

In the context of data reduction, the mean value can be used to simplify the data set. A collection of values in a sample or population can be simplified into a value, i.e. its measure of center. Mean is one measure of the data center. The simplification of such data is intended to make easy to memorize them.

The second context is a fair share. In this context, the values in a data are shared from one value to another in order to obtain the same value. This context is suitable for elementary school students because, for the first time, they will be introduced to the concept of sharing in division operation.

Mean as a typical value, assumes that the mean is a value that often appears in the distribution or value that is located in the center of data after the data are sorted. The mean here is considered to be the same as the mode or median. Although mean, median, and mode are generally three different values, they are all used to determine the center of data.

The last context that can be used to understand the mean is a signal in noise. Based on this point of view, each random observation has an estimated value of a specific value, but it is unknown. Each observation has an average that has a certain deviation to the true mean. The deviation is called as an error. The mean of each observed average is considered to be the actual data mean. This kind of context is perfect for building an understanding of distribution.

In addition to statistical reasoning, there are two other concepts that become the focus of current statistical learning called statistical literacy and thinking [8,9]. Therefore, the three concepts are first discussed in this paper and then will be explained by focusing on what statistical reasoning is.

Statistical literacy is the essential and basic skills used to understand statistical information or research results [10]. This skill consists of the skills to organize data, create and display tables, and represent data from multiple points of view. In addition, this skill also contains an understanding of the concepts, terms, and symbols in statistics, as well as an understanding of opportunities as a measure of uncertainty.

Statistical thinking is a person's understanding of the reasons why a statistical investigation is conducted, and how to do it [11]. With this understanding, one can choose the right method of analyzing data. In addition, with statistical thinking one can criticize and evaluate the results of statistical research.

Statistical reasoning is a way for a person to interpret statistical information and statistical ideas [12]. Reasoning stands for a person's understanding of the statistical processes and the person's interpretation of the statistical results.

From that sense, it is apparent that the concept of statistical thinking and reasoning are overlapped. But these two concepts are quite distinguishable. For example, someone who has knowledge of when and how to apply certain statistical procedures, then it can be said that the person has statistical thinking. And if the person is able to interpret the results of the statistical procedures he or she performs, then the person has statistical reasoning.

Based on the previous explanation, statistical reasoning is an ability that must be owned by school students. Students should not only be able to perform certain statistical procedures, but they must also be able to interpret the results of the statistical investigations they have undertaken.

To promote statistical reasoning, teachers should make students deal with real data around them, process and present the data, and ultimately draw conclusions from the data. Thus, they are able to give the reason why their conclusions are meaningful.

2. Methodology

To find out the statistical reasoning of students about the mean, the author analyzed data collected from 26 students. Data obtained in the form of student work result to the problem of mean. These problems can be shown as follows.

When asked to find the mean of the numbers 17, 43, 22, and 38, Joko turned in the computations illustrated below to show the mean was 30. Explain her method and illustrate it to find the mean of a set of different numbers.

17	43	22	38
20	40	22	38
20	30	32	38
28	30	32	30
30	30	30	30

(Adapted from Bennett, Jr., Burton, Nelson [13])

The problem is in addition to testing the students' understanding of the mean concepts and also for testing student pedagogies. In addition, the problem wants to show the mean concept as a fair share context.

After the data has been obtained, the authors analyze the types of student work. From the analysis, obtained three level of answers. The three types of answers are further described in the next section.

3. Results

The students' work on the mean problem given in outline can be divided into three categories, described as follows.

3.1 Fail in Understanding the Problem

Polya split the steps in solving the problem into four steps, such as understand the problem, devise a plan, carry out the plan, and look back. By using these Polya steps, it can be said that most students make mistakes in the first step, i.e. understand the problem.

Most students make mistakes when reading the numbers on the table. They assume all numbers are numbers to be searched for. So, they tend to add up all the numbers, then divide by 20, that is, the number of these values. Answers in this category are represented by student 1:

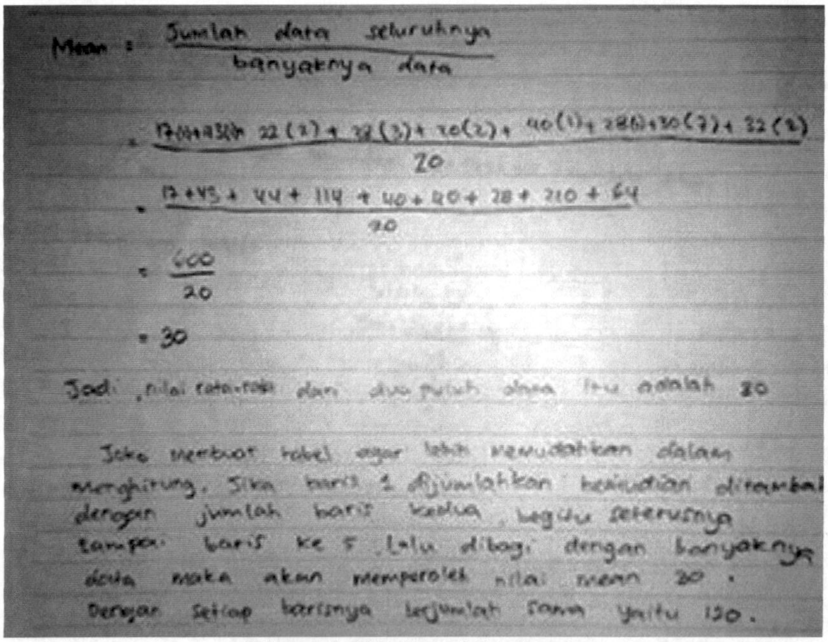

Figure 1 Solution of Student 1

The answer illustrates how students have not understood the problem as a whole. They have not analyzed in depth what relationship between the first, second, third, fourth, and fifth rows in the table. They are still fixated on the numbers that appear, and immediately assume that all of these numbers are numbers to be determined by the mean.

3.2 Success in Looking Pattern, Fail in Interpreting Method

The second category is the category where students are able to find patterns in the method of determining the mean, but they cannot interpret the method. This category is represented by the following 2 student work.

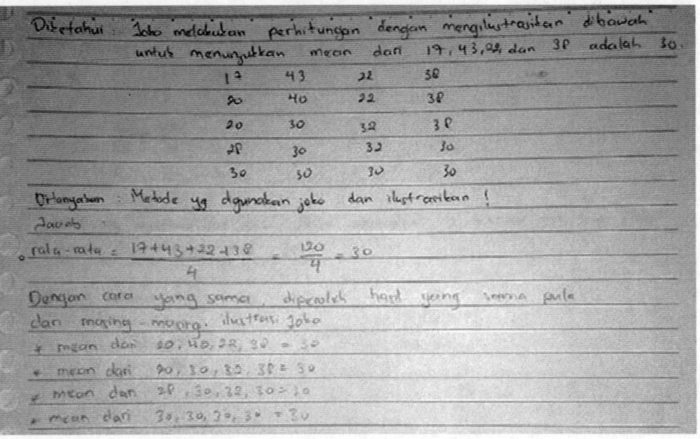

Figure 2 Solution of Student 2

The student's answer gives insight that they can already see the method of determining the mean in depth, i.e. by looking at the pattern of each of the steps taken, but they are still not able to interpret the method.

In the answer, it has been shown that each row has the same mean, but it has not been shown how the relationship between rows. In other words, the student looks at each line separately. He has not explained the relationship of the numbers in each line.

3.3 Success in Interpreting the Method

The third category is the category where the student can explain the steps or procedure of the method of determining the mean. The answers in this category are represented by the following 3 student responses.

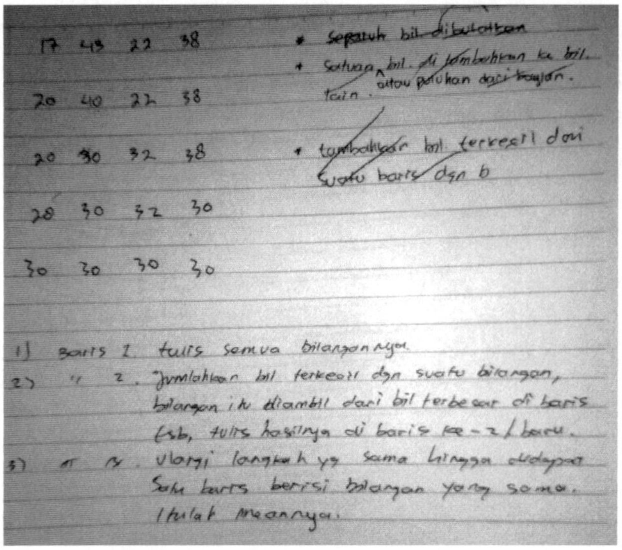

Figure 3. Solution of Student 3

From the answer, it can be seen that the student has explained the method used in detail. He has been able to find relationships between rows in the given steps. And the steps are rewritten in detail in order to be used to determine the mean in different cases.

4. Discussion

This paper wants to answer how students understand the concept of mean. Therefore, based on the data that has been collected the answers to these questions can be categorized as follows.

4.1 Ignoring Problem Context

From the data obtained, it can be said that most students still consider the mean is simply as the ratio between the sum of the overall value of the number of these values. Assuming this, they will be very sensitive to the numbers that appear, but instead, ignore the essence of the problem.

Students still cannot find the context of the given mean problem. Thus, they jump directly into the values given and look for what is asked. Once they find the given values and the ones being asked are the means, they instantly add up all those values and divide them by the number of the values. So, they do not realize that the numbers in each line are a representation of the method in the determination of the mean.

As a result, students still do not have statistical reasoning in interpreting the other method in finding mean. They have not been able to criticize the method and test its truth. Instead, they show the method they already know and used it on the given problem, even with a misconception of the context.

4.2 Not Seeing Relation Between Steps

The answer of the second category of students has shown that they have been able to see that the mean of each line is the same. Using the mean formula that they already know, they can show that the mean of each line is 30.

However, they can't find the relationship between lines. They think of one line with another separately. So, they still cannot find the "big idea" in the given method of determining mean.

If viewed from the statistical reasoning perspective, students in this category still cannot interpret the mean value. They are still fixated on the formula they've been using. Using this formula, they determine the mean of each row and find that the mean of all rows is the same. They have not found where the numbers in the second, third, fourth, and fifth rows come from.

4.3 Explaining the Method for Finding Mean

Students in this category can already find patterns in each row in the given method of determining the mean. They explain that the smallest number must be added with a certain constant, and the constant is "taken" from the larger number on the same line. In other words, they already associate the mean with a fair share context.

Based on data from this category, it can be said that some students already have enough reason statistical reasoning. They were able to analyze other methods of finding mean, by not gluing only to the methods they knew.

5. Conclusion

Based on the results of this study, it can be concluded that most students still have low statistical reasoning that in understanding the concept of mean. Most students still have not been able to analyze other methods in determining the data mean. Nevertheless, there are some students who have been able to interpret the steps in determining the mean by relating it to the fair share context.

Based on these conclusions, the authors provide recommendations in learning statistics at school or at the university level. Statistical learning should prioritize the achievement of students' statistical reasoning. This can be done by designing learning that requires students to collect data from their everyday life, then organize the data, and in finally they can make conclusions based on their own interpretations. This kind of learning can help students to connect the real data representation into a more abstract representation.

References

[1] Dani B Z and Joan G 2004 *Statistical Literacy, Reasoning, and Thinking: Goals, Definitions, and Challenges* (Netherlands: Springer)

[2] Kemendikbud 2016 *Kompetensi Inti dan Kompetensi Dasar Sekolah Menengah Pertama/Madrasah Tsanawiyah (SMA/MA)* (Jakarta: Kementerian Pendidikan dan Kebudayaan)

[3] Kemendikbud 2016 *Kompetensi Inti dan Kompetensi Dasar Sekolah Menengah Pertama/Madrasah Tsanawiyah (SMP/MTs)* (Jakarta: Kementerian Pendidikan dan Kebudayaan)

[4] Jacobbe T and Carvalho C 2011 *Teachers' Understanding of Averages* (Netherlands: Springer)

[5] Leavy A and O'Loughlin N 2006 Pre-service Teachers Understanding of the Mean: Moving Beyond the Arithmetic Average *J. Math. Teach. Educ.* **9** 53–90

[6] Pollatsek A, Lima S and Well A D 1981 Concept or computation: Students' understanding of the mean *Educ. Stud. Math.* **12** 191–204

[7] Watson J M and Moritz J B 2000 The Longitudinal Development of Understanding of Average *Math. Think. Learn.* **2** 11–50

[8] Garfield J and Ben-Zvi D 2009 Helping Students Develop Statistical Reasoning: Implementing a Statistical Reasoning Learning Environment *Teach. Stat.* **31** 72–77

[9] Garfield J and Ben-Zvi D 2004 *Research on Statistical Literacy, Reasoning, and Thinking: Issues, Challenges, and Implications* (Netherlands: Springer)

[10] Gal I 2002 Adults' Statistical Literacy: Meanings, Components, Responsibilities *Int. Stat. Rev.* **70** 1–25

[11] Greer B 2000 Statistical Thinking and Learning *Math. Think. Learn.* **2** 1–9

[12] Del Mas R C 2004 *[3] . (2004). A Comparison of Mathematical and Statistical Reasoning* (Netherlands: Springer)

[13] Bennet J and Burton N 2012 *Mathematics for Elementary Teachers: A Conceptual Approach* (New York: McGraw-Hill Companies, Inc.)

IOP Conf. Series: Materials Science and Engineering **296** (2018) 012038 doi:10.1088/1757-899X/296/1/012038

An analysis of Indonesia's information security index: a case study in a public university

W Yustanti, A Qoiriah, R Bisma, A Prihanto

Informatics Engineering Department, Universitas Negeri Surabaya, Indonesia

wiyliyustanti@unesa.ac.id

Abstract. Ministry of Communication and Informatics of the Republic of Indonesia has issued the regulation number 4-2016 about Information Security Management System (ISMS) for all kind organizations. Public university as a government institution must apply this standard to assure its level of information security has complied ISO 27001:2013. This research is a preliminary study to evaluate the readiness of university IT services (case study in a public university) meets the requirement of ISO 27001:2013 using the Indonesia's Information Security Index (IISI). There are six parameters used to measure the level of information security, these are the ICT role, governance, risk management, framework, asset management and technology. Each parameter consists of serial questions which must be answered and convert to a numeric value. The result shows the level of readiness and maturity to apply ISO 27001 standard.

1. Introduction

Currently, the implementation of Information and Communications Technology (ICT) governance is an obligation and demand for every organization of public service providers. The improving service quality as the realization of good governance is increasingly very important. The performance of organizational governance will be disrupted if information on ICT governance experiences security concerns that convey confidentiality, integrity, availability and non-repudiation. Guarantees to the security of ICT should refer to a standard that has been recognized by both national and international. There are various standards on information security, one of which is internationally recognized ISO / IEC 27001 in terms of the Information Security Management System (ISMS).

Public university as a government's organization must comply with government's regulations related to information security. This research investigates the completeness and the maturity level of information security in a public university in Surabaya. The methodology which used will explain the implementation of Indonesia's Information Security Index clearly. The results of the analysis of this study will be used to improve the organization's governance related to the security of information at a state university, especially the state university as a public service agency.

2. Information Security Management System (ISMS)

The Information Security Management System (ISMS) includes of rules, processes, guiding principle, and related assets and actions, which are accomplished communally by an organization, in order to guard its information stuff. The ISMS is a well-ordered methodology for building, applying, operating, checking, appraising, sustaining and refining the security of organizational information to accomplish

IOP Conf. Series: Materials Science and Engineering **296** (2018) 012038 doi:10.1088/1757-899X/296/1/012038

business goals. It is constructed by risk valuation and acceptance level of organizational risk considered to effectively handle and manage the risk. The success of an ISMS's implementation depends on analysis of requirements for applying proper controls to assure the protection of the information assets. The following basic rules also support the successful implementation of ISMS [1]:

- Awareness of the importance of information security
- The assignment of who is responsible for information security
- Management and stakeholder commitment to information security
- Improvement of social values
- Risk assessment that decides proper controls to gain acceptable levels of risk
- Security becomes an important part of information and systems networks
- Prevention and detection of active information security incidents
- Assure an overall approach to information security management
- Continual review of information security and the creation of proper modifications

2.1. ISO/IEC 27001

ISO (*International Organization for Standardization*) and IEC (*International Electrotechnical Commission*) have published International Standards for the management system. The standard describes the steps to be followed in setting up and operating the management system. This procedure is a combination of features in which field experts have reached consensus internationally.

One of the standards issued related to information security is ISO / IEC 27001: 2013, published in September 2013 which is a replacement of ISO / IEC 27001: 2005. ISO / IEC 27001 is the result of a discussion of the Joint Technical Committee of ISO / IEC JTC 1 (*Information Technology*) and subcommittee SC 27 (*IT Security Techniques*). Members of ISO or IEC participate and collaborate in the development of International Standards through technical committees organized by each organization with specific topics of common interest [2].

2.2. Indonesia's Information Security Index (IISI)

In Indonesia, the government adopted the ISO/IEC 27001:2005 to protect the public organization from information security risk. Ministry of Communication and Informatics of the Republic of Indonesia has issued the regulation number 4-2016 about Information Security Management System (ISMS). The law states that ISMS is an obligation for all organization which are processing their transaction using an electronic system based on risk principles.

The rule defines information security is a maintenance of confidentiality, integrity, and availability of information. The tool which is used to evaluate and analyse the degree of information security preparedness in an organization is mentioned as the Information Security Index (ISI).This method is published officially by National Standardization Agency with number SNI-27001.There are five domains to describe the level of information security based ISI standard[3]:

- *Information Security Governance*. This part evaluates the readiness of information security governance procedures and the agencies/functions, duties and responsibilities of information security executives.
- *Information Security Risk Management*. This part evaluates the readiness of implementing information security risk management as a basis for implementing an information security strategy.
- *Information Security Framework*. This part evaluates the completeness and readiness of the information security management (policy & procedure) framework and its implementation strategy.
- *Management of Information Assets*. This s part evaluates the completeness of safeguards against information assets, including the overall cycle of use of those assets.
- *Information Technology and Security*. This part evaluates the completeness, consistency and effectiveness of technology use in securing information assets.

The assessment process is then done through 2 (two) methods:

- The number (completeness) of the form of security. This method will evaluate the extent to which the respondent's institution has applied safeguards in accordance with the completeness of controls required by ISO / IEC 27001: 2009 standards.
- The maturity level of the information security management process. This method is an extension of the completeness evaluation and is used to identify the level of maturity of safeguard implementation with categorization that refers to the level of maturity used by the COBIT (*Control Objective for Information and Related Technology*) or CMMI (*Capability Maturity Model for Integration*) framework [4, 5].

Before the assessment process is done quantitatively, the role of ICT in the organization or scope must be evaluated first. The respondent will be asked to describe the existing ICT infrastructure in the organization briefly. The purpose of this process is to cluster the instance to a certain "size" i.e. Low, Medium, High and Critical. The categories of ICT Roles can be described as follows:

- *Minim.* The use of ICT in defined scope is insignificant, and its presence has no effect on the working process.
- *Low.* The use of ICT supports ongoing work processes, although not at a significant level.
- *Medium.* The use of ICT is part of the ongoing work process, but the dependency is still limited.
- *High.* ICTs are an integral part of the working process.
- *Critical.* The use of ICT is the only way to run a work process that is strategic or national scale.

3. Results and Comments

The methodology used for this study can be explained by steps such as Figure 1. The first step to do is to define the scope of the assessment. The scope of this research is the ICT Service Centre at a public university in Surabaya.

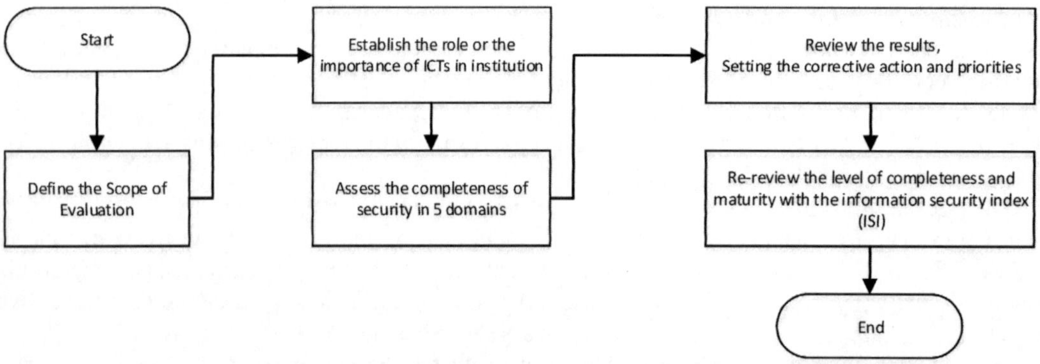

Figure 1. Procedure to evaluate the completeness and maturity level of information security indexes

The second step is to evaluate the role and the importance of ICT unit in the university. This activity will deliver a total score which describes how the dependence of organization on the ICT. The conclusions of the assessment can refer to table 1 that has been defined in the ISI index.

Table 1. Classification of ICT Role based on Total Score

Range	Classification
0 - 12	Low
13 - 24	Medium
25 - 36	High
37 - 48	Critical

IOP Conf. Series: Materials Science and Engineering **296** (2018) 012038 doi:10.1088/1757-899X/296/1/012038

By doing some interviews, the result of assessment can be described in table 2.

Table 2. Questions to evaluate the role of ICT in an organisation

No	Questions	Answers	Scores
1.1	The total annual budget allocated to ICT: • Less than IDR 1 Milliard = Minim • IDR 1 Milliard up to IDR 3 Milliard = Low • IDR 3 Milliard up to IDR 8 Milliard = Medium • IDR 8 Milliard up to IDR 20 Milliard = High • IDR 20 Milliard or more = Critical	Medium	2
1.2	The number of staff / users in institutions using infrastructure ICT : • Less than 60 = Minim • 60 to 120 = Low • 120 to 240 = Medium • 240 to 600 = High • 600 or more = Critical	Critical	4
1.3	The level of dependence on ICT services to run the duties and functions of your institution	Critical	4
1.4	The value of intellectual property owned and produced by your institution	Low	1
1.5	The impact of major ICT system failures used by your institution	Critical	4
1.6	The level of dependence on the availability of ICT systems to connect the work location of your institution	Critical	4
1.7	The impact of your agency's ICT system failures on the performance of other government institutions or on the availability of national government systems	Critical	4
1.8	The level of sensitivity of ICT system users in your institution	Critical	4
1.9	The level of compliance with laws and other legal instruments	Medium	2
1.10	The potential loss or negative impact of incidents penetrated information security ICT system of your institution	Medium	2
1.11	The level of dependence on third parties in running/operating ICT systems	Low	1
1.12	The level of classification/criticality of ICT systems in your institution, relative to threats of attack effort or information security breach	Medium	2
	The importance of ICT role:	High	34

The total score from Table 2 shows the number 34 if it refers to Table 1 can be said that the total evaluation score in the "High" category. It means that ICT is an integral part of the running work process. After knowing that the role of ICT in the university is very important then the next step based on figure 1 is to assess the completeness and maturity of information security. The assessment is done to 5 (five) domains. All questions in each area are grouped into 3 (three) categories of security, in accordance with the stages in the implementation of ISO / IEC 27001 standards. Questions related to

IOP Conf. Series: Materials Science and Engineering **296** (2018) 012038 doi:10.1088/1757-899X/296/1/012038

the basic information security framework fall under the category of "1", for the effectiveness and consistency of its application defined as categories "2", and those that refer to the ability to always improve information security performance are category "3". Table 3 shows the scoring of question answers.

Table 3. Classification of ISO 27001 Completeness

Type of Answers (*implementation status*)	Completeness Category 1	Completeness Category 2	Completeness Category 3
are not done	0	0	0
in planning	1	2	3
in deployment or partially applied	2	4	6
are applied wholly	3	6	9

Assessment is then done by analysing the number in each area and analysing whether the number has reached or exceeded the threshold of achieving a certain level of maturity. The calculation is done by applying the principle [4]:

- Achievement of maturity Level is done in accordance with the completeness and (*consistency and effectiveness*) of its implementation.
- A higher level of maturity requires the completeness, consistency and effectiveness of security at the lower levels.
- There is an additional level of maturity to provide a more detailed description, namely I +, II +, III +, and IV +, so that there are a total of 9 levels of maturity (figure 2).

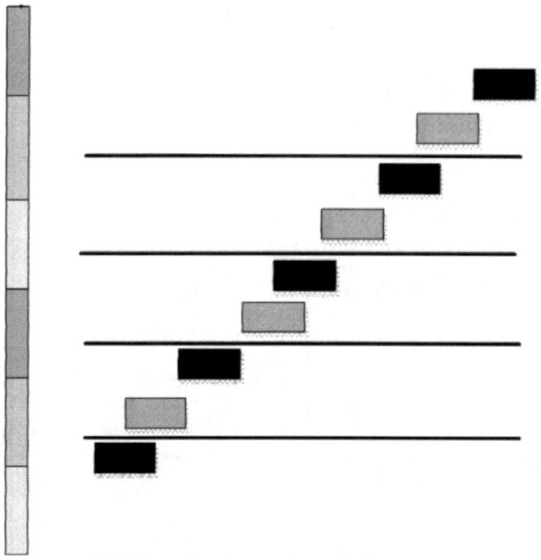

Figure 2. Levels of Maturity

In the information security scale based on the Information Security Index, the level of maturity is defined as follows:

- Level 0 : Unknown (*passive*)
- Level I : Initial Conditions (*reactive*)
- Level II : Implementation of the Basic Framework (*active*)
- Level III : Defined and Consistent (*proactive*)
- Level IV : Managed and Measurable (*continued*)
- Level V : Optimal Level (*optimal*)

Overall, the results of the assessment of the areas can be described in table 4.

Table 4. A Public University ICT Centre's Assessment Result

Domains	Total Score	Maturity Level	Completeness Status	The number of questions based on completeness category		
				1	**2**	**3**
Information Security Governance	58	II	Valid	8	6	6
Information Security Risk Management	40	II	Valid	9	4	2
Information Security Framework	27	I+	Not Valid	11	8	7
Management of Information Assets	94	II	Valid	21	9	4
Information Technology and Security	75	II	Valid	13	10	1
Total	294					

The data in table 4 will be compared to ISO 27001 control in a radar chart.

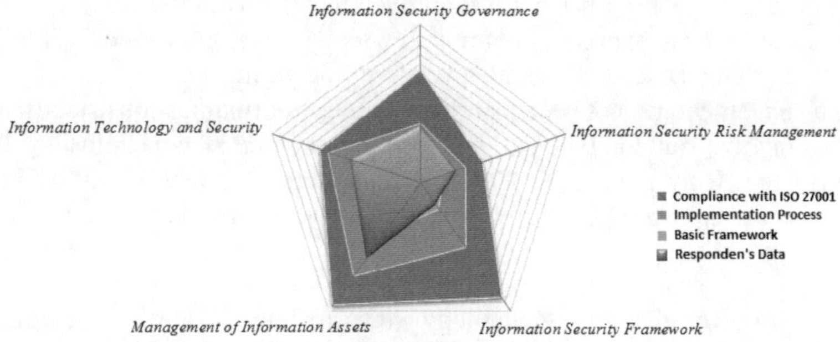

Figure 3. Completeness Level of 5 Domains in a Public University ICT Centre

In the radar diagram, the area background indicates the threshold level of completeness (category) 1 to 3 (light green to dark green).

Table 5. Correlation Matrix between Importance Level and Readiness Status

ICT Role Score	ICT Role Classification	Total Score of 5 Areas IS Index	Readiness Status to ISO 27001 Standard
0 - 12	Low	0 - 124	Not Feasible
		125 - 272	Need Improvement
		273 - 588	Good/Enough
13 - 24	Medium	0 - 174	Not Feasible
		175 - 312	Need Improvement
		313 - 588	Good/Enough
25 - 36	High	0 - 272	Not Feasible
		273 - 392	Need Improvement
		393 - 588	Good/Enough
37 - 48	Critical	0 - 333	Not Feasible
		334 - 453	Need Improvement
		454 - 588	Good/Enough

The value of each area is shown in the red area. In this diagram, can be seen the comparison between the conditions of readiness as a result of the evaluation process with reference to the level of completeness. The resulting quantities are then mapped in accordance with ICT's importance to the

IOP Conf. Series: Materials Science and Engineering **296** (2018) 012038 doi:10.1088/1757-899X/296/1/012038

scope of the organization based on the relationship matrix between the ICT level of importance and the total score (table 5). The analysis result from the respondent's data shows that the relationship between the total score (294) and the ICT's dependency (high category) conclude that the University ICT Centre still need improvement relate to Information Security. Table 5 illustrates that the higher the level of organizational dependence on ICTs or the more important the role of ICTs against the organization's work, then more forms of security are required and applied until it reaches the highest stage.

4. Conclusions

The conclusion that can be drawn from this study is that the level of readiness and maturity of the ICT unit at a state university in Surabaya in category II with a high level of dependence on ICT. This means:

- Security has been implemented even though most are still in the technical area and there is no linkage of security measures to obtain an effective strategy.
- The security process runs without documentation or an official recording.
- The operational safeguard measures applied depend on the knowledge and motivation of the implementing individual.
- The overall form of security has not been proven its effectiveness.
- Weaknesses in security management are still widely found and cannot be resolved thoroughly by the executor and the leader causing a very significant impact.
- Security management has not been prioritized and is not running consistently.
- The parties involved are likely to still not understand brand responsibility

For further study, the University ICT Centre needs detail correction in information security documentation, policy, and procedure in order to improve the maturity level to optimum status.

References

[1] ISO/IIEC 2700, Information technology--Security techniques -- Information security management systems -- Overview and vocabulary, 4th edition, 15th Feb 2016

[2] ISO/IEC 27001:2013, Information technology -- Security techniques -- Information security management systems – Requirements, https://www.iso.org/standard/54534.html, retrieved on 29th April 2017.

[3] Tim Direktorat Keamanan Informasi , Panduan Penerapan Tata Kelola Keamanan Informasi bagi Penyelenggara Pelayanan Publik (in Bahasa), 2011

[4] Covernace Institue 2017, Cobit 4.1 Framework Control Objectives Management Guidelines Maturity Models, , ISBN 1-933284-72-2 US.

[5] Junttila, Juho 2014. A Business Continuity Management Maturity Model, University of Turku

IOP Conf. Series: Materials Science and Engineering **296** (2018) 012039 doi:10.1088/1757-899X/296/1/012039

The effect of thematic learning model, learning style, and reading ability on the students' learning outcomes

L Nurlaela, M Samani, I G P Asto and S C Wibawa

Home Economics Department, Universitas Negeri Surabaya, Indonesia

luthfiyahnurlaela@unesa.ac.id

Abstract: This study aims at gaining empirical findings of the effectiveness of thematic instructional model as compared to conventional instruction; and the potential capacity of thematic instructional model in accommodating different learning styles and reading abilities. This is an experimental research design with 140 elementary students as research subject. The data were collected by using achievement test, learning style questionnaire, and reading comprehension test, and analyzed by using Anava. The results indicate: there is a significant difference in achievement between students who use thematic instructional model and those using conventional model; a significant difference in achievement between students with visual learning style and those having auditorial learning style; a significant difference between students with high reading ability and those with low reading ability. Student's achievement is influenced by the interaction between instructional model and student's learning style. Student's achievement is not influenced by the interaction between instructional model and student's reading ability, the interaction between student's learning style and student's reading ability, and the interaction among instructional model, learning style and student's reading ability. The conclusion is thematic instructional model was more effective than conventional instruction and thematic instructional model had a capacity in accommodating different learning styles and reading abilities.

1. Introduction

Integrated learning is believed to confer a benefit on the achievement of instructional and nurturant effects [18]. By using a more student-focused approach, integrated learning provides opportunities for involving a wide range of students' potentials in the learning process, and it will make learning more meaningful. The integrated learning model provides enough opportunity for the involvement of various students' experiences because the raised themes are selected from the things students put forward, which may be originated from previous experience, and based on the perceived needs of students [3].

There have been a great number of studies concerning the thematic integrated leanring. It has proved to be reliable not only to improve students learning outcomes but also for the student-student and student-teacher interaction. In addition, the thematic learning is also able to foster the collaborative activity, giving rise the nurturant effect by increasing the responsibility and ability to cooperate with students.

A study conducted by Mavropoulos, et al [11] addressed the application of interdisciplinary (integrated) approach in the learning of chemistry. In this study, the researchers concentrated on the teaching model applying an interdisciplinary approach to the topic of "food". The results indicated that there was an increase in student-instructor interaction and the similar application of interdisciplinary models to another topic was recommended.

Ruth [13] investigated the comparison between the scores of students in elementary reading that implemented an integrated thematic learning in the experimental group with the students' scores in the control group using conventional learning. During the two-year period, the experimental school students' scores showed an increase of 16%, while the control school only achieved an increase of 3%.

In addition, learning will run well when students' differences and cognitive processes are well understood by the teacher [15]. Students with different learning styles (visual or auditory) will be studied in accordance with the type. How they learn will affect their academic achievement [16]. In the thematic learning presenting the lessons based on themes that connect the various related subjects might provide different learning outcomes for each student with different learning styles.

The integrated learning facilitates children with visual learning styles. Rose and Nicholl state that more often a person uses visual models, easier to understand and remember because he uses the largest part of the human brain, the part that is dedicated to visual processing. Concept mapping and visual representations help learners think about a subject globally and allow flexibility in their thinking [12]. Based on these descriptions, it can be said that the learning outcome of the child with visual learning style is likely to be better than the one with auditory learning style.

Another factor that also must be considered on students, especially first-grade students, is the early reading ability since they are generally still in the early stages of learning to read or early reading. Activities of reading and writing will empower students to conduct exploration, researching, and enjoy the contents of knowledge according to the needs and interests of their own as independent learners [17].

In connection with the model of learning, the thematic models can accommodate differences in students' reading ability. The thematic learning models make all students active and engaged, regardless of reading ability. By using a more-focused student-centred approach, the thematic learning provides opportunities for a variety of potential involvement in the learning process, as well as accommodating students' individual differences including their reading ability.

Based on this background, this study focused on the effect of learning model, students' learning styles, and students' reading abilities on their learning outcomes in the 1st grade in Surabaya city. The learning model included thematic and conventional models, while the students' learning styles were categorized as 'visual' and 'auditory'. Furthermore, the students' reading skills are categorized as 'high' and 'low'. The models of learning, student learning styles, as well as students' reading skills, are what are regarded as factors affecting the students' learning outcomes.

2. Method

The present study employed an experimental research with a factorial design (2x2x2). This experimental study involved two groups of subjects (70 students in the experimental group, and 70 students in control group). The experimental group received the thematic learning model as the treatment and the control group had the conventional model. At the beginning of learning, a pretest was administered to measure the students' early ability of to determine whether or not the two groups have the same ability. Furthermore, the reading ability test, in which students read a text one by one, and then two people who had been trained as the observer, assessed their reading skills. A questionnaire on learning was completed by the class teacher, assuming that he knew each student well, considering they had been taught in the classroom since the first semester. The reason why it was the teachers who completed questionnaires student's learning style because it is assumed the 1st-grade students did not quite have the ability to understand the items on the learning style questionnaire developed in this study if it was completed by the students themselves, it was possible to cause bias. The independent variable was the learning model, which fall into two categories: thematic learning and conventional learning models. The moderator variable was the students' learning styles and reading abilities. Learning styles can be classified into two, namely, visual and auditory. The reading ability was also divided into two, namely, high and low. The dependent variable was the learning outcomes, i.e. cognitive learning outcomes, covering the aspects of knowledge and understanding (comprehension).

IOP Conf. Series: Materials Science and Engineering **296** (2018) 012039 doi:10.1088/1757-899X/296/1/012039

The influence of background variables was controlled by the subject to subject randomization in multi-stage cluster random sampling or random sampling stages. The achievement test administered for the conventional learning included the Indonesian, Mathematics, Science and Social Sciences subjects. While the test for the integrated learning refers to the themes developed, covering: Food and Animal. Such tests before being used for real research were first tested in the 1st grade of Wonokromo 1 Elementary School, to measure the content validity and the reliability of the instrument

To measure the validity, the Pearson correlation formula was used, while the reliability testing technique used the split-half Spearman-Brown formula [1]. The learning style instrument was developed based on the characteristics of each learning style, namely visual and auditory as well as modified from the instrument for learning styles by Bradway and Hills. The test for reading skills of the students used the proposed criteria by Kartika (2004), in the form of technical reading skills tests (reading aloud). The reasons for selecting the tests were that the tests were in accordance with the first-grade students' expected competencies putting an emphasis on the technical reading ability. The test for instrument reliability of teaching skills used the technique of inter-observer agreement [6]. An instrument is regarded as reliable if the reliability values obtained is ≥ 0.75 [2]. The data analysis regarding the research hypothesis testing used the three-way Analysis of Variance using the SPSS.

3. Results and Discussion

3.1. Requirement Test Analysis

The results of normality test found that almost all the group data are normally distributed with Kolmogorof-Smirnov and Shapiro-Wilk statistical values showing $p > 0.05$, except the conventional data-visual-low, for its Kolmogorof-Smirnov statistical value show $p < 0.05$; but the Shapiro-Wilk statistical value indicates $p > 0.05$. Therefore, the overall data of the dependent variable for both groups are of a normal distribution.

The results of the homogeneity test found that the p-value of all the variables is above 0.05. It is therefore concluded that all the variance between groups is homogeneous. Thus, the data can be further analyzed by ANOVA.

3.2. Results of Data Analysis

The summary of the results of ANOVA is shown in Table 1. The null hypothesis (Ho) is rejected if $p < 0.05$. Based on the table 1, it reveals that the results of the hypothesis testing by df = 1 and a significance level of 5% obtained are: (1) F = 9.651 and significance level or p-value = 0.002, meaning that the learning outcomes showed a significant difference between the variations of models of learning, (2) F = 13.844 and level of significance of p-value = 0.000, meaning that the learning outcomes showed a significant difference between variations in learning styles, (3) F = 6.428 and the significance level or p value= 0.012, meaning that the learning outcomes showed a significant difference between the variation the ability to read. (4) F = 6.768 and figures of significance p = 0.010, means that the learning outcomes are influenced by the interaction between the learning model and the students' learning styles, (5) F = 0.055 and significance level or p value= 0.814, meaning that learning outcomes are not influenced by the interaction between model of learning and the students' reading ability. Although individually, the learning model affects the learning outcomes, as well as the reading ability; but it turns out that the interaction between them has no effect, (6) F = 0.207 and significance level or p-value = 0.650, meaning that the learning outcomes are not influenced by the interaction between the learning style and students' reading skill, and (7) F = 0.977 and level of significance or pvalue = 0.325, meaning that the learning outcomes are not influenced by the interaction among the learning models, learning styles and students' reading ability.

Table 1. Summary of ANOVA Results

Tests of Between-Subjects Effects

Dependent Variable: POSTEST

Source	Type III Sum Square	df	Mean F of Squares		Sig.
Corrected Model	4096.543[a]	7	585.220	5.295	.000
Intercept	767132.858	1	767132.858	6941.462	.000
Learning Model	1066.533	1	1066.533	9.651	.002
Learning Styles	1529.920	1	1529.920	13.844	.000
Reading Ability	710.437	1	710.437	6.428	.012
Learning Model*Learning Styles	747.910	1	747.910	6.768	.010
Learning Model*Reading Ability	6.111	1	6.111	.055	.814
Learning Styles * Reading Ability	22.912		22.912	.207	.650
Learning Model*Learning Styles*Reading Ability	107.943	1	107.943	.977	.325
Error	14587.926	132	110.515		
Total	804130.810	140			
Corrected Total	18684.469	139			

a. R squared = .219 (Adjusted R Squared = .178)

3.3. Interpretation of Data Analysis Results

3.3.1. Thematic and Conventional Learning in Achieving the learning outcomes.

From the results of the hypothesis testing, it showed that the learning outcomes of students who took the thematic learning model were higher than the learning outcomes of students who took the conventional learning model. This proves that the thematic learning significantly and more highly influence the achievement of learning outcomes as compared to the conventional model of learning.

These findings are consistent with the findings of previous research, including (1) Research by Ruth (1989) found that the reading scores of elementary school students who applied the thematic learning during the two-year period showed an increase of 16%, while elementary schools applying the traditional model only reached an increase of 3%; (2) Buechler (1993) suggests one of the research results of the application of the thematic learning model called CLASS-a program in Indiana that used thematic models. The study analyzed the performance of 100 elementary schools in terms of testing the progress of learning called ISTEP (Indiana Statewide Testing for Educational Progress). His findings were that schools with ISTEP CLASS obtained a score higher than other primary schools in the state and that scores on CLASS schools continued to increase over time.

Benson (2005) suggested that the thematic learning involved a set of activities related and designed around topics or themes, and to reach some areas of the curriculum. The themes provide an environment that encourages learning process involving all students actively. The theme also builds the student interest and prior knowledge by focusing on topics relevant to their lives. The setting assists the students to relate to real life experience and develop what they know. The theme gives one of the best tools to integrate the content area in a way that makes sense for students and help them make connections to transfer the knowledge they have learned and apply it in meaningful ways. Another benefit of using the themes in teaching elementary school children includes: learning of factual information in depth, physically involved in the study, study skills processes, integrating learning in a holistic way, increasing closeness of the group, focusing on individual needs, and motivating students and teachers [9].

Numerous studies indicate the existence of the thematic learning advantages compared to the conventional learning. The advantages of thematic learning are also confirmed by the opinions of Shoemaker (1989) stating that the thematic learning allows students to learn in the most natural way. Teachers can make a deal to build a web of themes that attract students, and beneficial for all. These advantages include more coverage of the curriculum to learn naturally, to build students' interest, teaching skills in a meaningful context, and more flexibility.

3.3.2. Thematic and Conventional Learning in Accommodating Learning Styles.

Based on the hypothesis testing, it is known that the learning outcomes were influenced by the interaction between the learning models and the students' learning styles. From the results of the descriptive analysis, the learning outcomes of the students with visual learning styles were higher than the learning outcomes of the ones with auditory styles, both in the thematic and conventional learning groups.

The character of a child with visual learning styles is that he relies heavily on vision when absorbing information. By nature, the child is interested in the visual familiar sights and reminds visual signs such as motion, colors, shapes, and sizes. Most children have an excellent visual hand-eye coordination is. They are also visually superior in all fine motor activities, i.e. activities that require eyes and small muscles, such as those in the fingers. They enjoy reading and writing neatly remaining on the line. Relevant to the statement of Rose and Nicholl revealing the characteristics of children with visual styles in more detail, among other things are neat and orderly, meticulous to detail, good spelling and be able to see the words that are actually in their mind, considering what is seen rather than heard, having the visual association, usually not bothered by the commotion, read quickly and diligently, rather read than recite, and requires the views and overarching goals and be vigilant before mentally feeling sure about a problem [12].

Conversely, children prefer the auditory sounds and words on information provided to the views and touch. In general, children are very verbal auditory, therefore, are often warned for talking too much during the class. Unlike the visual children who love reading and writing neat, auditory ones would rather read out, or when he reads his own, his voice tends to be hard as giving instructions. Comparing to children with visual styles, auditory children's writing, in general, is less tidy. Rose and Nicholl state that auditory children are easily distracted by noise, find it difficult to write, like to talk, as well as have problems with jobs that involve visualization [12].

Based on these characteristics, it can be understood that in this study, the learning outcomes of visual children tend to be better than the ones of auditory children. This does not mean the visual learning style better than the auditory one. In this context, study materials which are designed in the form of Student Book are more favourable to the visual children. Rose and Nicholl cite that the visual children use a global or general format which allows the information indicated in a similar manner as the brain functions in multiple directions simultaneously because he uses the largest part of the human brain, the part that is dedicated to visual processing [12].

3.3.3. Thematic and Conventional Learning in Accommodating Reading Ability.

Based on the results of hypothesis testing, it turns out that the learning outcomes were not influenced by the interaction between the learning model and the students' reading ability. Although individually, the learning models affected the student learning outcomes, as well as the reading ability. The interaction between them had no effect on the learning outcomes. Based on the descriptive analysis, the learning outcomes of the students with higher reading ability turned out to be better than the students with low reading ability.

Studies on literacy have been reported by various sources. MacGilchrist suggested that there is one important finding of the role of reading, where the level of students' abilities at the age of 7 years is a good predictor of the top level of achievement in the future. A correlation was obtained between the reading ability at the age of 7 and the level of achievement exam at the age of 16 [10].

Reading ability is thus so important that it is necessary to provide a conducive learning environment, so as to create a literate generation [14]. The education system needs to be reformed in order to be able to develop the literacy skills of children from an early age. Learning should be directed towards developing students' creativity and thinking. Starting primary school, children should be accustomed to the reading assignments. In connection with this, the conventional learning encouraging the growth of interests and reading habits should be enhanced. The learning models that are better able to create a literate learning environment which can improve the students' reading ability should considerably be applied.

3.3.4. The Effects of Learning Models, Learning Styles, and Reading Ability on Student Learning Outcomes.

The thematic learning is more accommodating to differences in students' reading ability because it provides the conditions that encourage students to develop the ability to read. This learning makes students active and engaged no matter how their reading ability is. This leads to students' increased interest in reading, thus leading to increasing reading skills. As previously noted, theoretically, there is a relationship between reading and reading skills. The higher interest in reading is, the better a person's ability to read. In the thematic learning, learning strategies lead to increased students' interest in reading, which will then lead to increased reading ability, and ultimately lead to better learning outcomes.

This also applies to the interaction between the learning styles and the reading ability. The results showed that there was no interaction between the learning outcomes, both on the thematic and conventional learning. The learning styles individually affect the learning outcomes and the reading ability. The interaction between them had no effect on the learning outcomes. Research has shown that reading ability is more influenced by the interest in reading, and not one's learning style [4]. In other words, children with high ability to read and any learning style, whether visual or auditory, will achieve higher learning outcomes.

The Student books provided in this study was a material which was feasible to qualify, because it presented the concept and the structure and activities that the students had been identified. Then during learning, the teacher provided scaffolding to provide assistance to students. They were very beneficial for students to develop the reading ability, both for visual and auditory children.

Based on the previous discussion, it can be understood when it was later discovered that there was no interaction between learning models, learning styles and literacy on learning outcomes. Therefore, there was no interaction between the learning model with the reading ability, and there was no interaction between learning styles and the reading ability; then by itself, the interaction among these three (models of learning, learning styles and the ability to read) on learning outcomes was discovered.

4. Conclusion and Suggestion

Based on the findings, the conclusions of this study are summarized into seven points, they are as follows: 1) there is a difference in the learning outcomes for the students who use the thematic learning model and the students who use the conventional learning model. The learning outcomes of the students who take the thematic learning are higher than the students who follow the conventional teaching; 2) there is a difference between the learning outcomes of students who learn with visually style and ones with auditory learning style. The learning outcomes of students who learn with visual styles are higher than students with auditory learning styles; 3) there is a difference in the learning outcomes between the students with high reading ability and the students with low reading ability. The learning outcomes of students who have high reading ability are better than students who have low reading skills; 4) There is an interaction between the learning models and students' learning styles in the learning outcomes; 5) There is no interaction between the learning models and the reading ability in the students' learning outcomes; 6) There is no interaction between learning style and the reading ability in the students' learning outcomes .; and 7) There is no interaction between among the learning models, learning styles and reading ability in the students' learning outcomes.

Based on the conclusions, as an empirical implication, several suggestions can be put forward as follows: 1) The importance of the thematic learning to apply the learning in elementary school, especially the lower class; 2) The importance of attention to individual differences of students, in particular learning styles; 3) Reading ability should also be a concern, considering the elementary students, in general, are still in the early stages of learning to read or early reading.

References

[1] Ary D Jacobs LC & Razavieh, A 2002 *Introduction to research in education* Edisi keenam Stamford: Wadsworth Thomson Learning
[2] Borich G D 1994 *Observation Skills for Effective Teaching* (Englewood Cliffs: Merril Publishers)
[3] Brand,S R(ed) 1991 Integrating the curriculum: Educational leadership *Journal of ASCD* **49 (2)**
[4] Buechler, M 1993 *Connecting learning assures successful students: a study of the CLASS program* (Bloomington, IN: Indiana Education Policy Center) (http://wwwkovalikcom Retrieved on 10 April 2004)
[5] Diptoadi V L, Teopilus S and Tedjakusuma H 2003 The influence of learning style and learning strategies on the reading achievement of persons using English as a foreign language *Jurnal Teknologi Pembelajaran Teori dan Penelitian* **11 (1)** pp 27—36
[6] Grinnel Jr Richard M 1988 *Social Work Research and Evaluation* Third Edition (Itasca, Illinois: FE Peacock)
[7] Grisham,DL 1995 *Integrating the curriculum: The case of an award-winning elementary school* Paper presented at the annual meeting of the American Educational Research Association, Berkeley, CA http://wwwkovalikcom Retrieved on 4 September 2005
[8] Kartika E 2004 Memacu minat membaca siswa sekolah dasar *Jurnal Pendidikan Penabur* 3 **(III** pp 113—28
[9] Kostelnik M J, Soderman A K and Whiren A P 2004 *Developmentally appropriate curriculum: Best practice in early childhood education* (Upper Saddle River, N J: Merrill)
[10] MacGilchrist, B 2005 *Reading and achievement—some lesson for the future* www Literacytrustorguk Retrieved from 11 Juli 2006
[11] Mavropoulos A, Roulia M, and Petrou L 2003 An interdisciplinary model for teaching the topic "foods": A contribution to modern chemical education *Chemistry Education: Research and Practice* **5 (2)**
[12] Rose C and Nicholl M J 1997 *Accelerated learning for the 21 st century* (London: Judy Piatkus)
[13] Ruth N S 1998 *A comparative study of Integrated Thematic Instruction (ITI) and non-integrated thematic instruction* Doctoral dissertation, (Texas A&M University http://wwwkovalikcom Retrieved on 11 April 2005)

[14] Sammons P, Elliot K, Syiva K, Melhuish E, Siraj-Blatchford I, and Taggart B 2004 The impact of preschool on young children's cognitive attaintments at entry to reception *British Educational Research Journal* **30** pp 691—713
[15] Slavin R E 2000 *Educational psychology: Theory and practice* Sixt Edition (Boston: Allyn and Bacon)
[16] Soleste H R and Roland G T 2003 *Theoritical perspectives, research findings, and classroom implications of the learning styles of American Indian and Alaska Native Students ERIC Digest* http://wwwericdigestorg/2003-3/alaskahtm Retrieved on 15 April 2007
[17] Torgesen J K 1998 Catch them before they fail: Identification and assessment to prevent reading failure in young children *American Educator* pp 32—9
[18] Tudor L S 2013 Primary school skills development through integrated activities *Procedia - Social and Behavioral Sciences* **127** pp 722—27

IOP Conf. Series: Materials Science and Engineering **296** (2018) 012040 doi:10.1088/1757-899X/296/1/012040

The implementation of assessment model based on character building to improve students' discipline and achievement

Rusijono and K Khotimah
Educational Technology Department, Universitas Negeri Surabaya

Rusijono@unesa.ac.id

Abstract. The purpose of this research was to investigate the effect of implementing the assessment model based on character building to improve discipline and student's achievement. Assessment model based on character building includes three components, which are the behaviour of students, the efforts, and student's achievement. This assessment model based on the character building is implemented in science philosophy and educational assessment courses, in Graduate Program of Educational Technology Department, Educational Faculty, Universitas Negeri Surabaya. This research used control group pre-test and post-test design. Data collection method used in this research were observation and test. The observation was used to collect the data about the disciplines of the student in the instructional process, while the test was used to collect the data about student's achievement. Moreover, the study applied t-test to the analysis of data. The result of this research showed that assessment model based on character building improved discipline and student's achievement.

1. Introduction

In the National Education Day anniversary 2010, the President of the Republic of Indonesia launched the implementation of the National Policy for Character Building. Especially in National Education Departement, its main focus is on the school (students, teachers, education personnel), family, community, and the environment. Implementation of the National Policy for National Character Building is done gradually and continuously. In the future, the Ministry of National Education Departement will include character building through the strengthening of the curriculum from kindergarten to university as part of strengthening the national education system. The implementation of character building will be integrated into existing subjects matter, co-curricular and extracurricular activities, and involves the participation of the environment, families, and communities. [6].

The Minister of Education and Culture of Indonesia made a policy on Curriculum 2013. One of the principles in this curriculum is developed a balance between the development of spiritual and social attitudes, curiosity, creativity, cooperation with intellectual and psychomotor abilities. The objective of instructional in this curriculum includes attitudes, knowledge, and skills.

Anderson states the aspects that need to be assessed teachers on learners in the learning process are the behaviour of the learning process, the efforts of students to master competencies, and the achievements of the students [3]. Behavior in the learning process includes the attitude and actions of students against peers and teachers, particularly the attitude and actions related to the values in the course material and activities in the learning process. Efforts of students refer to the seriousness and discipline of students in the instructional process. Discipline and the seriousness associated with affective and psychomotor aspects of the learning process. Learner achievements of learners meant mastery level of competency to be achieved.

Koesoema Doni suggested that the criteria of assessment in the education based on character building, namely: the quantity of attendance, punctuality collect of the task, the number of violence behavior, the participation of student in program of cooperation between schools, student involvement in drugs, academic achievement, and non-academic culture [5].

Based on expert opinions in the paragraph before can be concluded that assessment model based on character building includes three components. Firstly, behaviour in the learning process includes the attitudes and behaviour of peers and teachers. Attitudes and behaviour in this component refer to the values in the material and learning activities. In the context of educational technology, the transformation of values can go through the material presented to the students or through learning activities. Secondly, student effort and it is referred to the seriousness and discipline of a student in the instructional participation. This component includes the following aspects: the presence of learning or lecturing process, the punctuality in instructional participation and collect the assignment, the frequency and quality of question, the frequency and quality of the opinion, the frequencies of consultation out of hours lectures in order enrichment, creativity, reasoning that shown in the activities or work-related lectures. Lastly, the achievement of students, including academic and non-academic.

The implementation of this evaluation model needs a punctuality and discipline both lecturer and student. The lecturer must be discipline to record the student's activity and the student must be punctual and active in the lectures process. The punctuality is assessed in this evaluation model is attendance in the lecture and collect the assignment. In addition, the student must also be active in the learning process because of their activities in the learning process also be assessed. The activity in the learning process comprises the frequency and quality of the question, the frequency, and quality of the opinion, the frequencies of consultation out of hours lectures in order enrichment, creativity, reasoning that shown in the activities or work-related lectures.

Finally, it is concluded that the discipline is a condition that is created and formed through a process of a series of behaviours that indicate the values of obedience, loyalty, regularity and or orderliness. At the beginning, discipline is a burden for their obligation to act in accordance with the rules. However, when the regulation went right to their soul the discipline of behaviour is no longer perceived as a burden, but on the contrary, would weigh on him when he did not act in accordance with the rules. The assessment model regulates any behaviour of a college student, from attendance at the lecture, activity in lectures, and the time of the collect assignment. Relevance to the discipline process, at the beginning the evaluation model is certainly burdening but when the regulation in the evaluation model went right to student's soul they are no longer a burden. Since all activities of the students assessed, so it is concluded that the implementation of this evaluation model can improve student discipline.

2. Research Method

This study used ADDIE model and the stages are Analysis, Design, Development, Implementation, and Evaluation. The subjects of this study were students of Departement Educational Technology year 2014, Faculty of Education, State University of Surabaya. The number students of Departement were 92 students, it is divided into two classrooms. The number of class A is 46 students and class B is 46 students. The distribution of the classes is not based on the quality of students, so it can be assumed that the quality of students of both classes is equal. So, the implementation of this assessment model used random static group comparison design.

Based on the theory that has been analyzed, assessment model developed includes all behaviour of student related to the learning process includes four components, namely: (1) participation, (2) assignments, (3) sub summation and (4) summative examinations. Student participation includes: the presence of the lecture, student activity in asking, present their opinion, creativity, reasoning that shown in the activities or work-related lectures, and student activity consults outside lecture hours in order enrichment lecture material. The assignment component includes all assignment from the lecture at one semester. The score of this component is mean of all assignment. Subsummative examination administrated two times. The score of the sub summative examination is mean of two times sub summative examination. Then, the final component of this assessment model is a summative examination.

The subject of study which used in the implementation was Epistemology and assessment. Determination of the subject of study was based on the consideration that the subject of study must be taken by all students of the Faculty of Education. Second, Epistemology was not liked by an undergraduate student, while the assessment was liked by an undergraduate student. Moreover, the characteristics of them are different, the material of epistemology is metacognitive knowledge and the material of assessment is procedural knowledge. Before implementation of assessment model begun, all of the component assessment model were socialized for the student.

The variables that observed during the implementation of the assessment model were applicability of assessment model, student discipline, and student achievement. The indicators of applicability in this study were applicability of participation assessment, assignment, sub summative examination, and summative examination. There was no difference sub summative and summative examination of this assessment model with the previous assessment model, so the component of applicability of the assessment model based on character building which studied just participation component. The indicators of student discipline are attendance punctuality in the lecture and collect the assignment. Then, the indicators of student's achievement were the total score of assignment, sub summative, and summative examination. Participation component is not included in the variable achievement because assessment at control group did not use components participation. Moreover, the study applied t-test to analyze the data.

3. Finding and Discussion

The first variable studied in this research is the applicability of the assessment models based on character building. The components of the assessment model are participation, assignments, sub summative and summative examinations. Sub-components of participation are: the presence of learning or lecturing process, the frequencies, and quality of the question, the frequency and quality of the opinion, frequencies of consultation outside the lecture hours for enrichment, creativity reasoning that shown in the activities or work-related lectures. Complete data on the student participation experimental group and the control group can be seen in table 1.

Table 1. The Participation Data of Students

Classes	Presence	Question	Opinion	Consult	Creativity	Score of Participation
Experiment	58,5	4,1	3,6	2,3	2,1	71,3
Control	55,7	2,4	1.9	1,7	1,4	64,2

Note: The data is the average of the Epistemology and Educational Assessment Courses

The data in Table 1 indicate that the presence of student between the experiment class and the control class was almost the same, the average of presence at experiment class is 58.5 and the average at control class 55.7. While the components of the activity of students in asking have significant differences, the average score of the student questions in the experiment class was 4.1 and the control class was 2.4. The average score of student expresses an opinion at the experiment class was 3.6 and the control class was 1.9. The average score of consultation outside the lecture hours for enrichment at experimental class was 2.3 and the control class was 1.7. While the average score of the students' creativity who appear in asking, questions, or expressing their opinions at experiment class 2.1 and the control class was 1.4. Overall score of student participation in the experiment class was 71.3 and the control class was 64.2.

The results of t-test analysis of the participation of the experiment class and a control class were 4.78. This result is greater than the t_{table} = 2,021 (df 44 and the significance level 0.05).

The second variable which studied in this research was the discipline of students in the lecture process. There were two components of discipline, namely: attendance punctuality in the lecture and collect the assignment. The data of the two components can be seen in table 2.

Table 2. The Discipline Data of Students

Classes	Punctuality		Average Discipline Score
	Attendance at the Lecture	Collect the Assignment	
Experiment	97,5	93.7	95.6
Control	92,8	76.8	84.8

Note: The data is the average of the Epistemology and Educational Assessment Courses

The results of t-test analysis of the discipline students between the experimental class and a control class were 5.21. This result is greater than the t_{table} = 2,021 (df 44 and the 0.05 significance level). Based on the results of this analysis it can be concluded that there was a significant difference of students' discipline between a class that implemented the assessment model based on character building and it is not implemented the assessment model based on character building. Variable discipline is a variable that it is embedded in the lecture so it is not possible to do a pretest. Therefore, as a substitute for pretests, it was used data on experimental class student discipline in the course of research methodology. The average score of student discipline in the course of research methodology was 84.4. The average score is slightly lower than that of the control student class average.

The results of t-test analysis of the experiment class score student discipline in the course of research methodologies and the two subjects tested (Philosophy of Science and Evaluation) was 5.43. This result is greater than the t_{table} 2,021 (df 44 and the 0.05 significance level).

The third variable was studied in this research is student achievement. The component of achievements includes score assignments, a score of subsummative and summative examinations. The data of the three components can be seen in Table 3.

Table 3. The Achievement Data of Students

Classes	The Component of Achievement			Score of Achievement
	Assignment	Sub summative	summative	
Experiment	78	81	80	79,7
Control	76	74	73	74,2

Note: The data is the average of the Epistemology and Educational Assessment Courses

The results of t-test analysis of student achievement between experimental class and control class 2.71. This result is greater than the t_{table} 2,021 (df 44 and the 0.05 significance level). This study does not do the pretest, as a substitute for pretests was used student achievement index in the previous semester. The results of t-test between achievement student on epistemology and assessment in a control class with student achievement index in the previous semester were 1.99 ($< t_{table} = 2.021$).

4. Conclusion

Based on the analysis that has been described, it can be drawn three conclusions. Firstly, a model-based assessment of character education can be implemented properly. The implementation of this assessment model requires the lecturer to inform the students that the lecturer had to record every student participation and opinion. If the number of students in the class is too large it is difficult to be able to carry out this assessment model. Therefore, this assessment model only suitable for a small class (20-30 students).

Secondly, the implementation of assessment model based on character building can improve student discipline in the instructional process.
Lastly, the implementation of assessment model based on character building can improve student achievement.

References

[1] Abdullah. M 2010 *Character Education, Character Building Children from Home* (Yogyakarta: Pustaka Insan Madani)
[2] Anderson, Lorin W 2003 *Classroom Assessment, Enhancing The Quality of Teacher Decision Making*. (Marwah: Lawrence Erlbaum Associates, Publishers).
[3] Anderson, O.W. dan Krathwohl, D. R 2001 *A Taxonomy for Learning, Teaching, And Assessing* (New York: Addison Wesley Longman, Inc).
[4] Departemen Pendidikan Nasional Republik Indonesia 2005 *Republic of Indonesia Government Regulation No. 19 of 2005 on National Education Standards* (Jakarta: Badan Standar Nasional Pendidikan)
[5] Doni K.A 2010 *Character Education, Strategy to Educate Children in The Global Age*. Jakarta: Gramedia.
[6] Edi D.W 2010 *Building The Character of The Nation In 2010-2025* (Jakarta: Pemerintah Republik Indonesia)
[7] Gronlund, N. E 1985 *Measurement and Evaluation in Teaching* (6th ed.). (New York: Macmillan Publishing Company)
[8] Johnson D.W. dan Johnson R.T. 2002 *Meaningful Assessmen*t (Boston: Allyn and Bacon)
[9] Reynolds, C.R., Livingstone, R.B., & Willson, V 2010 *Measurement and Assessment in Education* Second Edition. (London: Pearson Education International)
[10] Rusdinal, Elizar 2005 *Classroom management in Kindergarten* (Jakarta: Direktorat Pembinaan Pendidikan Tenaga kependidikan dan Ketenagaan Perguruan Tinggi)
[11] Rusijono, et al 1999 *Development of Evaluation Result Learning System To Improve Student Achievement Of IKIP Surabaya* (Surabaya: Lembaga Penelitian IKIP Surabaya)
[12] Rusijono 2009 *Implementation of lecturing system of Universitas Negeri Surabaya in Faculty of Education*. (Surabaya: Lembaga Penelitian Universitas Negeri Surabaya)
[13] Taylor, G.R. & MacKenney, Loretta 2008 *Improving Human Learning in the Classroom, Theories and Teaching Practices* (Lanham: Rowman & Littlefield Education)
[14] Tilstone, C 2002 *Observing Teaching and Learning* (London: David Fulton Publishers)

IOP Conf. Series: Materials Science and Engineering **296** (2018) 012041 doi:10.1088/1757-899X/296/1/012041

Online test application development using framework CodeIgniter

S C Wibawa[1], Y Wahyuningsih[1], R Sulistyowati[2], R Abidin[3], Y Lestari[4] and Noviyanti[4], D A Maulana[5]

[1]Information Technology Education, Faculty of Engineering, Universitas Negeri Surabaya

[2]Ecomics Faculty, Universitas Negeri Surabaya

[3]Graphics Design Program, Universitas Negeri Surabaya

[4]Social Science Faculty, Universitas Negeri Surabaya

[5]Mathematics Department, Universitas Negeri Surabaya

setyachendra@unesa.ac.id

Abstract. The purpose of this study is developing application *an online test* for vocational students and to know the *user acceptance testing* on the application. The method used in this research is the *Research and Development* (R & D) only up to the pilot phase of the product. The stage of the procedure of the research namely: (1) Analyze the exam using paper compared to using web-based application *test* online. (2) Design the media in accordance with the design of the author. (3) To test the product by including a questionnaire instrument against the application that has been done. Researchers carried out tests on class X on the computer and network engineering Vocational High School (SMK) Darul Ma'wa Plumpang. It can be concluded that: (1) application *online test* was created gets the value of the validator with the percentage of lowest value and the highest value for the validation of products: 25% and 100%. With a total number of 14 questions, after validation of the products obtained from the three aspects of the assessment scale from 81.25 to 100 obtained from 2 different validators with the meaning of an application that has been developed and very suitable for use in school. (2) Based on *User Acceptance Testing* (UAT), applications can be very well received by the students and recommend to replay the final semester and others. With the successful acquisition of a category which means it's ready and qualified.

1. Introduction

The growth of information systems is currently very fast and rapid, so it offers a convenience and comfort for the user. It can be seen from various activities that are organized by an information system. Therefore, we try to create an information system and an application to explore the challenges and sophistication of existing application programs. This study begins with the relevant research conducted by [5] entitled "Application Exam Online at International School Stubs (RSBI) Public High School (SMAN) 2 Mojokerto" and [7] entitled "Application Design Simulation TOEFL (Test of English as Foreign Language).

With the problems and the relevant research mentioned above, the researcher attempts to implement Development of Test Online by Using CodeIgniter Framework in SMK Darul Ma'wa Plumpang on the subject of Basic Network for class X. The purposes of this research are: (1) Determining the validity of the online test application development using CodeIgniter framework in SMK Darul Ma'wa Plumpang on the subjects of Basic Network for class X (2) Knowing user acceptance testing on online test applications.

According to [4], *CodeIgniter* is a *web framework* for the PHP programming language, created by Rick Ellis in 2006, inventor and founder of Ellis Lab (www.ellislab.com) [4]. EllisLab is a work team which was established in 2002 and engaged in the manufacture of *software* and *tools* for web developers. Since 2004 until now, EllisLab has awarded ownership rights of *CodeIgniter* to the British *Colombia Institute of Technology (BCIT)* to further the development process. Currently, the official website of CodeIgniter has changed from www.ellislab.com to www.codeigniter.com.

Online test application or online exam application is an application used to perform the test without using the paper process or can be called as *Computer Based Test (CBT)* or *Computer Assisted Test (CAT).* All test process is done through computers, ranging from constructing questions, classroom settings, user settings which can take the exam, setting teachers teaching, until the examination process, by utilizing web engineering technology. The type of questions that can be included in the online exam application is a multiple choice questions type [10].

2. Method

Type of research used in this study is a research and development (*R & D*) stated that methods of research and development are a research method that is used to produce a specific product and test the effectiveness of the product [8].

Sugiyono divided the various stages of research and development into 10 stages: (1) the potential and problems, (2) data collection, (3) the design of the product, (4) design validation, (5) design revisions, (6) test products, (7) product revision, (8) utility testing, (9) the revision of the product and (10) the mass production [8].

2.1 Stage Potential and Problems

The first stages are to analyze the potential and problems. Potential is anything that if it is utilized, it will have an additional value. In this case, it is an online based test. SMK Darul Ma'wa Plumpang has been supported by laboratory capable enough to implement a computer-based test, but it still uses the conventional exam. Therefore, this potential needs to be raised.

2.2 Data collection phase

Once the problem can be demonstrated factually, the next step is to gather information that can be used as product planning. The steps are made by the author:

1. Perform initial observation. It has been conducted by the researchers by doing the interview. Conducted studies on various ways of making online-based exams from various books on web programming, especially the book on the CodeIgniter framework.
2. Information obtained from the Professional Teacher Program (PPP) and the study will be used to create media-based test online.

2.3 Stage Product Design

After seeing the potential or problems and gathering information, the next step is to create a product design that will be developed i.e. computer-based exam through *online* exam using multiple choice questions. With the allocation of the following values:

- Values Multiple Choice

$$pg = \frac{\sum jawaban\ benar}{\sum soal\ pg} \times 60 \quad (1)$$

2.4 Data Collection Techniques
The study used two data collection techniques that is by way of validation to the experts as a validator and evaluation of learning outcomes that perform tests.

2.4.1 Validation. Validation is the process of collecting data on the feasibility of teaching materials and applications obtained previously reviewed with SMK teachers and expert lecturers based on the assessment of the validator. With reference to the validity of the content and application test *online* developed.

2.4.2 Test. The test is a series of questions or exercises and other tools used to measure skills, intelligence knowledge, talents possessed by individuals or groups [13]. Students' answers obtained from the test will be corrected in two ways: manual (teacher-corrected) and automatic ways (corrected by the online system). The correction result will be converted to a value ranging from 0-100. Furthermore, the value of manual and automatic correction result compared with using t-test to know the significance of the difference of both. Thus, it can be written

1. Variable 1 = Automatic correction Result
2. Variable 2 = Manual correction Result

   That has been given treatment from the control class and experimental class.

2.5 Data Analysis Techniques
Data analysis conducted by researchers are as follows:

2.5.1 Expert Analysis Applications. From validation, sheet will be obtained by the researcher validity of online exam application done by experts by providing valid response criteria. For validation data analysis applications used the same calculation, i.e. using descriptive statistical rating as follows:

a. Determination of the size of the study along with the weight value. The determination is as follows:

Table 1 Scale Rating [8]

Category	Percentage (%)	Weight
Invalid	0-25	1
Less Valid	26-50	2
Valid Enough	50-75	3
Valid	75-100,	4

b. Specifies the maximum score validator
   The determination is conducted by multiplying the number of validator/respondent with the highest value weighted on the qualitative assessment. The formula used as table 3:

Table 2 Scale formula [10]

Valid	Nx4
Enough Valid	Nx3
Less Valid	Nx2
Invalid	Nx1
Score Validation	....

2.5.2 Validation of Grain Problem. The research instrument is filled by a material that teachers of subject basic network class X to obtain opinions and feedback on the initial media used during the test or quiz taking place, so we get a decent medium to be applied. Validation of the application used the same

IOP Conf. Series: Materials Science and Engineering **296** (2018) 012041 doi:10.1088/1757-899X/296/1/012041

calculation, i.e. using descriptive statistical rating for determining the size of the study along with the weight value.

2.5.3 Analysis of the application acceptance students (User Acceptance testing). User acceptance test (UAT) is a process of testing by users who are intended to produce documents used as proof that software which has been developed to be accepted by the user if the results of the testing can be considered to meet the needs of the user [13].

2.5.4 Algorithm Problem Scrambling. The algorithm used in the scrambling problem is the *order by rand* which is *syntax* contained in the *SQL* that is used to display the results of *record* randomly. The algorithm is placed in the *syntax quiz_model,* so as to make the matter in accordance with the desired randomization. Standard used in the manufacture of these online test is such a national test in the sense that students can work on the problems with randomization matter though selecting the same number.

Explanation of randomization question stems from the decision process *database* later in the *query* tables that *savesoft_category* and choose a matter on a question bank with a table named *savesoft_qbank* and choose the level of questions that are available (level 1, 2, 3). Level 1, 2, and 3, namely the level of easy, medium and hard. Then the results of *random* questions in the table stored *savesoft_result* in *fields* or columns r_qids. The contents of the column r_qids are all about exams.

Online exam can be used on each PC installed with *browser* exam. Students are unable to see or access another application unless the application online exam only.

3. Results and Discussion

3.1 Results Research

This aims to create an application to support the implementation of exams on the subjects of the basic network in Network and Computer Engineering Department and determine the feasibility of application *Test Online* created as a means to test. The results also present a description of the results of validation in the form of media validation and validation problems.

3.1.1 Online test. Application *online test* is an application used to perform the test without using the paper process.

- The first page (homepage). It is the initial page when entering sites *test online* that allow users to be able to do the exam through *logging in.*
- The main page. It is the initial page of the application after the user has logged in, beginning at page contains quiz that has been created by admin and has already done by the user.
- Account page. This page contains the biographical data that has been registered account. These users can change the bio-data and fill as desired.
- Page quiz. Page quiz is a page for users to work on the problems or exams that have been made by the admin. This page is not much different from the main page. Students can directly work on the exam that has been set.
- The results page contains student experiments or users who have completed the exam. In this menu, there is a facility to view all test results that have been completed.
- A page out contains a *log out* so the user can be out of the application or exam. Then, the user will be on the first page.
- Main page admin is a page that contains users who are already registered.
- Admin Page Sub Additional User Menu. This page to add new users.
- Home Admin Menu Sub Users List is useful for the admin page to add a list of users.
- Page Admin Menu Question Bank Sub Add New
- Page Admin Menu Question Bank Sub Questionnaire

- Page Admin Menu Quiz Sub Add New. This page serves to create a new quiz
- Home Admin menu sub-register quiz serves to change quiz that was ready to be done.
- Results page Admin Menu contains test results of students.
- Exit page serves to get out of the admin page.

3.2 Discussion

Making online test application through several stages in research methods, validation research instrument used has a scale of 1-4 values, then the data obtained. Calculation analysis application quality according to product experts. Data have been obtained will be analyzed with the following provisions.

Table 3 Quality Criteria based on tables researchers

Answer Scale	Criteria	Assessment Scale	Scale Ratings (%)
Strongly Good	1	14 to 24.5	25 to 43.75
Less Good	2	24.5 to 35	from 43.75 to 62.5
Good Enough	3	35-4.5	62.5 to 81.25
Very Good	4	45.5 to 56	from 81.25 to 100

3.3 User Acceptance Testing (UAT)

From the table below, it shows that the application can be used in schools and applied to the test. The results of the assessment indicate success validator passed *user acceptance testing*. Assessment is done in accordance with the steps and procedures that have been arranged by the author.

Table 4 ratings based UAT, modified source [13]

No.	Use Case Process	Success / Fail
1.	Test Name: Log Description Testing: Verify the access rights can only be accessed by registered users Test Case - Username - Password Expected results: - If successful will display the main page of the application. - If it fails displays an error message	*Successfully*
2.	Name Test: Working Quiz Description Testing: The execution of a matter can be done by the user/admin Test Cases: - Working Quiz Expected results: - Users can complete the provided quiz	*Successfully*
3.	Name Test: Results List Description of the test: See results of workmanship exam. Test Case: Seeing Results Expected results: - Users can view the exam that has been done.	*Successfully*
4.	Test Name: Edit and delete Test Description: Can modify and delete users who have registered. Test Case: - Edit and delete Expected Results: - Admin can change and delete users who have registered	*Successfully*
5.	Test Name: Add New User Testing Description: Registration of new users. Test Case: - Entering personal data users	*Successfully*

IOP Conf. Series: Materials Science and Engineering **296** (2018) 012041 doi:10.1088/1757-899X/296/1/012041

	Expected Results: - If successful, the user will be registered as a member / new user.	
6.	Test Name: Add New Questions Test Description: Add Questions Test Case: - Can create a new question. Expected Result: - If successful, there are new issues	*Successful*
7.	Test Name: Questions Bank Testing Description: The matters: Test Case - Admin can delete the questions and change matter. Expected Result: If successful, item will change and will be deleted	*Successfully*
8.	Test Name: Add new quiz Test Description: Creates a new quiz Test Case: - Admins can create new quizzes Expected Results: If successful, quiz will appear on the page a list of quizzes.	*Successfully*
9.	Name Test: Quiz Description Testing: Experiment quiz Test Case: - Trial work on the problems Expected results: If the question, successfully done it can be seen in the results pages	*successfully*
10.	Name Test: Results Report Description Testing: Seeing Results Reports Test Cases: - View results - Delete users have been working on the exam expected Results: If successful, the user will be deleted	*successfully*

4. Conclusions

The research conducted in developing online test application using CodeIgniter framework in SMK Darul Ma'wa Plumpang on the subjects of the basic network of class X results in a conclusion as follows: Research produces products in the form of online test applications that can be used in SMK Darul Ma'wa Plumpang. From the validator, it can be obtained from the media expert value of 81.25% which means the application is very valid and can be used. From the validation expert about the aspect of Validation, Content get a *valid* value which means can be used, while the aspect of Language and Writing Problem gets a *very understandable* value which means can be used in testing. According to research by *User Acceptance Testing* (UAT), applications can be received very well by the students and can be used to do the final semester or midterm's exams. With the acquisition of the category, the application is successfully ready and qualified to be tested.

References

[1] Adha Muhammad 2011 *Applications Online Exam on Training Employee Classification Method Using Computerized Adaptive Test with Feedback* (Jakarta: Indonesian University)
[2] Ahfas Imam 2012 *Design Build Test Online Simulations Candidate Civil Servants (CPNS)* (Maria Kudus University: Holy)
[3] Alexander 2013 *Web Programming Power Pack* (Jakarta: Mediacom)

IOP Conf. Series: Materials Science and Engineering **296** (2018) 012041 doi:10.1088/1757-899X/296/1/012041

[4] Budi Raharjo 2015 *Self-taught Learn CodeIgniter Framework* (Bandung: Informatics Bandung)

[5] Chandra Ahmad 2012 *Applications Online Exam in RSBI SMAN 2 Mojokerto* (Surabaya: UPN Veteran)

[6] Hadi Diki 2016 *Understanding and How to Use Bootstrap* (www.malasngoding.com May 4, 2017)

[7] Purwanto, Toni. 2011. *Application Design Simulation TOEFL (Test of English as Foreign Language*) (Jakarta: University Syarif Hidayatullah)

[8] Sugiyono 2016 *Methods of* Education (Bandung: Alfabeta)

[9] Tutdianah Husnah 2012 *Alademik Potential Test Application Online Learning Process As A Tool For CandidateExam* (Surabaya: UPN Veteran)

[10] Riduwan 2015 *Scale-scale measurement of research variables* (Jawa Barat: Member of the Association of Indonesian Publishers (IKAPI))

[11] Wibawa, Setya Chendra and Rau Julia Ton. Without Year *Vocational Students' Creative Portfolio: Using Computer Aided Design (CAD) Course on Innovative Decorative Design.*, Available: https://www.academia.edu/14059261/VOCATIONALSTUDENTS_CREATIVE_PORTFOLIO_USING_COMPUTER_AIDED_DESIGN_CAD_COURSE_ONINNOVATIVE_DECORATIVE_DESIGN (Accessed 27 November 2017, 15:00 pm).

[12] Wibawa, Setya Chendra. 2013 *Creative Industry Collaboration for The Creative Portovolio* pp 55-56 Available: https://www.academia.edu/29578478/Creative_Industry_Collaboration_for_The_Creative_Portfolio (Accessed 27 November 2017, 15:15 pm).

[13] Betha 2006 *Understanding User Acceptance Testing* www.betha.wordpress.com/2006/05/01/user-acceptance-tes 27 November 2017

Using historical perspective in designing discovery learning on Integral for undergraduate students

Abadi and S Fiangga

Mathematics Department, Universitas Negeri Surabaya, Ketintang, Surabaya, 60231, Indonesia

Corresponding author: abadi@unesa.ac.id

Abstract. In the course of Integral Calculus, to be able to calculate an integral of a given function is becoming the main idea in the teaching beside the ability in implementing the application of integral. The students tend to be unable to understand the conceptual idea of what is integration actually. One of the promising perspectives that can be used to invite students to discover the idea of integral is the History and Pedagogy Mathematics (HPM). The method of exhaustion and indivisible appear in the discussion on the early history of area measurement. This paper study will discuss the designed learning activities based on the method of exhaustion and indivisible in providing the undergraduate student's discovery materials for integral using design research. The designed learning activities were conducted into design experiment that consists of three phases, i.e., preliminary, design experimental, and teaching experiment. The teaching experiment phase was conducted in two cycles for refinement purpose. The finding suggests that the implementation of the method of exhaustion and indivisible enable students to reinvent the idea of integral by using the concept of derivative.

1. Introduction

The fact that lecturers, in teaching calculus, tend to focus only on the procedural ability is undeniable. Calculus is treated as a tool in which the materials only cover symbolic manipulation of function [2–4]. In addition, calculus textbooks do not support fully about the conceptual understanding of calculus [2]. Students' understanding of calculus may appear as a procedural method only. This understanding could restrict the concept of calculus developed from the idea of the infinitesimal.

Misunderstandings on the concept of calculus lead to students' difficulties in working in integral calculus problems. In general, students would encounter difficulties in learning calculus. Limit and infinitesimal concept produce several cognitive difficulties, such as; (1) terminology conflict of terms involved ("approaching", "arbitrarily small", "reach infinity") and (2) inability in estimating what happen in infinity [5]. In detail, the students' difficulties can be stated as (1) wrong interpretation of notational meaning – Leibniz notation, (2) difficulties in translating real world by using suitable representation into calculus formulation that restrict imagery of a function mentally, and (3) lack of algebraic manipulation that makes the application of procedural method override the conceptual understanding.

Several approaches are proposed to help students in the beginning of their calculus class. Several correctives strategies can be proposed such as: (1) avoiding early reference to the language of limit, derivative, and integration; (2) confronting students with the continuous properties and infinitesimal concept as a new concept for the students; (3) emphasizing only in procedural aspect; (4) using formal definition axiomatically; etc. [1,6]. One of the perspectives that can be used to support the students in calculus is what is called History and Pedagogical in Mathematics [7–10]. Using historical perspective to support the students' mathematics learning development has been spreading in the recent decades. In this perspective, the relationship between ontogenesis (development of the individual) and phylogenesis (development of mankind) in the context of mathematics concept development has been proposed lately [11]. Table 1 shows how the development of mathematics through the millennia is parallel with the cognitive development in students.

Table 1 Comparison between Mathematics Development in History and Individual Progression [11]

Development of Mathematics through History	Mathematical Content	Development of Mathematics in Individual Progression
Primitive Man, Primary and derivative occupation	Practical Geometry Separate unit represented by symbols Counting (Gesture or Language Symbols)	Infancy
Geodesy Egyptian and Babylonian	Origin of Rhetorical Algebra Experimental Geometry (Inductive and approximate – empirical and intuitional evidence) e.g. Pythagorean Theorem Arithmetic (Development of symbols) Additive Principle for collective units e.g	Childhood
Astronomy and Mechanics Greek, Hindu and Arabic	Rhetorical Algebra Practical Trigonometry (Hipparcus - reckoned a branch of Astronomy and Surveying by the Greek) Deductive Arithmetic Deductive Mechanics (Archimedes) Deductive Geometry (Theory)	School
Physic, Europe	Rhetorical Algebra Practical Trigonometry Symbolic Algebra Arithmetic (Zero and Place Value)	College
	Trigonometry, Decimals (Arithmetic Development) Logarithm, Symbolic Algebra	
	Analytical Geometry	
	Differential Calculus and Theoretical Mechanics, Integral Calculus	
	Higher Arithmetic, Function Theory Universal Algebra, Probability	

The development of mathematics through ages can be seen similar to the mathematics developed in individual progression. Table 1 shows that the early development of mathematics started as practical geometry and its basic counting. That mathematical content used by primitive man. The same mathematical content also appeared in the infancy stage of development of mathematics in individual progression. In the next stages, the development of mathematics became more complex especially after Renaissance, Europe Era. This is in line with the individual development as they enter college stage. This point of view considers several objections to the implementation of historical perspective in the teaching of mathematics [11].

The history of integral calculus is evolved on the idea of measuring non-rectilinear figure. The first important and famous problem in the beginning of calculus was the circle squaring problem. The aim of this problem was to find a square that has an equal area with the circle. For the time being, the problem developed into the problems of finding the area of cycloid and even arbitrary figures. The progress of the method, which was started by the method of exhaustion and the method of indivisible, can be adapted and implemented to the learning activities of learning integral calculus. In designing learning activities using historical perspective, ones need to link a mathematical concept with the corresponding authentic experience. One of the approaches that can be used in designing such activities is Realistic Mathematics Education (RME). The development of the activities using RME should involve the learning activities by the designed parts which are constructed from guided reinvention, emergent modelling, and didactical phenomenology [12]. By experiencing those stages, students would gain conceptual understanding as the activities of learning.

This paper proposes an alternative approach in teaching integral using historical perspective for undergraduate students of the mathematics department. In fact, in higher education, calculus course had been taught persistently in a non-meaningful way. Hence, after analyzing the history of the development of calculus, we would suggest the use of appropriate big ideas that they can help the students understanding of integral calculus based on the historical perspective of integral. RME proposes a guided reinvention as one of a fundamental idea on how history can be integrated into the teaching of mathematics along with didactical phenomenology and emergent modelling [4]. The designed learning activities for the first implementation is elaborated in 3 parts which are the goals of learning. Therefore, this paper discusses how to design discovery learning using historical perspective on Integral Calculus course for undergraduate students.

2. Methods

The main concern of the researchers is to help students to overcome their difficulties on the topic of integral. Here, we not only can elicit the way of the methods work well, but also we can understand why they work well. The design would be carried out in the two different classes of undergraduate students of mathematics department who take the Integral Calculus course as two consecutive cycles, which are preliminary teaching phase and teaching experiment phase. The preliminary teaching was carried out on the class of eight students, while the teaching experiment was carried out on the class of 24 students.

In designing meaningful learning activities, the use of historical perspective could help the students to focus more on the underlying concept rather than procedural manipulation symbols. The big idea on indefinite integral concept emerges from missing function context into anti-derivative context. While the big idea in the definite integral is to develop from the relation between the area under the curve and the function of the curve to Riemann Sum that is formalized of the Fundamental Theorem of Calculus.

The implementation of the learning activities would be applied and revised accordingly. The resulting outcome of this study was a sequence of learning activities which are crucial to the learning of integral [13]. There are three phases of conducting design experiment [14]. These phases are (1) preliminary phase, (2) design experimental phase, and (3) teaching experiment. The preliminary phase had a goal to formulate hypothesized local instruction theory in the form of a sequence of learning activities. These learning activities were then carried out and tested during the design experiment [14]. The design experimental phase was aimed to carry out the learning design. In this phase, the learning sequence of 4 lessons that were developed in the previous phase was carried out in teaching experiment.

The teaching experiment was conducted in 2 cycles. In the first cycle, the experiment was implemented in a class of eight students with one of the researchers himself as the teacher. The researcher's observations of the lessons and the written works of the students were documented in

IOP Conf. Series: Materials Science and Engineering **296** (2018) 012042 doi:10.1088/1757-899X/296/1/012042

reports. The reports emphasized on classroom discussions, reflections and interesting individual contributions by students to support whether the conjectures really fulfilled or some new issues come out and were used for revising the learning activities in some extent. The second cycle was the actual classroom for the design experiment. It was a class of 24 students. The learning activities sequence was implemented by the teacher. In the class, researchers try to capture the classroom activities of the whole class. At the same time, the researchers also have a focus group to be analysed along with the whole class situation. The data collected from this second cycle will be analysed and used as the final revision of the learning activities to answer our research question.

3. Results and Discussion

This study had designed the historically based learning activities on the course of integral calculus for undergraduate students. The design was implemented and revised during the preliminary teaching and the teaching experiment. The result will be considered as historical based learning activities and the analysis on how the activities can help the students' difficulties is elicited.

In designing meaningful learning activities, the use of historical perspective could help the students to focus more on the underlying concept rather than procedural manipulation symbols. Table 2 summarizes the design of the learning activities in the course of integral calculus. The big idea on indefinite integral concept emerges from missing function context into anti-derivative context. While the big idea in the definite integral is to develop from the relation between the area under the curve and the function of the curve to Riemann sum which is formalized of the fundamental theorem of calculus.

Table 2. Historical Based Learning Activities on the Topic of Integral

Nu.	Subtopics	Learning Activities
Indefinite Integral		
1	**Initial Value Problem**	**The problem discussed in this phase is about finding the function which is given as anti-derivative and a point.** **The students should find the correspondence function then determine the function whose graph passing through the given point**
2		**The problem discussed in this phase is about Free Fall Movement.** **The students are given an initial condition of where the object starts to fall and how it would fall. By finding the anti-derivate of the function of free fall movement, the students would find the initial function.**
3		**Using previously activity, the students will try to:** **Find the anti-derivative function and Initial Position in Free Fall Movement** **Determine the velocity in a given point, And so on**
4	**Derivative of linear and quadratic function**	**The problem discussed is determining graph and its given functions (derivatively related).** **The students then are asked to compare the area under the given linear function and its anti-derivate function values in the interval values.** **The area under linear curve is equal to the difference between the anti-derivative function value in the edges of the interval (state1)**
Definite Integral		
5	**function and its anti-derivative**	**The problem discussed in this phase is still about the area under a graph of the linear function and how to extend this idea to more general function.** **The students will be asked to discussed what would happen if the function is not linear, whether the relation still holds**
6	**method of exhaustion and the method of indivisible for finding the**	**The students would be introduced the method of exhaustion and the method of indivisible.** **The demonstration of the method of exhaustion would be presented by the teacher.** **The students, then, try to experience the method of indivisible in determining the area of the given ellipse**

	area	**The idea of infinity and small partition are introduced**
7	**Riemann Sum Partition method**	**The problem discussed is about determining the area of the given function graph.** **The students are asked to make equal partition using rectangular grid and determine the approximation area from each partition** **Various solution from different students should be discussed on how small we can use in the partition**
8		**The problem discussed is various type of graphs and the area under the graph.** **The students would be asked to apply limit in approximating the value of area** **In the end, the relation (*The area under linear curve is equal with the difference between the anti-derivative function value in the edges of the interval*) is carried out as the concept of fundamental Theorem of calculus**

The historical aspects of mathematics used in the learning activities, which become the big ideas, are the Method of Exhaustion, the Method of Indivisible, and Riemann Sum. The activities illustrated in Figure 1 and 2 are aimed to reinvent the idea of integral as using the approach of infinitesimal concept. Following these activities, a reinvention activity on Riemann Sum approach. In designing meaningful learning activities, the use of historical perspective could help the students to focus more on the underlying concept rather than procedural manipulation symbols. Table 2 summarizes the design of the learning activities on the topic of integral. The development of the big ideas was necessary for the reinvention of the Fundamental Theorem of Calculus.

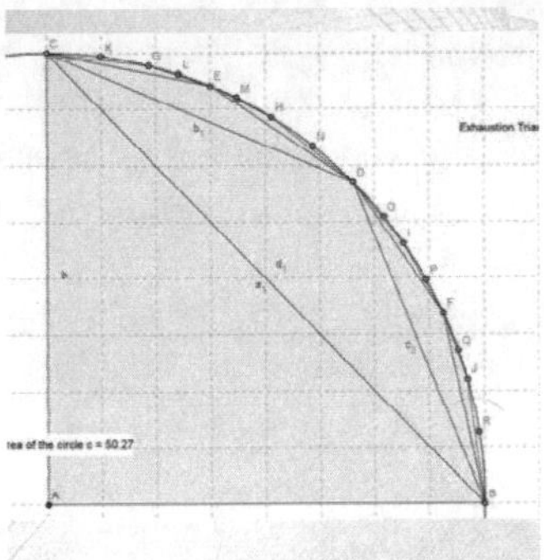

Figure 1. Activities on the Method of Exhaustion

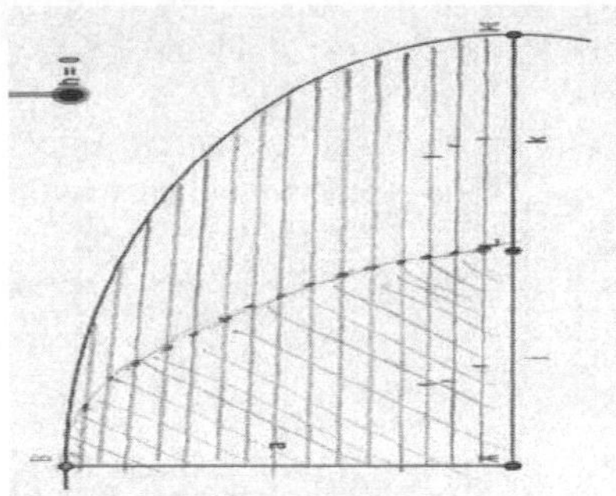

Figure 2. Activities on the Method of Indivisible

The Riemann Sum activity invited the students to discover the idea for finding the area under the given curve using partitions of equal width by applying the method of indivisible into the area (see Figure 3).

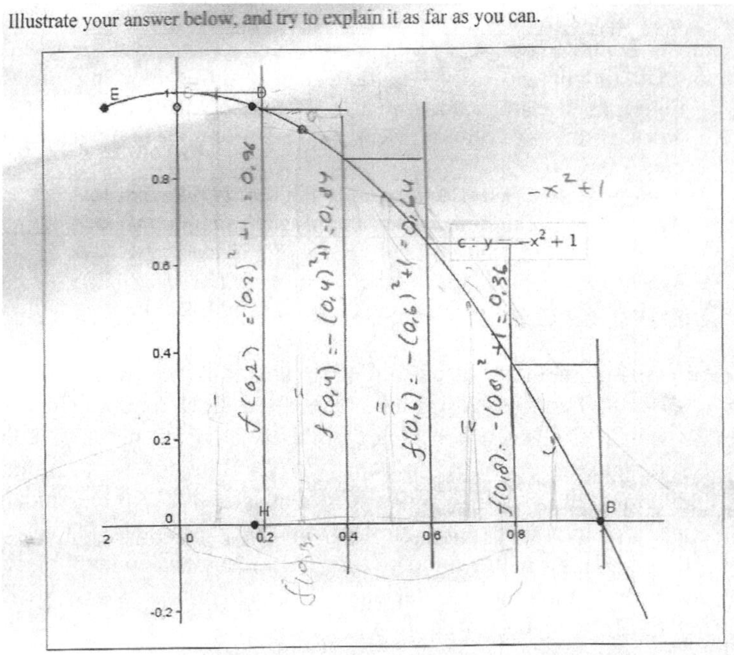

Figure 3. Using partitions of equal width

Moreover, making the partition width to be as small as possible (approaching 0 as applying the method of exhaustion) lead the students to the development of the definite integral concept resulted from Riemann Sum (see Figure 4)

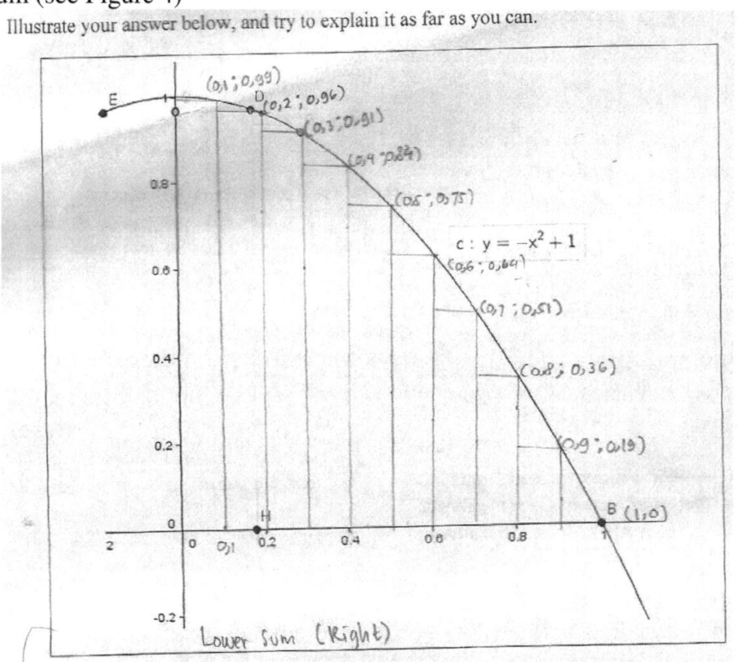

Figure 4. Taking partition width smaller

In short, from the discussed results, big ideas on the topics of indefinite integral and definite integral in the learning activities were adjusted accordingly. The context used in the lesson is made clearer to avoid students' misunderstanding and difficulties in using the sign of object movement orientation. The big idea of relation area under the curve and the introduction to Riemann Sum were improved by the use of Cavalieri's Principle. This improvement was required to bridge the gap where the students failed to understand the role of the method of indivisible in the reinvention of Riemann Sum. The students used the Cavalieri's Principle as an assisting material that leads the students in understanding the idea of the infinitesimal.

4. Conclusions

The learning activities in the integral course emerged from its relation to the derivative concept. In addition, the developed learning activities are parallel with the finding of Dubinsky and Schwingendorf [15]. The implementation of graph-based learning activities helped the students in developing their understanding of the concept of integral calculus [15]. This big idea involved the development of indefinite integral and its relation to function. Given the derived function as the known fact here that must be used for finding the function itself. One critical big idea was that the stage when the students must determine the connection between the area under a certain curve and the derivative of the curve. The conceptual learning happened when the students extended that big idea to a non-general function. Finally, the formulation of the Fundamental Theorem of Calculus involved the use of two big ideas of the method of exhaustion and the method of indivisible. In addition, the use of historical perspective could motivate the students during the course and could give an insight on how the mathematics concept emerged and used. The historical perspective could not be used in the classroom as it is. An adequate modification is needed based on the learning objectives. This view is in line with Savizi claims [16]. The correct interpretation is required to present the former historical problems into classroom materials and activities.

References

[1] Tall D 1992 Current difficulties in the teaching of mathematical analysis at university *Zdm* **92** 37–42

[2] Collignon P 2015 Using Calculus In Economy – Learning From History In Teacher Education *9th Congress of European Research in Mathematics Education*

[3] Gravemeijer K and Doorman M 1999 Context problems in realistic mathematics education: A calculus course as an example *Educ. Stud. Math.* **39** 111–29

[4] Doorman M and van Maanen J 2008 A historical perspective on teaching and learning calculus *Aust. Sr. Math. Teach.* **22** 4–14

[5] Tall D and Rashidi Razali M 1993 Diagnosing students' difficulties in learning mathematics *Int. J. Math. Educ. Sci. Technol.* **24** 209–222

[6] Tall D 1997 Functions and Calculus 289–325

[7] Fried M N 2001 Can Mathematics Education and History of Mathematics Coexist ? *Sci. Educ.* **10** 391–408

[8] Jankvist U T 2009 A categorization of the "whys" and "hows" of using history in mathematics education *Educ. Stud. Math.* **71** 235–61

[9] Katz V J 1993 Using the history of calculus to teach calculus *Sci. Educ.* **2** 243–9

[10] Tzanakis C, Arcavi A, de Sa C C, Isoda M, Lit C, Niss M, de Carvalho J P, Rodriguez M and Siu M-K 2000 Integrating History of Mathematics in the Classroom: An Analytic Survey *Hist. Math. Educ. ICMI Study* 201–40

[11] Radford L 1997 On psychology, historical epistemology, and the teaching of mathematics: Towards a socio-cultural history of mathematics *Learn. Math.* **17** 26–33

[12] Gravemeijer K 2004 Creating opportunities for students to reinvent mathematics *Proceedings of the 10th Conference of the International Congress of Mathematics Education* pp 1–17

[13] Bakker A and Van Eerde D 2015 An Introduction to Design-Based Research with an Example from Statistics Education *Approaches to Qualitative Research in Mathematics Education* ed A Bikner-ahsbahs (Bremen: Springer Berlin Heidelberg) pp 429–66

[14] Gravemeijer K and Cobb P 2006 Design research from a learning design perspective *Educational Design Research* pp 18–51

[15] Dubinsky E and Schwingendorf K 1997 Constructing calculus concepts: Cooperation in a computer laboratory. *Readings in cooperative learning for undergraduate mathematics.* pp 251–72
[16] Savizi B 2007 Applicable problems in the history of mathematics: practical examples for the classroom *Teach. Math. its Appl.* **26** 45–50

The impact of research-based learning on student's academic performance and motivation

M S Sumbawati and Y Anistyasari

Department of Informatics, Universitas Negeri Surabaya, Indonesia

meinisondang@unesa.ac.id

Abstract. Undergraduate students often have difficulty in finding the updated research topic for their undergraduate thesis. This is due to students have less knowledge about research in a particular field because they are less familiar with scientific articles. This work, therefore, proposed an implementation of research-based learning and measured students' academic performance and motivation. The study is conducted in four stages which are the plan, do, check, and act. Research-based learning is conducted based on its syntax. Paper and pencil test is then performed to measure students' academic achievement and a survey is devoted to determining students' academic motivation. The results show that most students obtain satisfied score and they feel comfortable to join a class with research-based learning.

1. Introduction

Inquiry-based learning (IBL) can be generally defined as a range of learning methods that are determined by a process of inquiry [1]. IBL generates other Problem-based learning, project work, field-work, and case studies can be classified as IBL. The characteristics of IBL includes learning particular disciplinary research methodologies [2]. On the other hand, undergraduate research is a part of the curriculum which also comprises learning research methodologies and producing research that imitates the forms of knowledge invention and dissemination in a specific field. Integrating research-based learning in undergraduate's curriculum is, therefore, becoming increasingly embraced. In this work, this integration is further defined as research-based learning (RBL).

The benefits of RBL for undergraduate students are increasing (1) student interest in knowledge, (2) confidence, (3) learning spirit, (4) learning achievement, (5) management, communication, and organizational skills, (6) motivation and enthusiasm, (7) interest in continuing studies to a higher level, (8) developing leadership skills, and (9) English proficiency. Students who actively participate in the learning process in encouraging research ideas which, if acted upon, will result in publication [3]. In contrary, some students are less motivated to participate in RBL.

In addition, undergraduate students have difficulty in finding the updated research topic for their undergraduate thesis. Therefore, students come to faculty adviser thesis without preparation research ideas. Some students submit thesis titles around topics that have been done by senior students. This is due to students have less knowledge about research in a particular field because they are less familiar with scientific articles.

Students are less interested in reading articles in English due to limited language skills. This can be performed by providing "coercion" to students in the form of the task of reading journals / English articles from accredited Journal. Then, students are asked to report the results of reading activities of the journal/article in the form of oral and written.

This study is devoted to undergraduate students of Information Technology Education in Learning Strategies subject. Students are divided into small groups of two persons, given a journal to read and report. When some students made presentations, others listen carefully and ask some questions for presenters.

2. Related Works

The aim of research-based learning (RBL) is that lecturers help students to build strong intelligent and applied connections between research boundaries and the students' own learning [4]. The benefits of RBL are:

- Introducing students to the disciplines' values, practices and ethics
- Confirming course content comprises advanced research findings
- Increasing students' understanding of how their field contributes positively to society
- Developing and enhancing students' skills and capabilities such as critical and analytical thinking, information retrieval and evaluation; and problem-solving
- Offering better opportunities for teaching and learning approaches such as inquiry-based and experiential methods that have been associated with constructive learning outcomes for students.

RBL is a theory referring to a variety of learning and teaching strategies that link research and teaching [4]. The best- practice in research-based learning includes:

- Research outcomes informing the curriculum
- Research-process based methods of teaching and learning
- Acquiring to use the tools of research
- Exploiting a complete research context

In RBL, research is regarded as an idea which supports teaching at various levels. In addition to integrating outcomes of research into the curriculum, it includes evolving students' knowledge of processes and methods of inquiry and making a comprehensive research.

3. Methodology

This work is quantitative research and is performed in four stages which are the plan, do, check, and act as depicted in Figure 1. The details of each step are described as follows.

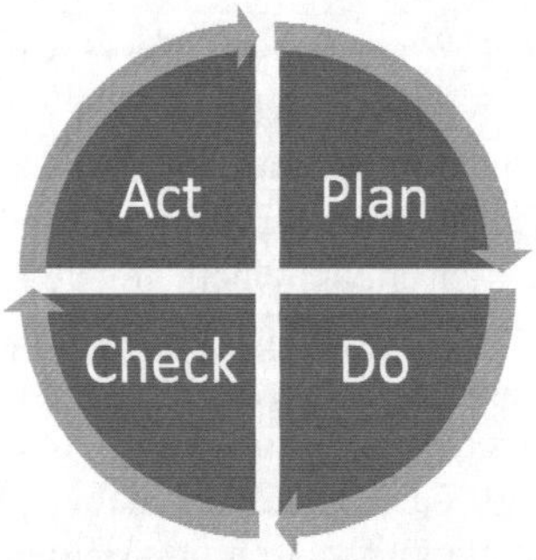

Figure 1. Research steps

3.1. Plan

In detail, the preparation of the study includes the actions undertaken to prepare the conducting research. The types of activities undertaken are as follows [5].

- Study literature on research-based learning, measurement of student response, and learning strategy learning materials.
- Initial data collection on student characteristics, then formulates indicators to measure the success of learning in accordance with the basic competencies that have been set.
- Preparing learning tools which consist of scenarios or lesson plans, student worksheets with the contextual approach, learning media, and test.
- Preparing and developing research instruments, which consist of observation sheet of students' activity during lecturing and questionnaire of student responses to the implementation of lectures.
- Validation and revision of instruments. Validation is performed by a competent learning expert. Based on expert input, the research team subsequently revised the instrument.

3.2. Do

This stage is a description of the action that will be conducted by researchers as an effort to apply research-based learning in the course of Learning Strategy. Research is conducted by applying the syntax of research-based learning.

- Plan learning activities which are combining the trending research topics into curriculum and providing examples to students about problems, solutions, and research results based on the teaching experience
- Choose the latest research publications which are conditioning the research topic with the theories being discussed in the lecture and providing an example of the dynamics of research on how the problem is done before and after research
- Conduct learning activities related to the research topic which comprises investigation and explanation about problem on which the research is conducted, analysing methods, arguments, and results presented in journal articles, giving conclusions in accordance with journal articles read, providing advice according to the results of the analysis and evaluation of students' understanding [6-7].

3.3. Check

In the check stage, we evaluate each process of aforementioned research-based learning. The results are compared against the expected targets or goals from the planning stage.

3.4. Act

This stage is also defined as an adjusted stage. It describes what improvements should be performed to the plan for the next cycle based on check stage results.

4. Results and Discussion

4.1. Student's Academic Performance

After attending lectures with research-based learning, students' knowledge is measured by written tests at the last meeting. Students are given about 25 items and asked to answer on answer sheet which has been provided within 50 minutes. Students' answers are corrected and the results obtained in Figure 2. 13% students gain score 51-60, 11% of them gain score 61-70, and 39% of them get score 71-80. The rest of them get 81-90 which is 21% of them and the remaining students gain 91-100. This implies that research based learning can be implemented to help students receive high academic performance.

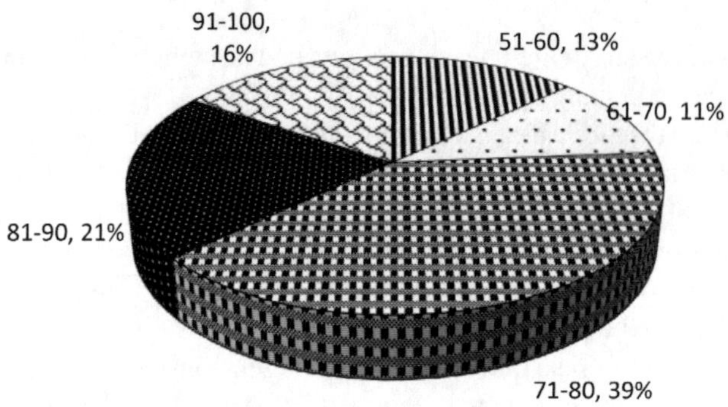

Figure 2. Student's academic performance

4.2. Student's Academic Motivation

Furthermore, we conduct a survey to measure student's academic motivation during the implementation of research-based learning. The results of students' academic motivation are described as follows.

4.2.1 Is the atmosphere of teaching and learning activities conducive to learning?

The student's responses are 50% stated occasionally, 27% stated always, and 23% stated frequently.

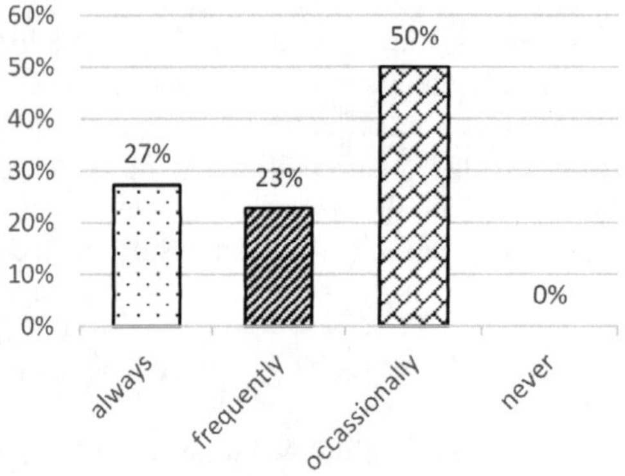

Figure 3. Response to question #1

4.2.2 Do you ask when you do not understand the material that the lecturer has taught?

The students' responses are 85% stated sometimes, 14% stated frequently, and 4% stated always.

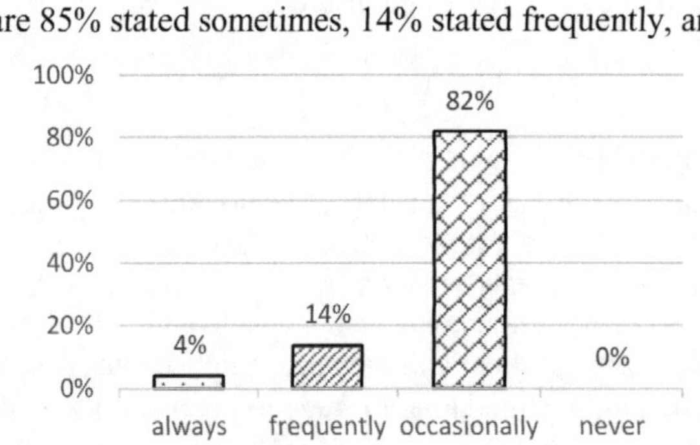

Figure 4. Response to question #2

4.2.3 Are you proactive in following the teaching and learning activities?

The students' responses are 68% stated occasionally, 27% stated frequently, and 5% stated always.

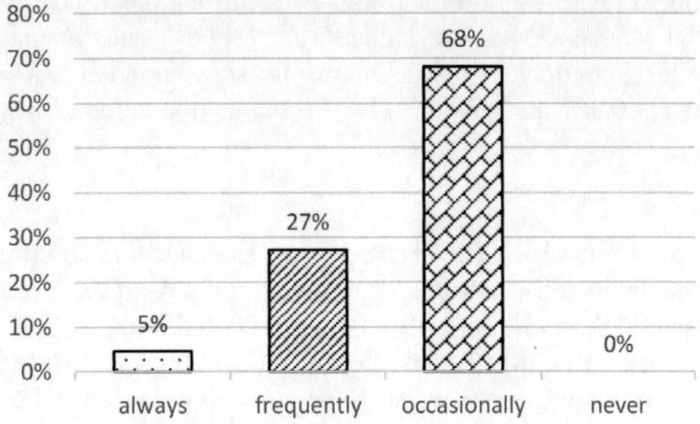

Figure 5. Response to question #3

4.2.4 Do you always do what the lecturer asks?

The students' responses are 41% stated sometimes, 32% stated frequently, 4% stated always, and 23% stated never.

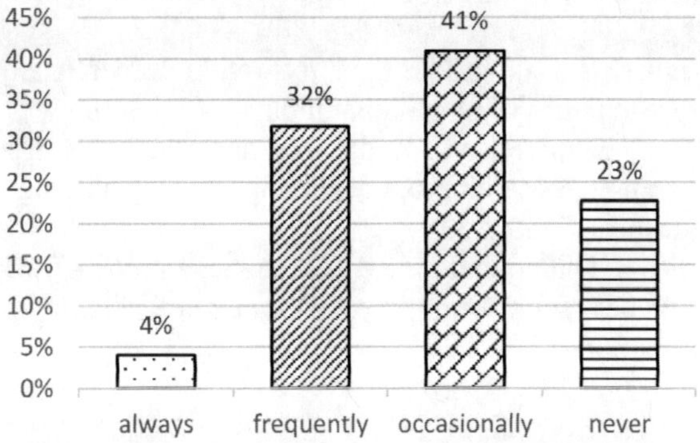

Figure 6. Response to question #4

4.2.5 Are you comfortable with learning through international journals?

The students' responses are 45% stated comfortable, 45% said less comfortable, 5% stated very comfortably, and 5% said uncomfortably.

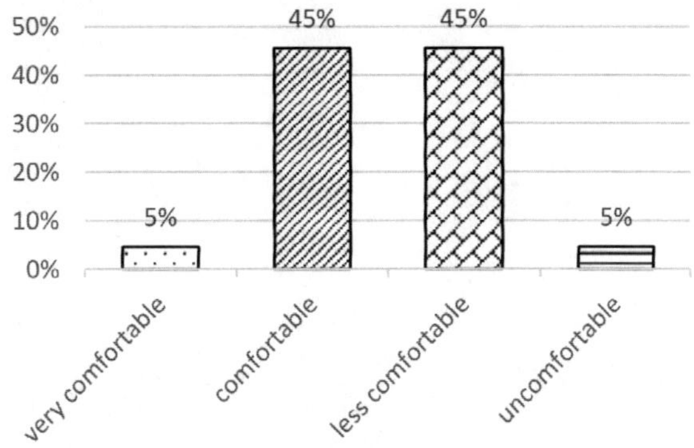

Figure 7. Response to question #5

IOP Conf. Series: Materials Science and Engineering **296** (2018) 012043 doi:10.1088/1757-899X/296/1/012043

5. Conclusion

From the results of data processing concluded that the research-based learning has been implemented in accordance with the syntax of research-based learning. Students are mostly (68%) proactive students in the learning process in the classroom. 85% of students actively ask questions and 41% always do the task. From the results of tests conducted to students to measure knowledge about Learning Strategy, obtained 39% of students get the score between 71-80, 21% get the value of 81-90, 16% get the value is 91-100, and the rest gets a value between 51-70.

References

[1] Calvo, L., & Prieto, C. (2016). The Teaching of Enhanced Distillation Processes Using a Commercial Simulator and A Project-Based Learning Approach. Education for Chemical Engineers, 17, 65–74. http://doi.org/10.1016/j.ece.2016.07.004

[2] Francese, R., Gravino, C., Risi, M., & Scanniello, G. (2015). Project-Based Learning in a Mobile Application Development Course — An Experience Report. Journal of Visual Language and Computing, 31, 196–205.

[3] Han, S., Capraro, R. M., & Capraro, M. M. (2016). How Science, Technology, Engineering, and Mathematics Project Based Learning Affects High-Need Students in The U. S. Learning and Individual Differences, 51, 157–166.

[4] Ionescu, C., Keyser, R. De, Keyser, R. De, & Ionescu, C. (2016). Interdisciplinary Project-Based Learning at Master Level: Control of Robotic Mechatronic Systems. IFAC-PapersOnLine, 49(6), 314–319.

[5] Kemmis, S., & McTaggart, R. (2007). Participatory Action Research: Communicative Action and The Public Sphere. Strategies of Qualitative Inquiry, 271–330.

[6] Lavero, A., Cura, J. M., Ezquerro, M., Rodrı, J., Lapuerta, V., & Cordero-gracia, M. (2015). Project Based Learning Experiences in The Space Engineering Education at Technical University of Madrid, 56, 1319–1330.

[7] Ríos, I. D. L., Cazorla, A., Díaz-puente, J. M., & Yagüe, J. L. (2010). Project – Based Learning in Engineering Higher Education: Two Decades of Teaching Competences in Real Environments, 2(2), 1368–1378.

IOP Conf. Series: Materials Science and Engineering **296** (2018) 012044 doi:10.1088/1757-899X/296/1/012044

Development of biology student worksheets to facilitate science process skills of student

Y S Rahayu, R Pratiwi and S Indana

Biology Education Program Universitas Negeri Surabaya, Indonesia
Jalan Ketintang, Surabaya 60231

yunirahayu@unesa.ac.id

Abstract. This research aims to describe development of Biology student worksheets to facilitate science process skills of student, at the same time to facilitate thinking skills of students in senior high school are equipped with Assesment Sheets. The worksheets development refers to cycle which includes phase analysis (analysis), planning (planning), design (design), development (development), implementation (implementation), evaluation and revision (evaluation and revision). Phase evaluation and revision is an ongoing activity conducted in each phase of the development cycle. That is, after the evaluation of the results of these activities and make revisions at any phase, then continue to the next phase. Based on the test results for grade X, XI, and XII in St. Agnes Surabaya high school, obtained some important findings. The findings are as follows. (1) Developed biology student worksheets could be used to facilitate thinking ability of students in particular skills integrated process that includes components to formulate the problem, formulate hypotheses, determine the study variables, formulate an operational definition of variables, determine the steps in the research, planning data tables, organizing Data in the form of tables/charts, drawing conclusions, (2) Developed biology student worksheets could also facilitate the development of social interaction of students such as working together, listening/respect the opinions of others, assembling equipment and materials, discuss and share information and facilitate the upgrading of skills hands-on student activity. (3) Developed biology worksheets basically could be implemented with the guidance of the teacher step by step, especially for students who have never used a similar worksheet. Guidance at the beginning of this need, especially for worksheets that require special skills or understanding of specific concepts as a prerequisite, such as using a microscope, determine the heart rate, understand the mechanism of specific indicators.

1. Introduction

According to Lai [1], the necessary skills to face the challenges of the 21st century is the critical thinking skills to be able to analyze the problem objectively. Duran and Dökme [2] state that critical thinking skills students can be trained by inquiry-based learning. Inquiry learning in science related to the effort to understand the various natural phenomena systematically, which includes attitudes, processes, products, and applications of science. Associated with the process, the science regarding the troubleshooting procedures using the scientific method to discover the concept of science. Even at Indonesia National Standard confirms that learning process organized to motivate students to participate actively in the learning process, and provide enough space for the creativity of the students [3].

Test TIMSS (Third International Mathematics and Science Study) indicates that the test measures five performance: (1) Understanding SimpleInformation,(2) Understanding complex information,(3) theorizing, Analyzing, and solving problems,(4) Using Tools, routine Procedures, and science processes,(5) Investigating the Natural World. Performance (2), (3), (4) and (5) it seems in line with the capabilities demanded in standard of competency above. It can be concluded that Indonesian students rank very low, it also means lower rankings in thinking ability so that it can be said also the competitiveness of Indonesian students in thinking the ability is also low.

This is supported by previous research results with a mapping exercise in one of senior high school in Surabaya Indonesia, showed that students' knowledge of process skills across school levels measured by the Science Process Skills Test relatively homogenous results are low. Based on these data can be interpreted that most likely from elementary, junior high and high school, students of this school has not yet obtained the maximum opportunity to learn science process skills. Or it can be said that science lessons in this school is not taught in their entirety. The students learn science as a product, not learn science as a process. This means that in this school, the science has not taught as products and processes. Master himself was also not mastered how to formulate an operational definition. Data skill experiment planning and formulation of low operational definition is a strong hint that during this experiment planning science in the classroom or in the laboratory carried out in a way that is not true [4].

Meanwhile, the results of the Test of Logical Thinking of students and teachers in this school showed that the majority of primary school students (66%) level of reasoning is concrete, while the junior high school students who think tangibly less (39%), high school students even fewer who think concrete (25 %). Based on these facts can be interpreted to mean that the student, either elementary, junior high, or high school is not maximized in the chance to engage in logical-mathematical experience, or social experience, or do a hands-on activity. Although the results showed that teachers in this school has the potential to teach their students to reach the level of formal thinking. Considering only the teacher who has the ability formal logic which can plan and carry out the teaching and learning process based on formal logic. In addition to the teacher, this school has a very good learning atmosphere that is supported by infrastructure and good management at all levels, elementary, middle, and high school [4].

Based frameworks already explained, the purpose of this research is to develop tools of learning science, especially biology high school that can provide ease teachers to teach and students learn science and thinking skills that, in this particular case is that it teaches process skills in the subjects of biology in high school through student worksheets and student worksheets key, Assessment Sheet and Assessment Sheet keys that have been developed. The three principal components of the main aspects in the development of student worksheets in this study are aspects of mathematical logic, social interaction aspects, and aspects of the hands-on activity.Aspects of mathematical logic which may reflect the ability to think conceptually recall is defined as the ability to think logically, critically and creatively, the ability to analyze and solve problems that are components that exist in science process skills.

Students worksheet and assesment sheets were developed by adopting and adapting biological learning device used in developed countries. Expected Students worksheet and assesment sheets has a quality equivalent to the quality and Students worksheet and assesment sheets used in developed countries. Standard grain assessment of the assesment sheets and developed worksheets that refer to future education and conform with global standards. The strategy adopted for it is adapting or adopting a learning device owned by developed countries and their grains assessment contained in the reference cutting-edge science learning device developed countries. This strategy is pursued in an effort to fulfill the main requirement of reference benchmark grains, namely content validity. It's believed that student worksheet and grains assessment contained in the latest reference has been developed with standard procedures by the relevant experts.

IOP Conf. Series: Materials Science and Engineering **296** (2018) 012044 doi:10.1088/1757-899X/296/1/012044

2. Method

According to Fenrich [5] measures the development of these devices can be visualized as in Figure 1. The development cycle includes a phase analysis (analysis), planning (planning), design (design), development (development), implementation (implementation), evaluation and revision (evaluation and revision). Phase evaluation and revision is an ongoing activity conducted in each phase of the development cycle. That is, after the evaluation of the results of these activities and make revisions at any phase, then continue to the next phase.

In the phase analysis to identify the components of any process skills that must be mastered high school students, according to SK and KD content standards that exist in biology to be a learning tool developed is expected to be used in high school. While in the phased planning by planning format Student Worksheet and Assessment Sheet developed, as well as the necessary KIT designed to complete the worksheet. Meanwhile, in the phase design do drafting 1 Student Worksheet and assesment sheets are equipped with an instrument and key of Student Worksheet and Assessment Sheet. At this stage, it is also prepared KIT.

In the phase of development, the study conducted a draft 1 student worksheet, assesment sheet by expert reviewers and review of the operation of the equipment and materials KIT. Expert reviewers' suggestions and input for consideration to revise worksheets, and KIT assesment sheet to a draft 2 were prepared to investigate. At the time of testing in other senior high schools conducted a formative evaluation. Based on input from teachers and students, to revise the draft 2 student worksheet, and KIT assesment sheet to a draft 3 which is also the master worksheets, print-ready assesment sheet and KIT are ready for use.

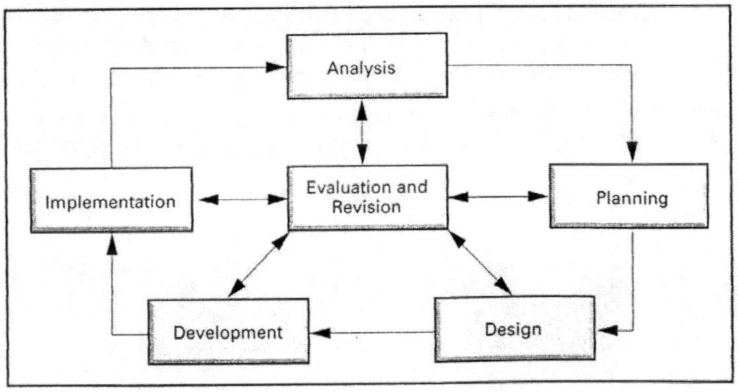

Figure 1. *Cycle Model Development* [6].

3. Results and Discussion

In this research, successfully developed 10 key of student worksheet and worksheets included, with 8 student worksheet tested. Accompanying for student worksheet, evaluation sheet was also developed for each of the keys along with the student worksheet and student sheet. On the development of student worksheet is also equipped with the KIT is a tool, materials, and infrastructure needed by students and teachers when using worksheets and assestment sheet are developed. The detailed results of the development of student workshett as Table 1 include details of process skills component drilled on each worksheet.

Table 1 shows that the mapping component of process skills in general at 10 student worksheet developed, highly variable. This is in accordance with the characteristics of the materials developed by KI and KD. But in general, the process developed skills component is a component of integrated process skills that better facilitate students to perform the steps of the scientific method that is substituted with a concept tailored to KI and KD. This is expected to further provide a learning science (especially biology) intact from the three aspects of mathematical logic, the social aspect and the aspect of the hands-on activity. Considering the aspects of mathematical logic to reflect the ability to think to remember is conceptually defined as the ability to think logically, critically and creatively, the ability to analyze and solve problems, then the demands of the standard competency of Junior High School / High School will be met through learning to use worksheets that were developed which facilitates students to process skills science.

Table 2 shows a summary of the results of tests on eight point in student worksheet being investigated. Three aspects that were recorded during the tests which aspects of mathematical logic, social interaction aspects, and aspects of the hands-on activity. Generally indicates that to aspects of mathematical logic, at first, the students find it difficult when student worksheets was released without the guidance of a teacher. But when students are given the guidance of the teacher step by step in doing worksheets that require a coherent mind-set of students, students can complete worksheets well. The teacher's role is as a facilitator to provide a learning environment that allows students to be actively involved in learning as well as a guide to steer students on the tasks that will be done [6]. When done mentoring students working on worksheets given but still found some difficulties of students in terms of operationally defining variables of the study, planning procedures, interpret the data and make the data tables/graphs are correct by using the notation size of recognized standards.

Social interactions showed quite good results considering the worksheets developed gives a chance to work in groups so that social interaction is to talk, ask questions, listen to the opinion of friends, give opinions, respect the opinion of friends, mutual trust, mutual cooperation can arise for students working on worksheets. In addition, it can be observed the joy and enthusiasm of students for doing worksheets, meaning that students in this student worksheet working with did not feel forced. Therefore, the worksheet test results show relatively good results, especially when students are asked to do the assesment. Rahman et al. [7] state that learning involves social interaction as a method of discussion to make students more active in the learning process and the students are able to process information received from the information rather than just listening to the teacher. In addition, Kruea-In et al (2015) stated that teaching and learning of science process skills are not only served as basic for the scientific method but also valuable opportunities for learning about the nature of science [8]

IOP Conf. Series: Materials Science and Engineering **296** (2018) 012044 doi:10.1088/1757-899X/296/1/012044

Table 1. Components of Science Process Skills are trained in Biology Worksheets on High School

No	Science Process Skills component	Biology Worksheets on High School									
		01	02	03	04	05	06	07	08	09	10
1	Observations	√	√	√	√	√			√	√	√
2	Scientific Figure		√								
3	Problem Formulation						√		√		
4	Formulation of hypothesis	√		√	√	√	√	√	√	√	√
5	Controlling the variables	√	√		√	√	√	√	√	√	√
6	Planning Experiments	√		√	√	√	√	√	√	√	√
7	Defining variables Operations				√	√	√		√	√	√
8	Conducting Experiments	√	√	√	√	√	√	√	√	√	√
9	Measuring and Counting	√	√				√	√			√
10	collection and Data Recording	√	√	√	√	√	√	√	√	√	√
11	Analyzing Data	√	√	√	√	√	√	√	√	√	√
12	Penginferensian		√	√	√	√			√	√	√
13	Pemrediksian								√		
14	Making Tables or Graphs	√	√		√		√	√	√	√	
15	Modeling										√
16	FormulationConclusion	√	√	√	√	√	√	√	√	√	√
17	Group Work	√	√	√	√	√	√	√	√	√	√

Worksheet 01: Seed germination
Worksheet 02: plasmolysis cell
Worksheet 03: Photosynthesis Require Chlorophyll
Worksheet 04: Photosynthesis Require CO_2
Worksheet 05: Photosynthesis Need Light
Worksheet 06: Heart Rate
Worksheet 07: Growth
Worksheet 08: phototropism
Worksheet 09: ENZYME
Worksheet 10: Respiration

Table 2. Aspects of mathematical logic, social interaction aspects, and aspects of the *hands-on activity* on the tested of students worksheets

No	Worksheet Number	Aspects of		
		Mathematical logic	**Social**	**Hands-on activity**
1	Biology Student Worksheet 01	With the guidance and direction of the teacher of high school students can finish the task on student worksheet 01 very well, although initially they should be guided step-by-step. Despite the students' answers indicate several answers were less than perfect on the identification component variables, operationally variable formulate and determine the procedure	In doing student worksheet 01, students to interact socially as well. Because of this student worksheet facilitate social interaction such as cooperation, mutual help, and mutual respect, be a good listener, active and able to be responsible for helping the group complete the work on time.	At first, the students are less skilled in conducting. This is because students are not familiar with student worksheet given pattern. However, after being guided students can perform well. Student worksheet and KIT BIO 01 SMA gave students the opportunity to engage in *hand on activity* overall for all students involved in the design and execution of experiments.
2	Biology Student Worksheet 02	Worksheet 2 were designed based on skill development in an integrated manner indicates that with the guidance and direction of teachers of high school students can finish the task on worksheet, although at first they should be guided step-by-step, especially in the use of the microscope as a prerequisite in the student worksheet this. But from the students' answers indicate several answers were less than perfect.	In doing these student worksheet, students interact well socially because this student worksheet facilitate the social interactions such as cooperation, mutual help, and mutual respect.	Obstacles in the student worksheet are demanding prerequisite that the students were able to use a microscope. Therefore, at first the students are less skilled in using the microscope, but after being given the guidance of students become more skilled. Students are also still difficult to distinguish between normal cells and cells that plasmolyzed and making incisions great epidermis so that the cells can be seen clearly.
3	Biology Student Worksheet 03	LKS High School BIO 3 were designed based on skill development in an integrated manner indicates that with the guidance and direction of	In doing these student worksheet, students interact socially as well as worksheets facilitate social interaction such	Initially, students are yet to test their skillful starch. But with guidance, students more easily work on worksheets as the tools used in this student

IOP Conf. Series: Materials Science and Engineering **296** (2018) 012044 doi:10.1088/1757-899X/296/1/012044

No	Worksheet Number	Aspects of Mathematical logic	Aspects of Social	Aspects of Hands-on activity
		teachers of high school students can carry out student worksheet well, although at first they should be guided step-by-step, especially in a test for the presence of starch as a criterion in this student worksheet. This means that students are able to carry out the steps of the scientific method to prove that requires chlorophyll photosynthesis. Despite the students' answers indicate several answers were less than perfect them being the conclusion, applying the results of experiments in explaining everyday phenomena, and interpret data. But in general, the students' answers are still within reasonable limits.	as cooperation, mutual help, and mutual respect.	worksheet, students very familiar so they are quick to skillfully use the tools and materials are there to test their hypothesis.
4	Biology Student Worksheet 06	BIO In doing LKS High School 6 at first the students had difficulty when working on worksheets that demands skills determine/measure the heart rate by counting the pulse as a prerequisite in this LKS. Therefore we need the guidance and direction of the teacher. This means that it requires the guidance of teachers LKS gradually at first so that students are able to carry out the steps of the scientific method or scientific activities to prove the factors that affect heart rate. From the students' answers indicate several answers imperfect include the formulation of a hypothesis, identify variables, and write procedures.	In doing this student worksheet, student interact socially as well as worksheets are designed to facilitate social interactions such as cooperation, mutual help, and mutual respect.	Students during work on worksheets initially less skilled in using a stopwatch, and determine the location of the artery to calculate the heart rate, but after being given the guidance of the students become more adept at calculating the number of pulses in a specified unit of time.

No	Worksheet Number	Aspects of		
		Mathematical logic	**Social**	**Hands-on activity**
5	Biology Student Worksheet 07	Worksheet 7 is designed based on the development of skills in an integrated process. student worksheet test results showed that in general beginning students find difficulty when working on worksheets that require thinking skills sequentially. However, because the BLM does not require complex skills, student worksheet can be followed easily by the students when they finish given guidance. This means that it requires the guidance of teachers LKS gradually at first so that students are able to carry out the steps of the scientific method or scientific activities. Despite the students' answers indicate several answers imperfect include the formulation of a hypothesis, identify variables, and write procedures.	In doing this student worksheet, student interact socially as well as worksheets facilitate social interaction such as cooperation, mutual help, and mutual respect.	At first, the students are not yet skilled measure the height of the root and trunk sprouted but after the guidance, students more easily work on worksheets are there to test their hypothesis.
6	Biology Student Worksheet 08	Student worksheet trial results show that in general as well as worksheets-worksheets that other students initially find it difficult when working on worksheets that require thinking skills sequentially. These worksheets can be followed easily by the students when they finish given guidance. This means that it requires the guidance of teachers, student worksheet gradually at first so that students are able to carry out the steps of the scientific method or scientific activities with good conduct. This is evident from the analysis of the work carried out	In doing this 08 BIO LKS High School students to interact socially is good for this student worksheet facilitate social interaction such as cooperation, mutual help, and mutual respect. Student worksheet was designed to work in groups. Students are a lot of discussions when planning research activities and analyzing the study data.	In the beginning, students lack the courage to write ideas at BIO LKS 08. This is expected because the students are not accustomed to taking steps and write down their ideas on the worksheet provided. However, after being guided students can perform well. student worksheet and KIT BIO 08 SMA gave students the opportunity to engage in *hand on activity* overall for all students involved in the design and execution of experiments.

No	Worksheet Number	Aspects of Mathematical logic	Aspects of Social	Aspects of Hands-on activity
		worksheets that students demonstrate that all components can be answered correctly by students.		
7	Biology Student Worksheet 09	In common- student worksheet. Student worksheet others, initially, the students find it difficult when working on demanding LKS sequentially thinking skills, especially when students are required to connect between the way work the indicator with the results of their analysis. In addition, they are also required skilled use of laboratory equipment which, although simple, but requires a systematic mindset to understand the mechanism of action of the enzyme. However, from the work of student worksheets indicate the general ability of the students has been good although still have difficulties in terms of identifying variables and write the procedures.	In doing this student worksheet, student interact socially as well as worksheets facilitate social interaction such as cooperation, mutual help, and mutual respect. Students work in collecting saliva in an amount sufficient to obtain amylase enzyme used in the experiments.	At the start of work on student worksheets rather difficult to collect saliva, students are less skilled in using a pipette to bring the solution. But after the training and guidance of their skills increase.
8	Biology Student Worksheet 10	worksheet 10 developed based on the development of skills in this integrated process, demanding the student's skills to skillfully assemble simple respirometer tool modification results. If that is not owned by the students as a prerequisite, the students will have difficulty in doing this worksheet. Worksheet test results showed that in general beginning students feel difficulties. But after being given a step by step guidance they were able to accomplish	In doing this student worksheet, student interact socially as well as worksheets facilitate social interaction such as cooperation, mutual help, and mutual respect. In doing this LKS student very often discuss the procedures that must be compiled.	At first, the students are less skilled in assembling the tools and experimental materials in particular about putting cotton wool soaked in NaOH. But after getting guidance ultimately more skilled students.

No	Worksheet Number	Aspects of		
		Mathematical logic	**Social**	**Hands-on activity**
		this student worksheet well to prove that the process of respiration requires oxygen. However, from the work of student worksheets still shows that students are still difficulties in terms of identifying variables and write the procedures.		

For the aspects of the hands-on activity, student worksheet that was developed generally develop these skills. Therefore, students will experience obstacles in doing worksheet when basic skills in using certain tools as a prerequisite cannot be satisfied students. This is evident when students do worksheet 02, which requires students to use a microscope, as well as for worksheet 03-skill test starch, worksheet 06 demanding to determine pulse to calculate heart rate, worksheet 09-skill test starch with amylase of saliva and use the pipette correctly, and worksheet 10 which requires students to compose / assemble models simple respirometer. However, when teachers provide guidance gradually students can have a hands-on activity skills that can be improved so that the work can be done well. According to Ekwueme et al. [9], hands-on an approach where students are guided to learn from experience. The use of hands-on learning to be able to effectively teach a concept which is supported by other findings [10 - 12].

4. Conclusions and recommendations

Based on the test results of Biology Student Worksheet in High School for grade 10th, 11th, and 12th obtained some important findings which are also conclusion in this study. The findings are as follows.

4.1 Biology worksheet that have been developed in this study can be used to train thinking ability of students in particular skills integrated process that includes components to formulate the problem, formulate hypotheses, determine the study variables, formulate operational definitions of variables, determine the steps in the research, planning data table, organize Data in the form of a table / graph, drawing conclusions.

4.2 Biology worksheet developed also facilitate the development of social interaction of students such as working together, listening/respect the opinions of others, assembling equipment and materials, to discuss work procedures and share information and facilitate the improvement of skills of hands-on students' activity.

4.3 Biology worksheet developed basically be carried out with the guidance of teachers' notes with step by step, especially for students who have never had similar worksheets. Guidance at the beginning of this need, especially for worksheets that require special skills or understanding of specific concepts as a prerequisite, such as using a microscope, determine / measure the heart rate, understand how / mechanism specific indicators.

4.4 The use of integrated process skills worksheets substituted giving the meaning to study biology as a whole.

Based on the findings above anything that could be recommended by the researchers that the teaching skills of the process should not be separated from the subject of biology but should be based on activity. In order for the student activities directional, it is necessary to develop skill-oriented process with accordingly. In order for students, logical thinking skills honed then student worksheet presented must be loaded with science process skills through the guidance gradually when execution

References

[1] Lai E R 2011 *Critical Thinking: A Literature Review* (Research Report: Pearson)

[2] Duran M and Dökme I 2016 The effect of the inquiry-based learning approach on student's critical-thinking skills *Eurasia Journal of Mathematics, Science & Technology Education* **12 (12)** pp 2887—908

[3] Regulation of the Minister of Education and Culture of the Republic of Indonesia Number 22 The year 2016 About Standard Process Unit of Primary and Secondary Education.

[4] Nur M Abadi, Siswono T E Y, and Rahayu Y S 2008 *Measurement Capabilities Activity Report Logic Student and Teacher YLPI Al-Hikmah*

[5] Fenrich P 1997 *Practical Guidelines for Creating Instructional Multimedia Applications* (Fort Worth: The Dryden Press Harcourt Brace College Publishers)

[6] Abdi A 2014 The Effect of Inquiry-based Learning Method on Students' Academic Achievement in Science Course *Universal Journal of Educational Research* **2 (1)** pp 37—41

[7] Rahman F, Khalil J K, Jumani N B, Ajmal M, Malik S, and Sharif M 2011 Impact of Discussion Method on Students Performance *International Journal of Business and Social Science* **2 (7)** pp 84—94

[8] Kruea-In C, Kruea-In N, Facharoenphol W 2015 A Study of Thai In-Service and Pre-Service Science Teachers' Understanding of Science Process Skills *Procedia - Social and Behavioral Sciences* **197** pp 993—7

[9] Ekwueme C O, Ekon E E, and Ezenwa D C 2015 The Impact of Hands-On-Approach on Student Academic Performance in Basic Science and Mathematics *Higher education studies* **5 (6)** pp 47—51

[10] Kruea-In N and Thongperm O 2014 Teaching of Science Process Skills in Thai Contexts: Status, Supports and Obstacles *Procedia - Social and Behavioral Sciences* **141** pp 1324—9

[11] Jeenthong T et al. 2014 Promoting integrated science process skills through betta-live science laboratory *Procedia - Social and Behavioral Sciences* **116 (2014)** pp 3292—6

[12] Hodosyováa M et al 2015 The Development of Science Process Skills in Physics Education *Procedia - Social and Behavioral Sciences* **186 (2015)** pp 982—9

IOP Conf. Series: Materials Science and Engineering **296** (2018) 012045 doi:10.1088/1757-899X/296/1/012045

Literacy Infrastructure, Access to Books, and the Implementation of the School Literacy Movement in Primary Schools in Indonesia

K Laksono and P Retnaningdyah

Faculty of Languages and Arts, Universitas Negeri Surabaya

kisyani@unesa.ac.id

Abstract. Literacy Infrastructure and access to books are the foundation of literacy activity. Indonesia has regulations from the Ministry of Education and Culture requiring that 15 minutes should be used each day before the learning begins to read books other than textbooks. However, many schools are not yet obeying this requirement. The purposes of this study are to describe the literacy infrastructure in primary schools in Indonesia, to analyze access to books in primary schools, to explain the School Literacy Movement implementation, and to identify issues around the implementation of reading strategies in a context in which there is limited access to books. The questionnaire and interview study were conducted in 30 primary schools in East Java, Indonesia. The study concluded that the literacy infrastructure and access to books in 30 primary schools are below standard, but the school community enthusiastically implements the objectives of the School Literacy Movement. Many primary schools are already implementing good many reading strategies although there are some problems related to teacher competence.

1. Introduction

Indonesia is listed as one of the countries that are succeeding in reducing the illiteracy rate. United Nations Development Programme noted that the level of public literacy in Indonesia reached 92.8% for the adult group, and 98.8% in the category of teenagers [19]. This achievement shows that the literacy crisis is Indonesia has passed. Nevertheless, challenges remain, and one of these currently is a widespread lack of interest in reading. In addition, the availability of books throughout Indonesia has been inadequate; the government is also facing low motivation for reading among students, teachers and the wider community.

The results of the Progress International Reading Literacy Study or PIRLS, which evaluates the reading ability of students in grade IV, ranks Indonesia at the 45th place out of 48 participating countries with a score of 428, below the average value of 500 [17]. In the Programme for International Student Assessment (PISA), a survey which evaluates the ability of learners aged 15 years and includes reading, maths, and science, Indonesia which was originally in PISA 2009 ranked 57th with a score of 396, while the average score of Organisation for Economic Cooperation and Development or OECD is 493, showed a decline in rank in 2012, when it was ranked 64th with a score of 396 (the mean score average OECD is 496) [13]. This data is consistent with findings of UNESCO concerning the reading habits of the people of Indonesia, that only one in 1,000 people of Indonesia is reading for pleasure.

Reflecting on this fact, the Ministry of Education and Culture (MEC) as the stakeholders feel responsible and that they need to make changes strategically and systematically. Hence the Regulation of the Minister of Education and Culture (RMEC) No. 23 of 2015 on "Growth Character" published in mid-July 2015. In this rule, literacy became one of the principal concerns. This concern was articulated in the obligation "to use 15 minutes before the day of learning to read books other than textbooks (every day)."

Fifteen minutes reading activity in schools is not a new program in Indonesia. Some primary schools (PS) have taken the initiative to create reading corners, they require students to read literary books, and they publish books by students and teachers. The publication of RMEC No 23 was intended as a legal protection for good practices that have been implemented in these PS and at the same time to become a reference for emerging regulatory regimes at the local level. However, despite the fact that the RMEC was published in July 2015, still many schools have not implemented it. Therefore, to support the RMEC No 23 and to create a safe atmosphere for literacy, a task force was set up to develop a School Literacy Movement (SLM). This was established in January 2016 by the Directorate General of Primary and Secondary Education, Ministry of Education. SLM Task Force members come from academia, practitioners, literacy activists and non-governmental organization. The Task Force has four tasks: (1) socialization and coordination to introduce the movement, (2) implementation, (3) promotion of the SLM's purposes to the wider public, and (4) monitoring and evaluation.

Literacy infrastructure and access to books are very crucial to understanding state literacy policy in Indonesia. All states maintain a state literacy education office headed by an official designated by the state director. In Indonesia, the state director is Minister of Education and Culture. RMEC No 24 of 2007 identifies the fact that the number of books, as one element of the literacy infrastructure, is very crucial in the development of literacy. The number and types of books for all schools have been set in the Ministerial Regulation or RMEC No. 24 of 2007 on Infrastructure Standards, including enrichment books in primary schools.

There are 13,466 islands that make-up Indonesia, and one of the most heavily populated islands in Indonesia is Java. Java is relatively well equipped regarding infrastructure. Therefore, schools in Java (including primary schools in East Java) can be one barometer for judgments related to school infrastructure and the implementation of SLM strategies.

This study is important because RMEC No 24 of 2007 has been in effect for almost ten years and RMEC No 23 has been in effect for more than one year, but the result of PIRLS and PISA are still low. Therefore, based on the problems, the purposes of this study are to describe the literacy infrastructure in East Java primary schools in Indonesia; to analyze access to books in East Java primary schools and SLM implementation, and to identify and discuss how to implement SLM with limited infrastructure.

2. Method

This study was conducted in 30 primary schools (PS) in East Java (9 cities/districts): 14 PS in Surabaya; 3 PS in Sidoarjo; 3 PS in Lamongan; 2 PS in Gresik; 2 PS in Mojokerto; 2 PS in Bangkalan; 1 PS in Madiun; 1 PS in Trenggalek; 1 PS in Bojonegoro. Respondents are teachers. Some of them are now studying further while still working.

Data were collected in September 2016 by questionnaire (related to access to enrichment books and about SLM implementation). The survey was followed by an interview to investigate answers more thoroughly. The questionnaire asked about the number of students and enrichment books in PS; phases in the development of SLM (habituation, development, and learning); strategies used such as reading aloud and sustained silent reading. Regarding SLM implementation, respondents were asked to answer "already" "not yet", "yes" or "no" to indicate whether a particular strategy had been put into action. Data were calculated in percentages (the number of PS choose "already" and "yes"). Therefore, according to the Guttman scale [22], the results are presented in one of two categories: 0-49% is "less" and 50--100% is "good". Interviews conducted for exploration of questions about access to books strengthen the questionnaire data, especially if there are questionable data or empty responses. Furthermore, the interviews also explore the reasons for the answers in the questionnaire including a respondent's solution or an idea.

3. Findings and Discussion

3.1 Literacy Infrastructure and Access to Books

RMEC No. 24 of 2007 on Infrastructure Standards for School in Indonesia cited a variety of book that should be used in PS. One of the important types of book is an enrichment book. All PS (100%) should meet the standards, especially concerning the number of enrichment book titles. Regarding enrichment books, the standard number that should be available at each PS is 840 titles, composed of 60% nonfiction (504 titles) and 40% fiction (336 titles).

However, although the regulation has been in effect almost ten years, PS in East Java -- which include schools which have received much attention and many amenities compared to schools outside Java-- almost nothing meets the required standards. The number of students in PS studied ranged from 62--850 students. Based on the data, from 30 PS, there are only two PS that meet the total number of titles. One PS has 1000 titles of enrichment books consisting of 500 titles of fiction and 500 titles of nonfiction. Regarding the number of enrichment books, it meets the standards. However, regarding details, the number of nonfiction books does not reach the minimum standard (504 titles). A second PS has 845 titles of enrichment books consisting of 225 fiction and 620 nonfiction. Regarding the number of enrichment books, it meets the standards. However, regarding details, the number of fiction books does reach the minimum standard (336 titles). The third PS has 625 titles of enrichment books consisting of 375 fiction (meets the standard) and 250 nonfiction (does not meets). The respondents from these three schools said that their primary schools had been getting a block grant for books of fiction and nonfiction and a special allocation fund for the purchase of books since 2007. The other PS did not receive this grant. There are 7 PS that do not have enrichment books: one PS that only had one enrichment book, and another had only two enrichment books and 10 PS have no fiction at all.

The data shows that many PS do not meet the number of enrichment books. Based on the data, from 90 entries, only five (5%) meet the standards (check mark), and this indicates that many students have poor access to books. In addition, respondents said that there are two PS that have no library. Some of PS do not have a record of the number of books in their schools. The issues than, for PS, are funding, poverty amongst their students, limited numbers of books and no libraries. Therefore, many students have very poor access to books.

Many PS lacked enrichment books but turned out to have various ideas as follows:

(1.) To laminate some stories from the newspaper.
(2.) To empower students to bring books from home to read together and exchange books among students when they have finished reading (although unfortunately, many poor students have no books because of the terrible cost)
(3.) Every few weeks, exchange books between classes.
(4.) Transfer books or borrow from each other between PS.
(5.) Invite a government car library to visit the school.

Respondents from Surabaya's PS said that many schools in Surabaya have even been aided by the city library to enter book details online and have the books delivered to their schools. In this scheme, each school has one or two librarians assigned to it.

The rules on the standard of infrastructure were issued in 2007. Until now many schools that do not comply. Studies confirm that children of poverty have very little access to books at home and in their communities, with poor access to good public libraries and bookstores [12]. Children from impoverished households have access to fewer books and other reading materials than do their more financially stable peers. Not only do poor children have fewer books in their homes, but they also live in communities with fewer books in the classroom, school, and public library [16]. The study by Batool and Mahmood concludes that children will only become information literate with proper learning tools, resources, and flexible infrastructure in schools. In the U.S. schoolchildren, one of the lack of educational infrastructure is books [3].

If many PS in Java (East Java) do not meet the standards of facilities and infrastructure, how about the school in the other islands? There needs to be more attention from the government, community, business, and industry to increase access to books, especially enrichment books on the other islands. It appears that the funds from the government (Operational Aid to School Program), minimum 5%, which were allocated for the development of libraries, have not been used to procure books. This is the case even though studies confirm that increased access to materials such as books can reduce the effects of living in poverty for children (Krashen et al. 2010; McCoy, 2013 and increased access to books is related to increased reading achievement [7].

3.2 SLM Implementation

Beers et al suggest that for schools to be at the forefront of the development of cultural literacy there are several strategies needed to create a positive culture of literacy in schools [2]. These are conditioning the physical environment to make it friendly, promoting the social and affective environment, and promoting the academic environment. These strategies were adopted in the three phases in SLM. The implementation phases of the SLM are habituation, development, and learning [23].

3.2.1 Phase 1 of SLM: Habituation. RMEC No 23 of 2015 required that 15 minutes a day should be devoted to reading in all schools. SLM was formed in January 2016. In this case, the Ministry through SLM requires all schools (100%) to be carrying out the habituation phase, development, and learning.

From the ten items in this habituation phase, five items (50%) were classified as good, and five items (50%) were classified as less than good. In the habituation phase, it seems that 67% of PS have implemented the 15-minutes reading, including PS which used the time to read Al-Quran (8%). PS that have not implemented a 15-minute reading said they did not know about SLM, yet they understood the importance of literacy and its implementation in PS. There are several answers from these respondents which included:

(1) Teachers and principals do not know the RMEC No23 of 2015.
(2) There were no orders from their direct superiors (in this case the district/city education offices) which was one of the reasons why the respondents have not started SLM. It also means socializing SLM is not spread evenly.
(3) The PS does not have enough books.
(4) The PS has been allocated 15 minutes before class to read Al-Quran.
(5) Teachers need to gain insight into the reading which is suitable for students.
(6) The working time has been filled with a load of teaching and administration.

Ninety-three percent of schools have libraries, but not all libraries have a particular room, there is sometimes just a bookshelf in the corner of the school. 70% of PS have a reading corner in each class. Respondents said that books in the reading corners turned out to be a book that students themselves had brought to PS to read together. The school should provide the enrichment books, but according to many respondents, the books belong to the students because the teacher asked them to bring books to the school. The item "rich text materials are posted in each classroom, corridors, and other areas in the school" which reaches 87%, although in the form of simple rich text material. However, with the limitations, the spirit of the school community to carry out literacy is good: "Respondents and the principal have commitments" reaches 73%. The others were still waiting for direct orders from the immediate supervisor (the education office).

Less encouraging responses can be seen in the item "Teachers, principals, and educators to be a model of in fifteen minutes reading activities" which reaches 4% because-- according to respondents-- reading comprehension is for students and not the teacher who is in charge of overseeing the students. "Students have a daily reading journal" reaches 43%. PS that have not applied daily journal reading appears to be still waiting for instructions. According to respondents, the important thing is students have to read. Of the 67% of PS implementing the program of reading 15 minutes, only 40% apply it for one whole semester of school. Respondents that implement reading for 15 minutes for less than one semester said that they would continue it.

Only 47% of school gardens, canteens, and student health units are clean, healthy, and rich text. There are still PS that do not have a library, a garden, and student health unit because of limited space and funds. Problems associated with an area of PS infrastructure became one of the causes have not been fulfilled SLM implementation. In addition, some respondents also said that their PS is a school with poor students.

Furthermore, the involvement of the public (parents, community, and the alumni) to develop literacy activities is also lacking (17%). Respondents also argued that the public was rarely involved in literacy activities at PS because they were not started and had not been involved. In addition, the tracking of alumni (treasure study) is still not happening. A few PS engage alumni because the alumni's homes are relatively close and there is concern from alumni themselves.

Regarding the physical environment, there are constraints on the limited number of books, space constraints (narrow class of PS and PS yard limited), and limited funds to create rich text material. Studies confirm that children of poverty attend schools with poorly supported classroom libraries and school libraries (Neuman and Celano, 2001). This is even though increased access to materials such as books can reduce the effects of living in poverty for children [15].

3.2.2 Phase 2 of SLM: Development. From the ten items in this development phase, four items (40%) were classified as good, and six items (60%) were classified as less than good.

The items which have been classified "good" consist of "Libraries used for various activities of literacy" which reaches 77%; "Reading corner is used for various activities of literacy" which reaches 63%. According to the respondents, the library and the reading corner have been used to promote literacy activities, ranging from reading, discussing, and creating works.

"There are various follow-up activities to the fifteen minutes of reading" reaches 63% and "There are oral and written demands that are used as a non-academic assessment", reaches 50%. "The variation of the follow-up fifteen minutes reading activities" reaches 63%. This is to prepare a resume, notes, draw, retell, discuss. Of the 63%, not all respondents were used to non-academic demands (13%) because they think there should be no assessment.

The less encouraging results can be seen in the item of "There is Team of School Literacy (TSL) formed by the principal", which reaches 43%. Furthermore, there are several reasons for PS that without TSL, student portfolios, and activity that promotes a culture of literacy schools, namely:

(1) PS have not implemented the TSL because they do not know about it,
(2) There is no clue about the TSL and student portfolios.

"Students have a portfolio that contains a collection of reading response journals" which reaches 43%; of the 43% of schools using reading response journals with students, only 33% were on display in the classroom or the school corridor. "There is an academic activity that promotes a culture of literacy school" was found in only 37% schools. The reward for the achievement of students in literacy activities at regular intervals" was found in only 30% school. According to the respondents, some PS can not provide rewards because many PS are constrained by funding and the school community are economically disadvantaged (students of poverty). Furthermore, "The celebration activities in specific days on the theme of literacy" was found in only 23% school.

3.2.3 Phase 3 of SLM: Learning. From the eight items in this learning phase, three items (40%) were classified "good", and five items (60%) were classified "less than good". The items that have been classified "good" consist of "There is rich text material related subject displayed in each classroom" which reaches 73%. When compared with the habituation phase in item "There are rich text materials are posted in each classroom, corridors, and other areas in the school", which reaches 87%, it means 14% of rich text material is not related to the subject. Furthermore, "Fifteen minutes each day reading followed by other activities by the academic demand" reaches 57%. The literacy activities related to subjects include a resume, storytelling, and discuss subjects related to Bahasa Indonesia. "Reading already entrenched and became school community needs," reaches 53%. According to respondents, this indicator came from students who love to read.

The less encouraging can be seen in the item "Team of School Literacy (TSL) arranges the planning, implementation, and assessment of school literacy program" which reaches 43%. These percentages are the same as those in the development phase, i.e. there is TSL formed by the principal. It means all TSL arrange the planning, implementation, and assessment of the school literacy program. Furthermore, "School networking with external parties for the development of literacy programs" reaches 43%; "There is a reward for the achievement of participants in the literacy activities based on academic demand" was found in only 43% school. It's more than "There is a reward for the achievement of students in literacy activities at regular intervals" which reaches 30% in the development phase. According to the respondents, rewarding the achievement of students in the literacy activities (based on academic demand) is done by the teacher during a lesson, through praise or a small gift such as a pencil. However, this is less than those rewarded in the development phase when the reward was given by the school rather than the teacher.

"There is show the work of students in celebration of certain days with literacy theme" was found in only 23% school. This percentage is the same as "there is a celebration activity in specific days on the theme of literacy" in the development phase and it means all celebration activities show the work of students. In addition, a show of the work of students (the results of the literacy activities) in celebration of certain days with literacy theme and students journal displayed in the classroom, and school corridors are not done by many PS because of limited funds.

Students' journals (from reading enrichment books and textbooks) displayed in the classroom and school corridors was found in only 23% school. It's less than "Student's reading response journals displayed in the classroom and school corridors" in the development phase which reaches 33%, and it means that not all students' response journals are the result of reading textbooks.

Based on the analysis performed, the recapitulation of the three phases can be seen in the following table.

Table 1. Recapitulation of SLM Phases in PS

Phase	Good (50--100%)	Less (0--49%)
Habituation	50%	50%
Development	40%	60%
Learning	40%	60%

The table appears that in the habituation phase, 30 PS in East Java is in a category that is balanced. But in the phase of development and learning are still classified in the less than good.

3.2.4 Read-Aloud and Sustained Silent Reading as Learning Strategies. Some PS was already implementing stages of read-aloud and sustained silent reading in the 15 minutes of reading and learning of subject [18] [11]. SLM wants all PS (100%) to carry out read-aloud and sustained silent reading as strategies.

Most PS was already implementing a read-aloud as one of the strategies (classified as good). However, this does not only apply to the fifteen minutes reading program but also in the learning of subjects.

Eighty-six percent of teachers selected books/stories that are useful and exciting, and after reading a story, the teacher conducted a discussion of the book; teachers who do not choose a book do so because of the limited books in their school. 83% teachers read the text with the pronunciation and intonation clearly and not too fast, and they asked a question on the sidelines of the story to arouse students' reaction and response. In this case, respondents, as teachers were felt, are still not right in pronunciation and intonation when they were reading aloud. They want to learn to be more confident when they read aloud, especially when reading to the public. 79% teachers asked things related to the story that would be read, and they gave freedom for students to choose books in accordance with their interests and pleasures. Teachers who could not give freedom to the students to choose books did so because of the limitations of existing books. In addition, many students do not have books, even though they have been asked to bring books to school. It is associated with the limited funds of the student's family.

Furthermore, 76% of teachers in PS had read the book that would be read aloud and could open a conversation about the reading material. Teachers who had not read the book that would be read aloud felt that the story was familiar and well-known.

Several sessions were allocated to reading the book to completion (if a readable book is thick) or they chose one section of a thick book to be read aloud. This reaches 76%. In addition, teachers who were not following with the discussion of books and other things related to books read did not do so because they feel the time is very limited.

Most PS was already implementing Sustained Silent Reading as one of the strategies (classified as good). However, this does not only apply to the fifteen minutes reading program but also in learning of subjects. Of the three steps beginning in sustained silent reading, it appears that only 69% students can choose his preferred place to read. In this study, some teachers cannot give freedom to the students to choose books because of the limitations of existing books. However, studies confirm that increased access to books makes sense, given findings that show that self-selected reading is a powerful predictor of reading achievement [5].

"Students may choose another book" reaches 79%. According to the respondents, in sustained silent reading, some students can not choose his preferred place to read because of space limitations, including in the classrooms. Students can only take the book and returned to their seat again for reading. The limited number of books including thick books makes it difficult to choose the books and teachers (respondents) felt no need to explain a thick book. In addition, "teachers explain that the students will read the book to completion within a particular time, depend on the thickness of the book" reaches 76%. Not many books are thick in PS because of the limitations of funds for the purchase of books. Many PS have proposed the procurement of books to the foundation or government, although it is recognized that the purchase of books is not a funding priority. In addition, the respondents admitted that they do not read many books, especially literary books. Reading habits have not become a part of the culture in their everyday lives.

Based on the analysis performed, the recapitulation of Read Aloud and Sustained Silent Reading as Strategies is good (50—100%). Although read aloud and sustained silent reading strategies have been implemented as well, there are some interesting findings, namely:

(1) Competence and reading habits teachers need to be improved
(2) The limitations of the book and place are the constraints of implementation, backed up by limited funds and poor students
(3) The short time constraints make it difficult for teachers to improve their creativity, including discussion about the book.

4. Conclusions

The standard of literacy infrastructure--including the number of enrichment books--is the basis of access to books and SLM development. The condition of schools in Indonesia, especially in East Java is not the same across all schools. The 30 PS in this study in East Java is still below standard (than RMEC No 24 of 2007), especially in the number of enrichment books. Therefore the access to books is poor. This problem arises from the lack of funding for the procurement of books and this is not a priority in school. Suggesting that "students must bring books from home" is not the right step because many of the students are poor.

Furthermore, the implementation of the three phases of the SLM is still less than good and this is due to the fact that the SLM is still relatively new. In this implementation, the factor of access to books is a major constraint. Nevertheless, the implementation of the read-aloud and sustained silent reading as a strategy (including in 15 minutes reading and learning) is "good" although there are some problems related to teacher competence.

In spite of the fact that the number of enrichment-books is still below standard, and there is poor access to books, the school community in 30 East Java PS enthusiastically implements the objectives of the School Literacy Movement. Even though there are some PS still in *wait and see* mode because there are no orders from the immediate supervisor (district education offices/city), the reverberations of SLM is increasingly clear, not only for schools but also for society more widely.

Growing the culture of literacy in schools in Indonesia has started. SLM is a continuous and long-term program. The result cannot be quick; we must proceed slowly but surely. SLM has begun to appear and is expected to be a virus that spreads among students and teachers. Local government support should continue to be pursued so that the literacy movement extends to the schools, families, and societies. Hopefully, SLM can help brighten the world of education in Indonesia.

References

[1] Batool Syeda Hina and Khalid Mahmood 2012 Teachers' Conceptions About Information Literacy Skills of School Children *Pakistan Journal of Library & Information Science* **13**

[2] Beers C S, Beers J W, and Smith J O 2009 *A Principal's Guide to Literacy Instruction* (New York: Guilford Press)

[3] Cohen David K and Monica Bhatt 2012 The Importance of Infrastructure Development to High-Quality Literacy Instruction Journal *The Future of Children* **22** **(2)**

[4] Krashen et al 2010 An analysis of the PIRLS 2006 data: Can the school library reduce the effect of poverty on reading achievement? *CSLA (California School Library Association) Journal* **34** **(1)**

[5] Krashen S 2004 *The Power of Reading* (Portsmouth: Heinemann and Westport)

[6] Krashen Stephen [No date] Protecting Students Against the Effects of Poverty: Libraries *New England Reading Association Journal* (in press)

[7] Lindsay, J. 2010. "Children's Access to Print Material and Education-Related Outcomes: Findings from a Meta-Analytic Review". Naperville, IL: Learning Point Associates. Available at: https://static1.squarespace.com/static/5731ee0840261d67c7155483/t/5767517d29687f6102b04f75/1466388897631/Lindsay_Children%27s+Access+to+Print+Material+and+Education-Related+Outcomes_2010.pdf (Accessed: 25 October 2016)

[8] McCoy Emily 2013 *Lost for Words: Poor literacy, the hidden issue in child poverty* A policy position paper. National Literacy Trust. Available at: http://www.literacytrust.org.uk/assets/0001/7863/Lost_for_Words_-_child_poverty_policy_paper.pdf (Accessed: 25 October 2016)

[9] Minister of Education and Culture 2007 RMEC No. 24 of 2007 on [Infrastructure Standards for Primary Schools / Islamic Elementary Schools (SD / MI), Junior High Schools (SMP / MTs), and High School/Madrasah Aliyah (SMA / MA)] (In Indonesian)

[10] Anonyme 2015 RMEC No 23 of 2015 on Growth Character (In Indonesian)

[11] Morrison V and Wlodarczyk L 2009 Revisiting Read-Aloud: Instructional Strategies That Encourage Students' Engagement With Texts *The Reading Teacher* **63** **(2)** pp 110—8

[12] Neuman S B and Celano D 2001 *Access to print in low-income and middle-income communities: An ecological study of four neighbourhoods* Reading Research Quarterly. Availabel at: http://www-personal.umich.edu/~sbneuman/pdf/AccessToPrint.pdf (Accsessed: 26 October 2016)

[13] OECD 2013 *PISA 2012 Results in Focus. Programme for International Student Assessment.*

[14] Pilgreen J 2000 *The SSR Handbook: How to Organize and Manage a Sustained Silent Reading Program* (Portsmouth, NH: Heinemann-Boynton/Cook Publishers)
[15] Schubert and Becker 2010 Social inequality of reading literacy: A longitudinal analysis with cross- sectional data of PIRLS 2001 and PISA 2000 utilizing the pairwise matching procedure *Research in Social Stratification and Mobility* ***29*** pp 109—33
[16] teacher.scholastic.com 2013 *Access to Book* Family and Community Engagement Research Compendium Available at: teacher.scholastic.com/products/face/pdf/research-compendium/access-to-books.pdf (Accessed: 26 October 2016)
[17] The International Association for the Evaluation of Educational Achievement or IEA 2012. *PIRLS 2011 International Results in Reading*
[18] Trelease J 2013 *The Read-Aloud Handbook* (New York: Penguin)
[19] UNDP 2014 2014 *Human Development Report* United Nations Development Programme Available at: http://hdr.undp.org/sites/default/files/hdr14-report-en-1.pdf (Accessed: 12 February 2016)
[20] UNESCO 2012 "Youth and Skills Putting Education to Work *Education for All Global Monitoring Report 2012* Available at: unesdoc.unesco.org/images/0021/002175/217509E.pdf (Accessed: 12 February 2016)
[21] W Widhiarso 2011 *SKALO Program Analisis Skala Guttman* (Yogyakarta) Available at: widhiarso.staff.ugm.ac.id/wp/skalo-program-analisis-skala-guttman (Accessed: 25 October 2016)
[22] Wiedarti Pangesti and Kisyani-Laksono (eds) 2016 *Desain Induk Gerakan Literasi Sekolah* (Jakarta: Dikdasmen, Kemendikbud) Available at: http://dikdas.kemdikbud.go.id/wp-content/uploads/2016/04/Desain-Induk-Gerakan-Literasi-Sekolah.pdf (Accessed: 22 September 2016)

Line integral on engineering mathematics

L H Wiryanto

Faculty of Mathematics and Natural Sciences, Institut Teknologi Bandung, Ganesha Street Number 10 Bandung, Indonesia.

Email: leo@math.itb.ac.id

Abstract. Definite integral is a basic material in studying mathematics. At the level of calculus, calculating of definite integral is based on fundamental theorem of calculus, related to anti-derivative, as the inverse operation of derivative. At the higher level such as engineering mathematics, the definite integral is used as one of the calculating tools of line integral. the purpose of this is to identify if there is a question related to line integral, we can use definite integral as one of the calculating experience. The conclusion of this research says that the teaching experience in introducing the relation between both integrals through the engineer way of thinking can motivate and improve students in understanding the material.

1. Introduction

The line integral is one of the topics in engineering mathematics courses for electrical engineering department during the last three years in Bandung Institute of Technology. Other departments learn the line integral in advanced calculus. This study describes how to teach the line integral in engineering mathematics course [1].

Students learn line integral after studying definite integral in calculus [2,3] (first year). As we know, a definite integral is usually introduced from the large of area under a curve. Since the calculation using limit of the Riemann sum is not easy, the students are introduced to a tool called Fundamental Theorem of Calculus. We can see the theorem in references such as Varberg et. al. [4]. Therefore, the definite integral is calculated based on anti-derivative. It is compulsory for the first-year students being able to calculate the definite integral for simple functions.

Based on the knowledge in the first-year calculus, the students in electrical engineering department should study line integral as the continuation of the definite integral. What is it and how is it calculated? From mathematical approach, it can be introduced a similar process in definite integral: discretize, Riemann sum and limit, see for example in Kreysig [5] and Wiryanto [6]. But it is not interesting for engineering students. They require a physical explanation to motivate their study. It presents in this paper.

2. Line Integral

In explaining a line integral, there is prerequisite that is usually given in the previous section, i.e. line curves presented in vector function, scalar, and vector fields. Basically, there are two types of line integral written as

$$\int_C f(x,y)ds \text{ and } \int_C \bar{F}(x,y).ds.$$

Here, we discuss the integral along a curve C in a plane. f is scalar field and $\overline{F}$ is vector field [7]. In this paper, we focus only for the first type of integral.

Mathematically, the notation of the line integral is a number that is the result of the limit of Riemann sum [8,9]. The process to construct the Riemann sum and then calculate the limit are impractical, and are useful in analytical work. Practically, we can calculate the Riemann sum using the following formula:

$$\int_C f(x,y) = \int_a^b f\big(x(t),y(t)\big)s'(t)\,dt. \tag{1}$$

for $C{:}\,\boldsymbol{r}(t) = x(t)\boldsymbol{i} + y(t)\boldsymbol{j},\ \ a \le t \le b$ and relation $[s'(t)]^2 = \boldsymbol{r}'(t)\cdot\boldsymbol{r}'(t)$. The relation (1) indicates that we can write the line integral into definite integral [10]. In practice, we can describe the line integral as the area of a wall built above a curve C, and the height of the wall is various, following $z = f(x,y)$. In real life, it can be told as the same height of the wall, but the contour of the ground is up and down, so that the total height is $z = f(x,y)$. So, how to calculate the area of the wall? Engineers will answer by pulling the wall into straight (see Fig.1.), and then it is calculated such as we know in Calculus. We could formulate in (1) mathematically.

In case $f(x,y) = 1$, (1) becomes

$$\int ds = \int_a^b s'(t)dt \tag{2}$$

From the above explanation of wall area, (2) is the formulae to calculate the area of a wall built along a curve C with homogenous height 1. But, we can also explain the left-hand side of (2) as the length of a wire with shape following C, and the right-hand side is the way that we measure the wire by straightening. No-one measures a wire in curve form.

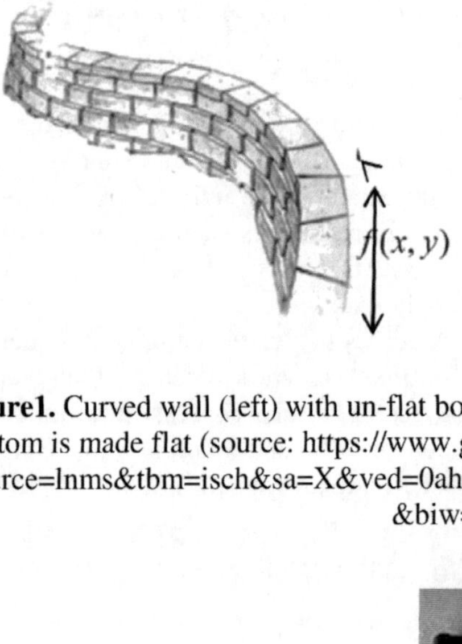

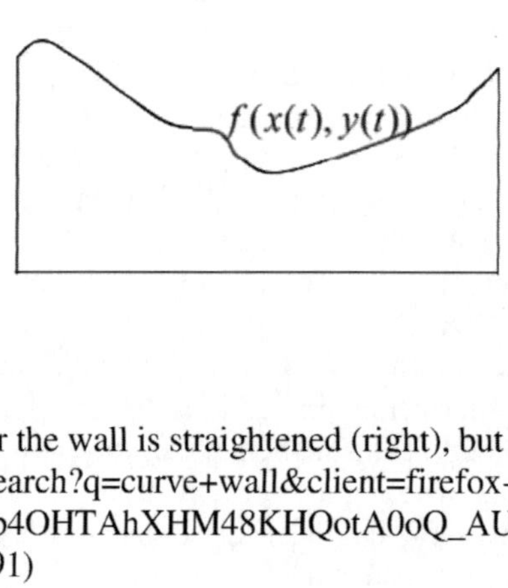

Figure1. Curved wall (left) with un-flat bottom and after the wall is straightened (right), but the bottom is made flat (source: https://www.google.com/search?q=curve+wall&client=firefox-b-ab&source=lnms&tbm=isch&sa=X&ved=0ahUKEwjA4cap4OHTAhXHM48KHQotA0oQ_AUICigB&biw=845&bih=491)

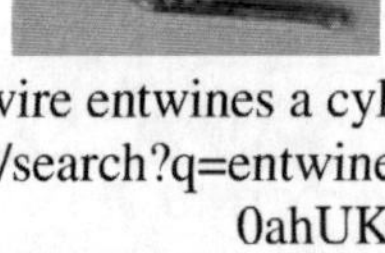

Figure 2. A wire entwines a cylinder (source: https://www.google.co.id/search?q=entwines&source=lnms&tbm=isch&sa=X&ved=0ahUKEwj-vrKa2OHTAhUQTo8KHcmxBykQ_AUIBigB&biw=1267&bih=736#tbm=isch&q=kumparan&imgrc =9wfGI-_dF5HIaM)

From the idea of wire length, we can extend the problem into C -curve in 3-D. If we have a horizontal cylinder with radius R, and we want to entwine n rounds of a wire having radius r from left to right, see Fig.2. How length of wire does we need? The first step is to construct the curve C as

$$C: \mathbf{r}(t) = R\cos t\ \mathbf{i} + R\sin t\,\mathbf{j} + t\,\mathbf{k}, \qquad \text{for } 0 \le t \le 2n\pi.$$

In one circle, each $2p$ of t, the wire must be shifted $2r$, so that the wire does not pile up each other. The next step is to calculate using (2), i.e.

$$\mathbf{r}'(t) = -R\sin t\,\mathbf{i} + R\cos t\,\mathbf{j} + \frac{r}{\pi}\mathbf{k}$$

$$s'(t) = \sqrt{R^2 + \frac{r^2}{\pi^2}}$$

and the length is

$$\int_C f(x,y)ds = \int_0^{2n\pi} \sqrt{R^2 + \frac{r^2}{\pi^2}}dt = 2n\pi\sqrt{R^2 + \frac{r^2}{\pi^2}}$$

the similar problem could be calculated the length of a rope rolled such as shown in Figure 3. To do that, we straight the rope into a line. The difficult thing was how to formulate the vector function representing the rolled rope.

Figure 3. The length of a rope in that form measured by integral. (source: https://www.google.com/search?q=rope&client=firefox-b-ab&source=lnms&tbm=isch&sa=X&ved=0ahUKEwiGopbK5OHTAhVMQo8KHQtbADgQ_AUICig B&biw=845&bih=491)

3. Conclusion

The definite integral is used as one of the calculating tools of line integral. The relation between both integrals through the engineer way of thinking, so that can motivate and improve students in understanding material.

References

[1] Strang G 2007 *Computational Science and Engineering* (Wellesley: Cambridge Press)
[2] Sealey V 2006 *Psychology of Mathematics Education*
[3] Ryzhik I M and Gradshteyn I S 2014 *Tables of Integral, Series, and Products*
[4] Varberg D, Purcell E J and Rigdon S E 2007 *Calculus* (Pearson Education)
[5] Kreyzig E 1999 *Advanced Engineering Mathematics 8-th edition ()* (John Wiley)
[6] Wiryanto L H *Handout matematika Teknik*
[7] Wegenkitt R and Groller E 1997 Fast oriented line convolution for vector field visualization via the Internet VIS '97 *Proceedings of the 8th conference on Visualization '97* pp 309–16

[8] Khattri S K 2009 Computing limits through Riemann sums electronic *J. Math. Technol.* **9** 301–7
[9] Ralph H 1988 A reimann-type integral of lebesgue power *Can. J. Math.* 79–87
[10] Noboru E and Kornilowicz A 2003 The definition of the Riemann definite and some related lemmas *J. Formaliz. Math.* **11**

Acknowledgment
The research of this paper was supported by P3MI Research Grant ITB 2017 and PUPT DIKTI 2017. Therefore, the author acknowledged.

IOP Conf. Series: Materials Science and Engineering **296** (2018) 012047 doi:10.1088/1757-899X/296/1/012047

Cultivating characters (moral value) through internalization strategy in science classroom

M Ibrahim[1], Abadi[2]

[1]Science Education of Postgraduate Program, Universitas Negeri Surabaya, Indonesia
[2]Mathematics Education of Postgraduate Program, Universitas Negeri Surabaya, Indonesia.

musliminibrahim@unesa.ac.id

Abstract. It is still in a crucial debate that characters play an important learning outcome to be realized by design. So far, most people think that characters were reached as nurturance effect with the assumption that students who are knowledgeable and skillful will have good characters automatically. Lately, obtained evidence that this assumption is not true. Characters should be taught deliberately or by design. This study was designed to culture elementary school students' characters through science classroom. The teaching-learning process was conducted to facilitate and bridge the students from the known (concrete images: Science phenomena) to the unknown (abstract ideas: characters: care, and tolerance. Characters were observed five weeks before and after the intervention. Data were analyzed from observation of 24 students in internalization strategy-based courses. Qualitative and quantitative data suggested that the internalization strategy that use of science phenomena to represent abstract ideas (characters) in science classroom positively cultivating characters.

1. Introduction

The main important goal of science education is to empower students in all level of education comprehensively in building a good character: such as spiritual and social attitudes, knowledge and skills [1], so they will be prepared for their future life. While most of the science classes focused on science concepts only [2], less attention has been paid to the characters. The reasons of this conditions: (a) we need more times, efforts, patience, and commitment to developing characters; (b) there is no complete information that can be used as a guide in order to develop characters; (c) there is no certain strategy of teaching that can be used to teach characters as well as cognitive domain.

This present study built upon past research on the use of internalization strategy of teaching in the science classroom [3]. In this strategy of teaching, students are encouraged "to learn" characters through modelling. The model of attitudes is not a human being but the scientific phenomena. In this kind of strategy, the lesson consist of two steps. Firstly, the teacher encourages the student to learn science content and make discoveries using a scientific approach to the first step of their lesson. Secondly, teacher use science phenomena that student already studied in the first step of their lesson as a model or prototype of a certain character [4] and then guides student "to learn" character.

The internalization strategy is a teaching strategy using science phenomena to represent the character and facilitate students to compares the similarities of the scientific phenomena within the character. The students were facilitated between two concepts "that are neither completely similar nor completely different" [5]. An internalization strategy has two domains: a base domain and a target domain. The base domain (model) is usually a familiar object that students possess from previous experiences, and the target domain is usually an unfamiliar object or new information. In this kind of process, students make an image transformation of the model/analog and transfer it into the target [6].

In the science classroom, for example, the model can be represented by the picture of a rice plant, and the target is characters (attitudes: good behavior). This relationship is analogous because both have similarities. People is symbolized by the rice plant with straight upward at the first time when it's fruit appear. Slowly, as long as the seed become content fully and fruit become go down. These processes are similar to the process in human attitudes development.

When models are used for instructional purposes, such as those which were examined in this study, the base domain is a common object that is well-known to students, and the target domain is a new object or concept that is to be learned by students. By associating attributes between the known (old) and the unknown (new), instructional model aims to help students gain a better understanding of the target concept. Modeling supports the instruction by relating "new information to knowledge and experience that students already possess, particularly if the new information is abstract and difficult to grasp" [7]. According to Orgill and Bodner [8], there are at least three classes of instructional modeling based on relationship, presentation, and level of enrichment. With respect to the relationship, modeling between known and unknown objects can be either *structural* or *functional*. The modeling relationship is structural when both objects share similar physical structures such as their external features. It is functional when both objects function or behave in similar ways. An instructional modeling can also represent a relationship that is both structural and functional.

Based on their presentation, modeling can be either visual or verbal [9]. In visual presentations, the instruction is enhanced by using the picture of a known object (concrete picture) that is analogous to an unknown object (target concept). These presentations have been called visual internalization strategy or *visual analogies* [10,11], *pictorial analogies* [12,6], *graphic analogies* [13], *Big Picture diagrams* [14], and '*new-look' representations* [15], In verbal presentations, analogous relationships are expressed in writing. These written presentations have been called *verbal Internalization Strategy or Verbal analogies* [9] and *text analogies* [16]. An internalization modeling can also be both visual and verbal. Regarding the level of enrichment, verbal internalization modeling can be simple, enriched, or extended [8].

This study was guided by the following research questions: How did visual internalization strategy of teaching affect positive attitude: Care and Tolerance of the elementary students?

2. Purpose

This study aims to determine the effectiveness of internalization strategy usage in a science classroom in cultivating positive attitudes: care and tolerance of elementary school students.

3. Method

3.1. Research design

To examine the effects of the internalization strategy through science lesson, we conducted a study using a time series design (Frankael, 2012).

$O_1\ O_2\ O_3\ O_4\ O_5 \qquad X\ \ O_6\ O_7\ O_8\ O_9\ O_{10}$

O1-O5: observation before intervention
O6-O12: observation after intervention
X: intervention using Visual Internalization Strategy (Figure 1)

Along five weeks period before the intervention, the observation was conducted to make sure that character: care and tolerance already exists on the subject or not. The intervention was conducted in three meeting time, using internalization strategy and visual modelling with @ 3 x 35 minutes time allocation for each meeting. Along teaching-learning period we make an observation concerning:

- Constraints

- Feasibility of the teaching materials
  After intervention, for five weeks, observation was conducted

3.2. Subject
This study is undertaken by the fourth-grade students at North Sangatta 1 Elementary School Student in the city of Sangatta, East Kalimantan Indonesia in 2013-2014 academic years. One class is treated as a group that consists of 24 students.

3.3. Instructional Intervention: Visual Internalization strategi
Five weeks before and after the intervention, the researcher makes a daily observation of characters (care and tolerance). The number of students that show their characters is recorded. The study consist of two phases: the first phase of the lesson, teacher facilitate students to learn in the area of science concept. Students learn about many things such as concept, procedures, principles, information, theory etc. They also observe several kinds of scientific phenomena, conduct an experiment and so on. At the second phase of the lesson, the teacher asks their students to observe widely and deeply about the certain science phenomenon and leads students to see the similarities about science phenomenon (base domain) with certain good characters of a human being (target domain) (Fig. 1). Frameworks of internalization strategy of teaching:

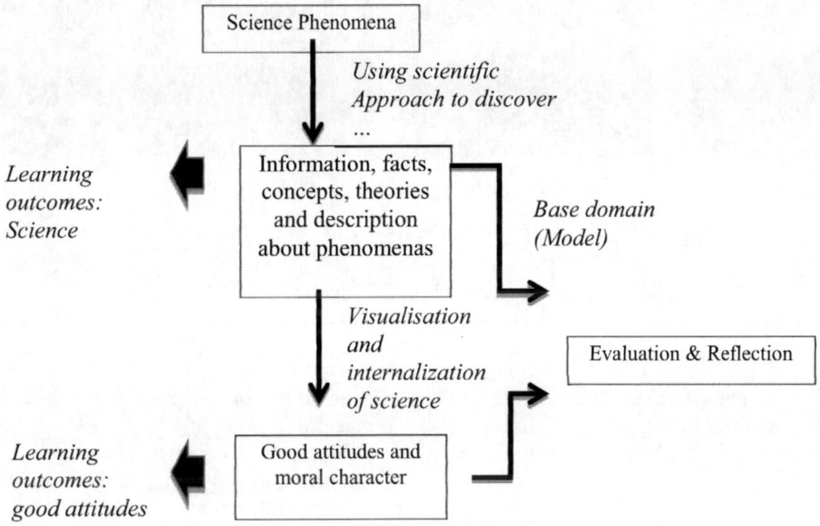

Figure 1. The framework of internalization strategy of teaching [5].

Here are the examples of positive attitudes toward scientific phenomenon as its model.
Example 1: Science topic and phenomena: Our body part
1. Hands (*fact, concept, procedures, etc.*)
   a. The name
   b. The function
   c. How to take care our body parts
   d. How our hands work, etc
2. Our hands as a model of attitudes

IOP Conf. Series: Materials Science and Engineering **296** (2018) 012047 doi:10.1088/1757-899X/296/1/012047

Table 1. The model of attitude and its target

Model	Target (Good attitudes)
We have two hands: right and left	Boys or girl (student with his/her friend
Right hand and left hands have Different function	All of us have certain role in our community
In conducting our activity, right and Left hands usually work together (synergetics)	Student dan their mate usually working collaboratively to reach their goal(s)

In this modeling processes (Table 1), students are encouraged to internalize the phenomena as a model of good attitude (left column) and the relevance good attitudes (right column).

Example 2: Science topic and phenomena: The Life Cycle of Butterfly

1. Life cycle of Butterfly (Fact, concept, procedures, etc.)
   a. The name of each step of life cycle
   b. The characteristics of each step.
   c. Comparison between caterpillar's behavior and the behavior of butterfly
   d. Explore what the caterpillar do along the phase of cocoon

A. Eggs B. Caterpillar C. Cocoon D. Imago

Figure 2. The steps of butterfly's life cycle (Amini, Siti: 2016)

Table 2. Life cycle of butterfly as a model of characters

Model	Target (Characters)
Caterpillar (Fig.2B), usually feeding on every Parts of plant, so they make farmer Upset---- bad behavior	The student understands and have knowledge about bad behavior and make them try to avoid
Cocoon: stop feeding and start to Control their behavior	Students are experienced through visualization how to control bad Behavior--- self-regulated
Butterfly: beautifully, feeding on food resources (nectar)	Visualization the model of good behavior beautiful: make anyone happy and usually use good food resources

Based on the processes, the students learn about characters bad and good behavior and how to change the bad behavior into good ones. Finally, observation has been done after the intervention for five weeks.

3.4. Instrument

In this study, an observation test with several indicators of positive attitudes: care & tolerance was developed and used. The observation sheets were piloted with 15 students. Inter-rater reliability of the test was to be 0.81. In the study, data were collected by classroom observation as long as teaching-learning process and school daily life observation before and after the intervention. The observation was conducted by two well-trained observers from postgraduate students.

4. Results and discussion

Teaching learning process using this internalization strategy can take place in accordance with the scenarios that have been developed in the lesson plan. The main obstacle was that the teacher facing the difficulty of making a positive attitudes model and how to internalize it to the students. To overcome this constraint, researcher designed lesson plans model of positive attitudes that relevance with the content, train teacher to operate the lesson plan in peer teaching forum.

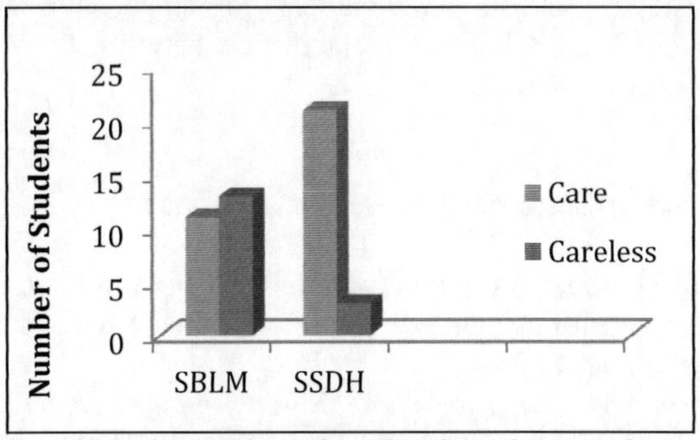

Figure 3. The distribution of student based on careless and care before (SBLM) and after (SSDH) teaching-learning process.

Characters (care and tolerance) were observed as follow: Both the characters were observed five weeks before and fives weeks after intervention. The growth of these characters develop form careless and intolerance (unfavorable) as shown in Figure 3 and Figure 4.

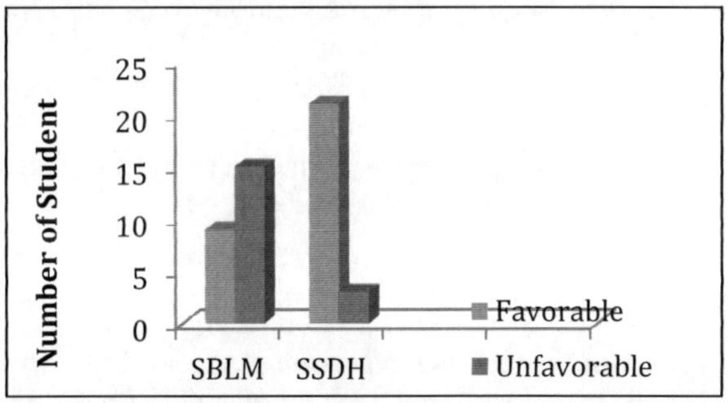

Figure 4. The distribution of student base on unfavorable an favorable before (SBLM)and after (SSDS) teaching-learning process.

In graphs 3 and 4, there were the differences in orientation of students towards tolerance and caring. This condition can be explained as follows. As shown in Figure 1 above, the internalization strategy had two phases. In the second phase, i.e after students learned science content, students learned positive attitudes through a variety of intervention strategies. In this phase, the teacher used a phenomenon of science as a model (prototype) positive attitude. The teacher influence students through *knowing, feeling, and acting.* Teachers try to give insight to students about the positive attitude that will cultivate through model, come up the feeling about those, subsequently affect student orientation towards positive attitudes

This results, which states that the internalization strategy can cultivate a character: especially care and tolerance. This result supported by another finding [3–5,17] on the moral sensitivity of biology class of Senior High School.

Learning something with complex characters can be easier when it is done through modelling [10]. Through such direct comparisons, visual and graphic analogies provide a shortcut to learning and help learners retain complex ideas [13].

Visual analogies that are used in the internalization strategy have been studied extensively for teaching science, from elementary through college [18]. One research focus has examined science analogies on the computer [15].

5. Conclusion

Characters especially care and tolerance can be cultivating through science classroom. We call it as Visual Internalization Strategy. Visual Internalization strategy is a teaching strategy that using scientific phenomena that represent characters and facilitate students to compare the similarities of the scientific phenomena with the characters. The students were facilitated between two concepts “that are neither completely similar nor completely different”.

References

[1] Partnership 2008 Partnership 21st Framework 2008 21st Century Skills Education and Competitiveness Guide
[2] Susetyowati E, Ibrahim M, Rahayu Y S and Bill A 2015 The evaluation of effectiveness of authentic tasks on students’ learning achievement of plant anatomy concept in Surabaya state university *J. Turkish Sci. Educ.* **12** 21–30
[3] Ibrahim M and Wahyu S 2014 *Penelitian Hibah Pascasarjana: Pemberdayaan Siswa Sekolah Dasar Untuk Berperilaku Positif Dan Berkemampuan Berpikir Melalui Pengembangan Perangkat dan Diseminasi Pembelajaran Berorientasi Pemaknaan*
[4] Ibrahim M 2009 *Penelitian Strategis Nasional: Pengembangan Perangkat Pembelajaran Sikap untuk mengajarkan Sensitivitas Moral*
[5] Ibrahim M and Sukartiningsih W 2014 *Pembelajaran Inovatif Melalui Pemaknaan: Belajar Sikap positip dari Alam* (Surabaya: University Press)
[6] Field C T and Graves R L 1981 The hint factor in pictorial analogies *Teach. English Two-Year Coll.* **7** 205–7
[7] Hargittai B and Hargittai M 2007 The use of artistic analogies in chemical research and education part 2 *MIT Press* **40** 357–61
[8] Orgill M K and Bodner G M 2006 An analysis of the effectiveness of analogy use in college-level biochemistry textbooks *J. Res. Sci. Teach.* **43** 1040–60
[9] Radford D 1989 *The Effects of Student Achievement of the Use of Extended Verbal Analogies in High School Biology Textbooks* (University of Georgia)
[10] Krieger P A 2008 *A Visual Analogy Guide to Human Physiology* (CO: Morton: Englewood)
[11] Krieger P A 2005 *A Visual Analogy Guide to Human Anatomy* (CO: Morton: Englewood)
[12] Issing L J 1990 Learning from pictorial analogies European *J. Psychol. Educ.* **5** 489–99
[13] Bailey T 2003 Analogy, dialectics, and life-long learning International *J. Lifelong Educ.* **22** 132–46
[14] Brightman H 2006 *Master Teacher Seminar sponsored by the Office of Faculty Development* (Birmingham: AL: University of Alabama)
[15] Perkins D N and Unger C 1994 A new look in representations for mathematics and science learning *Instr. Sci.* **22** 1–37
[16] Curtis R V 1988 When is a science analogy like a social studies analogy: A comparison of text analogies across two disciplines *Instr. Sci.* **17** 169–177
[17] Ibrahim M 2008 *Pengembangan model pembelajaran IPA inovatif untuk mengajarkan sikap Laporan Penelitian Inovatif Nasional* (Jakarta: Pusat Penelitian Kebijakan dan Inovatif, Kementerian Pendidikan Nasional)
[18] Brightman H 2007 Unlocking the gate to critical thinking: The meaningful-integrated learning level MBA Innovation 16–25

Acknowledgments

I would like to thank the Ministry of National Education and the Ministry of Research, Technology and

Higher Education of Indonesia, which has funded this research in a row starting from National Innovative Research for developing The Innovative models of teaching (2008 and 2009), National Strategic Research to develop the learning package as a tools to implement the internalization strategy in the class, and research graduate grant to disseminate of its results in 2012, 2013 and 2014. Special thanks to Mr. Bill Atweh, professor of Mathematics and Science at Curtin University, Perth, Western Australia on His suggestions to complete the research proposal. Thanks to the elementary students that participated in this research and last but not least to the Rector of Universitas Negeri Surabaya, Indonesia, Dean of the Mathematics and Science Faculty and staff of LPPM Unesa who have facilitated this research.

The Consortium of Asia-Pacific Education Universities (CAPEU) IOP Publishing
IOP Conf. Series: Materials Science and Engineering **296** (2018) 012048 doi:10.1088/1757-899X/296/1/012048

Learning from the coffee shop: increasing junior high school students' self-confidence through contextual learning based on local culture of Aceh

Sarmini[1], A Supriono[1], and Ridwan[2]

[1]Faculty of Social Sciences and Law, State University of Surabaya, Postal Code 60231 telephone/fax 0318281466
[2]Master SMP Negeri 3 Aceh Jaya, Indonesia

sarmini@unesa.ac.id

Abstract. Teachers should be able to provide meaningful learning, create a fun learning, and encourage the self-confidence of students. The reality is learning in Junior High School still teacher-centered learning that results the level of self-confidence of students is low. Pre-action showed 30% of students do not have self-confidence. The research aims to improve the self-confidence of students through contextual learning in the course from the social studies of Aceh based on the local culture. This type of research is classroom action research that conducted in two cycles. The research focus is the students' responses. The coffee shop is a source of learning social studies. Students Involved in the coffee shop interact with villagers who have made the coffee shop as social media. Students participate meetings to address issues of rural villagers. The coffee shop as a public share with characteristics of particularly subject as a gathering place for many people regardless of social strata, convey information, chat, and informal atmosphere that stimulates self-confidence.

1. Introduction

Honor students affected by several factors in the learning process, one of the key factors outlined in this article is confident. But most teachers use the conventional approach that weakens self-confidence. Confident referred to in this article were measured by questionnaire responses of indicators: (1) dare to ask the question, (2) express an idea, (3) respect the opinions of others, (4) cooperate, (5) open to criticism, (6) independent, (7) optimistic, and (8) bold presentation. The author raised the local culture based contextual learning as a solution. Contextual learning is student-centered, students closer to the experience, and meaningful life [1]. Local culture is used in this article is the social interaction in a coffee shop as a center of interaction. Since the first coffee shop for Acehnese serves as a gathering place, to learn together, develop self-confidence, and speech. Post-conflict and tsunami in Aceh functions of the coffee shop is growing rapidly. The coffee shop can be used as a learning resource [1].

This paper uses McClelland *the N-Ach* as the desire to achieve the highest achievement [2], and social learning theory Bandura which emphasizes the concept of *self-esteem,* personality, *behavior,* and *environment* influence each other [2]. The level of self-confidence only on the average score of 38.31% category (less good), even indicators of students work together in learning and daring expression stands at 0%. So far the school has made several attempts to develop a model of learning by involving teachers in studies of evaluation process and send teachers to train, even taking one day off duty at the school, for elaborating in the deliberation of subject teachers (MGMPs) in Aceh Jaya district. However, the reality of learning in the classroom, in general, is still using a conventional pattern.

The study of self-confidence in learning attracted the attention of many researchers from various fields of study. In general, the studies draw together confidently by increasing motivation, the satisfaction of learning, creativity, learning interesting, and behavior in the classroom [3,4]. Effect of self-confidence on the courage, responsibility, and reduce anxiety and doubts [3–10]. The influence of the learning environment, social and cultural rights against self-confidence [3–13]. The strategy for improving self-confidence in general use metacognitive, calibration, and semi therapy [3,4]. Analysis of the level of self-confidence in general use attitude scales, geriatric trials evaluating physical intervention [3], and psychometric testing of the Norwegian version of the questionnaire [4].

The study of variations in contextual learning in various countries, a new study to understand the contributions of visual stimuli to contextual fear conditioning, honed working memory, repetition [4,5]. Contextual facilitates learning through fast mapping [4,5]. Studies of social context and contextual interference effect in motor skill learning [4–6]. Studies of contextual models and collaborative aspects of information literacy [14–22]. Studies student response and significance of contextual learning [22,23]. The position of the subject of research in Kuta Tuha village of Panga District of ± 20 km from the center of Aceh Jaya district. This study focuses on increasing self-confidence using discovery learning models examines social interaction through public spaces Aceh coffee shop. Various forms of interaction ranging from individual interaction with the individual, the individual with the group and the group with the group packed in learning to improve self-confidence.

2. Methods

This study uses *classroom action research.* In connection with the class action, research can use the four principal components of action research by the following steps: (1) *planning,* (2) *acting,*(3) *observing* (4) *reflecting.* Data collected by student questionnaire responses and analyzed with measures of data reduction, presentation, and verification. Data reduction means summarize, choose the basic things, focus on the things that are important, and look for themes and patterns.

Questionnaire was made to determine the level of self-confidence. Each indicator consists of 3 positive statement item and 3 items of negative statements. This questionnaire every number has an answer: (a) always, (b) often, (c) sometimes, and (d) never. Criteria for scoring positive items: (1) answers an (always) will be given a score of 4, (2) answer b (often) will be given a score of 3, (3) answer c (sometimes will be given a score of 2, and (4) the answer d (never) will be given a score of 1. While the negative item will be given an opposite score to positive items. classification score acquisition of each of the respondents was classified into four classifications: (1) 156-192 confidently, (2) high 117-155 trust themselves, (3) 78-116 confident being, (4) low 39-77 confident themselves, and (5) 0-38 insecure.

The steps validity: (1) create table item's bullet analysis, (2) analysis of these scores are grouped into two items odd and even numbers, (3) score hemisphere odd correlated with the score hemisphere even using the formula Product Moment correlation, (4) seek reliability index using the formula Brown. From the data obtained known that the reliability of the questionnaire by the formula halved odd-even exist was 0.708, whereas N = 48 and r table 0.279, the questionnaire was declared unreliable and of a 48 item questionnaire confident everything were tested and the validity of this questionnaire has been used as many as one in class VII SMPN 3 Panga Aceh Jaya.

3. Results and discussion

3.1. Student response pre-action

Table 1. Student response pre-action

No.	Score Range	Classification Type Attitude	Percentage
1.	156-192	Very self-confidence	3.33%
2.	117-155	High self-confidence	10%
3.	78-116	Mid self-confidence	20%
4.	39-77	Low self-confidence	36.66%
5.	0-38	Not self-confidence	30%

(Source: student responses *pre-action*)

3.2. First cycle student response

Results of student questionnaire responses after the first cycle of treatment based on eight indicators can be seen in Table 2 below:

Table 2. Response cycle students I

No.	Score Range	Classification Type Attitude	Percentage
1.	156-192	Very self-confidence	10%
2.	117-155	High self-confidence	18.33%
3.	78-116	Mid self-confidence	33.33%
4.	39-77	Low self-confidence	26.67%
5.	0-38	Not self-confidence	11.67%

(Source: student responses post-cycle I)

Based on the results of student questionnaire responses after the first and second act of the first cycle can be observed that (1) the student confidently increased 6.67%, (2) high 8:33% self-confidence increases, (3) self-confidence was 13:33%, (4) low self-confidence decreased by 10%, and (5) lack self-confidence diminished 18:33%.

3.3. The student response cycle II

Results of student questionnaire responses after the second cycle of treatment based on eight indicators can be seen in Table 3 below:

Table 3. Student responses after the second cycle II

No.	Score Range	Classification Type Attitude	Percentage
1.	156-192	Very self-confidence	25%
2.	117-155	High self-confidence	40%
3.	78-116	Mid self-confidence	16.67%
4.	39-77	Low self-confidence	13.33%
5.	0-38	Not self-confidence	5%

((Source: The response after the second cycle students II)

Based on the results of student questionnaire responses after the second cycle can be observed that (1) the student confidently increased 21.67%, (2) high self-confidence increases 30%, (3) self-confidence is easing 3:33%, (4) low self-confidence is reduced 23%, and (5) are not confident reduced by 25%.

3.4. Comparison of student response cycle

Comparison of student questionnaire responses post-treatment cycle I and II, based on eight indicators can be seen in Table 4 below:

Table 4. Comparison of student response to cycle I and II

No.	Score	Type Classification	Cycle I	Cycle II	Average
1.	156-192	Very self-confidence	10%	25%	17.50%
2.	117-155	High self-confidence	18.33%	40%	29.17%
3.	78-116	Mid self-confidence	33.33%	16.67%	25%
4.	39-77	Low self-confidence	26.67%	13.33%	20%
5.	0-38	Not self-confidence	11.67%	5%	8.33%

(Source: student responses post-cycle I and II)

Based on the results of questionnaire responses students post-cycle I and II can be observed that: (1) students confidently achieve an increase of 15%, (2) high self-confidence increased 21.67%, (3) self-confidence was reduced 16.67%, (4) self-confidence low reduced 13:33%, and (5) lack self-confidence decreased 6.67%. According to the analysis the author an increased level of self-confidence of students in general because the teacher asks to deliver judgment criteria, reveals the idea of ideas, listen and cooperate. Based on these data we can conclude that the contextual approach can increase the self-confidence of students. Comparison of the level of self-confidence of students in learning between the conventional approach in cycle 1 and 2 in this study look very significant.

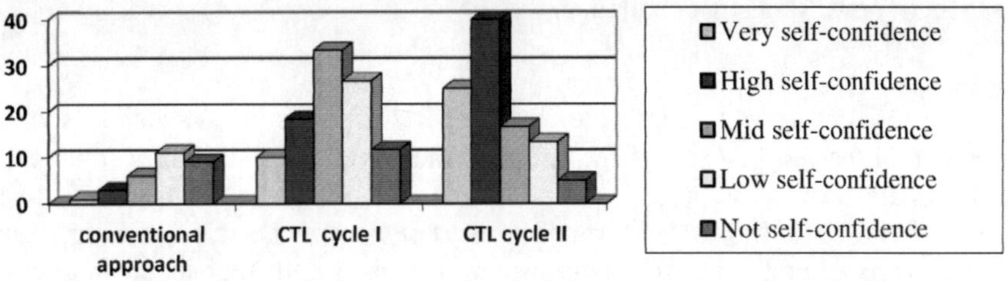

Figure 1. Comparison of the level of self-confidence among the students in learning to use the conventional approach, CTL approach in cycle 1 and cycle 2

Based on data from student questionnaire responses can be observed an increase in self-confidence between comparison with the conventional approach accumulation cycle I and II approach based contextual local culture Aceh showed a significant increase, namely (1) the conventional learning reached only 3:33% to a category confidently or one person, whereas in the first cycle of contextual approach to reach 10% or 3 people, an increase of 6.67% or 2 people. When compared with the second cycle reaches 25% or 8 people, an increase of 21.67% or 7 people. (2) Pre-action on the category of high self-confidence is 10% or 3 people, while in the first cycle reaches 18:33% or 5 people, an increase of 8:33% or 3 people. When compared with the second cycle reaches 40% or 12 people, an increase of 30% or 9 people. (3) The pre-action category was 20% self-confidence or 6 people, while in the first cycle stands at 33.33% or 10 people, it means the difference between 13:33% or 4 people. When compared with the second cycle reaches 16.67% or 5 people, meaning it has decreased 3:33% or 1 person.

Comparisons between students lack self-confidence based on questionnaire responses of students after learning of the pre-action cycle I and II, namely: (1) Classification of self-confidence lower in the pre-action 36.67% or 11 people, while in the first cycle stands at 26.67% or 8 people, meaning it has decreased by 10% or 3 people. When compared with the second cycle at position 13:33% or 4, meaning it has decreased by 23:33% or 7 people. (2) The classification of no self-confidence in the pre-action is 30% or 9 people, while in the first cycle stands at 11.67% or 4 people, means has diminished 18:33% or 5 people. When compared with the second cycle at 5% or 1, means that it has been reduced by 25% or 7 people.

According to the author analyzes, the conventional approach is a very narrow access for students to be active. While the local culture-based contextual approach would open wide active student access. Therefore, based on the results of research of class action is the second problem formulation in this study is based on questionnaire responses of students has proven 55% increase to higher self-confidence and reduced lower 48.33% of students who believe themselves through learning the local culture based contextual Aceh.

From this research, the authors can compare five different keys of using contextual method based on the local culture by conventional methods such as: (1) contextual learning based on local culture encourages students to be more active and creative, whereas conventional learning students are only required to listen and memorize, (2) particulars related learning materials constructed by the students through group work and tasks assigned by the teacher, whereas conventional learning through lectures students are watching and considering the information provided by the teacher, (3) using the immediate environment as a learning resource, while the conventional learning to use information of teachers and textbooks as a source of learning, (4) centered on the students through individual and group work, whereas conventional teacher-centered learning, and (5) a process of peer learning independently or in groups, while at conventional class disable in the learning process takes place between teachers and students only. The contextual approach encourages active student, the teacher must comprehend the different characteristics and capabilities. Teachers give students the opportunity to acquire and construct their own knowledge. From the results of the study, authors recommend that ccontextual approach provides a great opportunity for students to grow and develop self-confidence in participating in learning and have high achievement motivation, self-confidence, discipline higher time to meet the ambition and passion to learn new things that challenge.

4. Conclusion

Based on the research results, it indicates few important points in conclusion, such as: a local culture based contextual learning Aceh (coffee shop) at the center of community interaction in Junior High School 3 Panga Aceh can improve self-confident classified excellent, efforts to increase the self-confidence of the eight indicators through constructivism stage, find, asking, learning community, modeling, reflection, and the actual appraisal well successfully enhanced in 2 cycles performed by integrating the lesson plan and applied in learning activities in the classroom

Refferences

[1] Ordonez O, Carlos J, Stoller, Remmele F and Bernd 2015 2015 1 *Procedia Soc. Behav. Sci.* **171** 982–6

[2] Edelman M 2001 20 *Annu. Rev. Anthropol.* **30** 285–317

[3] McClelland D 1976 *The achievement motive* (New York: Irvington Publishers)

[4] Bandura A 1977 *Social Learning Theory* (New Jersey: Prentice-Hall)

[5] Franklin A, Burns P and Lee C 2014 2014 J. 3 *Nurse Educ. Today* **34** 1298–304

[6] Zvolensky M J, Miller M O B, Feldner M T, Feldner E L, McLeish A C and Gregor K 2016 2016 J. Add *Addict. Behav.* **31** 429–39

[7] Poon J C Y, Au A C Y, Tong T M Y and Lau S 2014 2-40 *Think. Ski. Creat.* **14** 31–40

[8] Chao H J, Chin L C, Yueh H M, Chun K Y, Chih W C and Ying C C Learning *Learn. Individ. Differ.* **42** 157–62

[9] Albrecht K, Essen E, Parys J and Szech N 2013 2013 J. 60 *Eur. Econ. Rev.* **60** 144–59

[10] Chan J C K, Fong D Y T, Tang J J, Gay K P and Hui J 2015 Clinical Simulation in Nursing **11** 278–83

[11] Rabaz F C, Castuera R J A, Arias G, Echeverria C F and Arroyo M P M 2014 2014 *Procedia Soc. Behav. Sci.* **132** 682–5

[12] Safta C G 2015 2015 203 *Procedia Soc. Behav. Sci.* **203** 341–7

[13] Woodman T, Akehurst S, Hardy L and Beattie S 2010 2010 *Psychol. Sport Exerc.* **11** 467–70

[14] Fallah N 2014 30 p140-7 *Learn. Individ. Differ.* **30** 140–7

[15] Hanton S, Mellalieu S D and Hall S 2004 2004 *Psychol. Sport Exerc.* **5** 477–85

[16] Filippin A and Paccagnella M 2012 2012 J. 31 p *Econ. Educ. Rev.* **31** 824–34

[17] Kleitman S and Gibson J 2011 2011 J. 21 *Learn. Individ. Differ.* **21** 728–35

[18] Zlata B V 2013 2013 86 p *Procedia Soc. Behav. Sci.* **86** 482–6

[19] Florack A, Rohman A, Palcu J and Mazziotta A 2014 2014 *Int. Intercult. Relations* **43** 278–88

[20] Marcen C, Gimeno F, Gomez C, Saenz A and Gutierrez H 2013 2013 *Procedia Soc. Behav. Sci.* **82** 750–4

[21] Hatzigeorgiadis A, Zourbanos N, Mpoumpaki S and Theodorakis Y 2009 2009 J. 10 p 186-192 *Psychol. Sport Exerc.* **10** 186–92

[22] Sar A H, Avcu R and Isiklar A 2010 201 *Procedia Soc. Behav. Sci.* **5** 1205–9

[23] Beattie S, Hardy L, Savage J, Woodman T and Callow N 2011 2011 J. 1 *Psychol. Sport Exerc.* **12** 184–91

IOP Conf. Series: Materials Science and Engineering **296** (2018) 012049 doi:10.1088/1757-899X/296/1/012049

Design of Intelligent Robot as A Tool for Teaching Media Based on *Computer Interactive Learning and Computer Assisted Learning* to Improve the Skill of University Student

M S Zuhrie[1], I Basuki[1], I G P Asto B[2], L Anifah[2]

[1]Department of Electrical Engineering, Faculty of Engineering, Universitas Negeri Surabaya, Indonesia.
[2]Department of Computer Engineering, Faculty of Engineering, Universitas Negeri Surabaya, Indonesia.

zuhrie@unesa.ac.id

Abstract. The focus of the research is the teaching module which incorporates manufacturing, planning mechanical designing, controlling system through microprocessor technology and maneuverability of the robot. Computer interactive and computer-assisted learning is strategies that emphasize the use of computers and learning aids (computer assisted learning) in teaching and learning activity. This research applied the 4-D model research and development. The model is suggested by Thiagarajan, et.al (1974). 4-D Model consists of four stages: Define Stage, Design Stage, Develop Stage, and Disseminate Stage. This research was conducted by applying the research design development with an objective to produce a tool of learning in the form of intelligent robot modules and kit based on Computer Interactive Learning and Computer Assisted Learning. From the data of the Indonesia Robot Contest during the period of 2009-2015, it can be seen that the modules that have been developed confirm the fourth stage of the research methods of development; disseminate method. The modules which have been developed for students guide students to produce Intelligent Robot Tool for Teaching Based on Computer Interactive Learning and Computer Assisted Learning. Results of students' responses also showed a positive feedback to relate to the module of robotics and computer-based interactive learning.

1. Introduction

The development of robotics in Indonesia has been very encouraging. The barometer is the success of the Indonesian Robot Contest. In the contest no less than 40 major universities in Indonesia took part. But the development of such robots is limited to the contest and has not been developed to address the issues more real, especially in the industrial world.

The focus of this research is the design of the device in the course of intelligent robotics through the firefighter home robot that is booming and will be implemented in annual of the Indonesia Robot Contest and the industrial in order to advance the world of robotics Indonesia. Robots of this type developed by developed countries such as America, Japan, the United Kingdom the cost is very high. Currently, the robot actually uses high tech, because the control is fully controlled by a microprocessor as a substitute for human and we strive to develop the Low-Cost Technology [1].

The focus of research in the first year is a teaching module manufacturing, planning mechanical design, control system through microprocessor technology and maneuverability of the robot. The first year of the study is expected to produce an intelligent robot that is reliable both in terms of technology and in terms of economy.

The principal issue is how the curriculum and the learning modules are implemented in accordance with the approach of industrial needs [2]. To support competency-based curriculum needs of the industry, it has also been compiled modules both manual and interactive, oriented to the achievement of competence as well as accommodate the working life skills [1,3–6]. With regard to the problems that will arise is what kind of learning modules appropriate to the characteristics of the attainment of the goal of learning the world of work as well as the achievement of competence in the field of life skills by using the main teaching materials in the form of modules and learning media [7–10].

To answer these problems in this study will be compiled in the form of learning devices in the course of intelligent robot tool for teaching based on computer interactive learning and computer-assisted learning to the needs of industry. Expected results of this study will be able to bridge the needs of energy, particularly of graduates majoring in Electrical Engineering Department, the State University of Surabaya which will work as a vocational teacher or plunge into the world of industry.

2. Theoretical framework

2.1. Computer interactive learning And computer-assisted learning strategy

Computer interactive computer and computer-assisted learning strategies is a teaching strategy that emphasizes the use or the use of computers and learning aids (assisted learning) in teaching and learning activity. The use of computers as a learning medium used for a university student to simulate the material learned [11]. The simulation results of these computers will be applied, or applied to the learning tools (assisted learning) that exist [9]. So through interactive learning strategies and computer-assisted learning easier for students to understand the material and apply the concept of the material on teaching aids directly.

To support the teaching and learning process and assisted interactive computer-based learning, the necessary infrastructure in the form of learning equipment consist of:

- Computer interactive learning is a computer device containing computer programs (software) that can be used to perform simulations of materials studied by a university student. Examples of simulation programs for learning Computer Interactive Learning [11].
- Computer Assisted Learning is a learning tool that can be used to apply the results of computer simulation on the appliance directly [9].

2.2. Development learning tool 4-D model

*2.2.1. Define stage.*The purpose of this stage is to determine and define the conditions of learning. This phase has five basic steps include front-end analysis, learner analysis, concept analysis, task analysis, and formulation of learning objectives [12].

2.2.2. Design stage. The design of prototype learning device is produced. The result of this phase is usually a preliminary design of the learning device depending on requirements. Device components used are very diverse, such as student book, modules, teacher guide, student activity sheets, lesson plans, tests of student learning outcomes, and learning media [12].

2.2.3. Develop stage. This stage aims to produce learning tools that have been revised based on the input of experts. The next step is a trial with an appropriate number of students in a class of real (not unlimited). This activity was conducted to determine how effective learning tools developed when applied to the learning process. The effectiveness of the learning device can be seen through the observation, for example, the activities of teachers and students, the ability to manage learning and student achievement test [12].

IOP Conf. Series: Materials Science and Engineering **296** (2018) 012049 doi:10.1088/1757-899X/296/1/012049

2.2.4. Disseminate stage. At this stage is the stage of deployment and use learning device. Learning tools that have been tested and revised earlier, duplicated and distributed for use in learning on a larger scale [12].

3. Research methods

This research was conducted by applying the research design development with the aim to produce a tool of learning in the form of intelligent robot modules and kit based on Computer Interactive Learning and Computer Assisted Learning at the Department of Electrical Engineering to improve the skills of university student [13].

Procedure development in research carried out through several stages as follows:

- Identifying the various issues surrounding the intelligent robot used in contests robot through the study of literature and data mining to be used as reference for the formulation of theory, simulation and applications approach;
- Analyze and formulate the identification of various issues surrounding the intelligent robot relevant to developed in the Department of Electrical Engineering Analyze and formulate the identification of various issues surrounding the intelligent robot relevant to developed in the Department of Electrical Engineering;
- Make analysis and draw up scenarios Content and Design of Learning and infrastructure needed for the development of equipment, teaching staff, laboratory personnel, and the establishment of appropriate evaluation system material needs robotics;
- Formulate indicators of success for teaching and learning process oriented to the achievement of the performance demands of the professional needs;
- Implementation of the manufacture of teaching modules and learning devices (kit) based on Computer Interactive Learning and Computer Assisted Learning as a means to integrate the theory, simulations, and applications suitable with various issues surrounding the learning needs and intelligent robot for contest;
- Conduct field trials on intelligent robot course tools of learning based on Computer Interactive Learning and Computer Assisted Learning in the Department of Electrical Engineering oriented to the mastery of the competency needs;
- Analyzing the results of field trials and make improvements to validate test procedures and test results;
- Doing the final revision of the intelligent robot course tools of learning based on Computer Interactive Learning, Computer Assisted Learning, the results of field test and validation [13].

4. Data collection and analysis techniques data

Information on the results of field surveys and discussions in order to implement the standards needs of the Indonesia Robot Contest and the learning device is analyzed with descriptive techniques. Likewise, data from focus group discussions in order to identify and formulate the essential topics as teaching materials for integrated robotic courses in the form of theory, simulation, and applications by using qualitative descriptive analysis techniques [11]. Qualitative descriptive analysis techniques emphasis on making the evaluation and synthesis of the conclusions resulting from the activities. Synthesis and conclusions on the outcome of this study were formulated through forums workshops and focus group discussions [14].

5. Results and discussion

5.1. Data from the Indonesia Robot Contest 2009-2015

From the data of the Indonesia Robot Contest during the period 2009-2015 can be seen that the modules that have been developed have reached the fourth stage of the research methods of development that disseminate methods. So that, the module is perfect, ready to be duplicated and distributed as learning devices subject to intelligent robot courses. In Indonesia Robot Contest, material-based teaching module compiled specifically for the contest is not only electrical or electronic engineering majors alone, but multi-disciplines, and the object of research being developed not to be the latest technology and appropriate technology alone, but include the development of learning tools.

5.2. Data socialization result in student learning tool

Based on the data about the socialization of the learning is done on students majoring in electrical engineering at the above resulted in the following data. For questions 1, 2, and 3 all respondents consisting of 10 students from representatives of various courses in the department of electrical engineering (100%) answered no understanding of the application of intelligent robot in the industry, or about intelligent robot equipment. Respondents also argued robotics competence that is not taught in the department of electrical engineering. This shows that the image of the competence of the department of electrical engineering is currently only limited to the classic electronic circuit, which ranges on the flip-flop, NAND gate, AND gate only, not offensive competence in the industrial intelligent robot. Yet today the development of machinery industry has many uses industrial intelligent robot.

To question no. 4 of intelligent robot teaching module of computer-based interactive learning and assisted learning of respondents who answered draw was as much as 9 students (90%) of the 10 student-vocational and vocational representatives argue unattractive as much as one student (10%). As for the question no. 5 and no. 6 on the use of computer tools and modules of respondents who think "interesting" as many as 10 students (100%) and argue "unattractive" as none (0%). This shows the positive response of the device in the form of learning modules and devices are developed lesson

To question no. 7 all respondents as many as 10 people (100%) found the module that was developed to facilitate the understanding of the matter. All respondents (100%) believed that they feel happy and motivated by learning to use teaching modules and tool of learning. This indicates that the module can be motivating university student and assist university student in understanding the material.

5.3. Data from the validation of learning by teaching staff (lecturers)

A module that has been created and then validated on 5 validators consisting of faculty learning experts, education experts, engineering experts, and grammarians. And the average assessment is 3.34 validator included in either category. So that the module can be used in trial 2 that experiments performed in the course of the intelligent robot.

6. Conclusions

Based on the results of data analysis and discussion, the researchers can draw conclusions as follows:

- This is a module developed research results and monitoring research officer with the team for 6 years continues to develop research-based robotics contest, with the hope of the future be able to a tool of learning in the form of intelligent robot modules and kit based on Computer Interactive Learning and Computer Assisted Learning;
- The results of student responses during learning tools developed socialization shows that the image of competence for the department of electrical engineering is currently still limited to the classic electronic circuit, which ranges on the flip-flop, NAND gate, AND gate only, only, not offend industrial robots. Yet today the development of industrial automation has many uses industrial robots. Therefore we need a learning device that can support learning robotics competence in the department of electrical engineering. From the results of student responses also showed a positive response to the module of intelligent robot based on computer interactive learning and computer-assisted learning developed;
- A module that has been created and then validated on 5 validators consisting of faculty learning experts, education experts, engineering experts, and grammarians. And the average assessment is 3.34 validator included in either category. So that the module can be used in trial 2 that experiments performed in the course of.

Based on the analysis of the data research and conclusions as well as the real conditions for research in the field, the researchers can provide advice as follows:

- Keep the learning modules other than the subject of intelligent robot modules such as sensor module, the module microcontroller, and others to support the course of robotic;
- The method of computer interactive learning and computer-assisted learning lectures should be done on any existing courses in addition to courses in the intelligent robot. Because of lectures that use computers and assistive devices as a medium of learning can help improve the motivation of students to attend lectures and to facilitate students
- to understand the course materials;
- Competence robotics needs to be taught in the department of electrical engineering, both for high school or university. Because intelligent robot competence is necessary for graduates majoring in engineering when it plunged into the world of industry, especially industries that the machines operate automatically.

References

[1] Greinert W D 1992 *The Dual System of Vocational Training in the Federal Republic of Germany POSTFACH 1 6101 6326* (Eschborn: Federal Republic of Germany)
[2] Brown S 1999 Reinventing the university *Assoc Learn. Technol* **6** 30–7
[3] Depdikbud 1993 *Konsep Sistem Ganda pada Sekolah Menengah Kejuruan di Indonesia* (Depdikbud RI: Depdikbud RI)
[4] DirektoratDikmenjur 1998 *Keputusan Menteri Pendidikan Menengah dan Kebudayaan RI No.323/U/1997 tentang Pelaksanaan Pendidikan Sistem Ganda pada Sekolah Menengah Kejuruan* (Jakarta, Indonesia: Direktorat Dikmenjur)
[5] Loftis H A 1980 *An Exemplary Model for Change, Vocational Instruction* (Columbus: Flarson Publisher)
[6] Harris R, Guthrie H, Hobart B and Ludnberg D 1995 *Competency-based Education and Training, Between a Rock and Whirlpool* (South Yarra, Australia: Macmillan Education)
[7] Bonk C J, Cummings J A, Hara N, Fischler R B and Lee S M 2000 *A Ten Level Web Integration Continuum for Higher Education: New Resource, Partners, Courses, and Markets, ed Abbey B* (University of Hindiana)
[8] Carr M M, Reznick R K and Brown D H 1999 Comparison of Computer-assisted Instruction and Seminar Instruction to Acquire Psychomotor and Cognitive Knowledge of Epistaxis Management (Otolaryngol Head Neck Surg) **121** 430–4
[9] Elves A W, Ahmed M and Abrams P 1997 Computer-Aided Learning Experience at the Bristol Urological Institute in the Teaching of Urology (BR J Urol) **80** 59–62
[10] Daniel J S *The knowledge media in Mega-universities and knowledge media Technology Strategies for Higher Education* (London: Kogan Page)
[11] Phillips R 1996 *Developers' Guide to Interactive Multimedia, A Methodology for Educational Applications* (Perth: Curtin University Press)
[12] Finch C R and John R C 1983 *Curriculum Development in Vocational and Technical Education* (London: Allyn and Bacon, Inc.)
[13] Norton R E 1985 *Dacum Handbook* (Columbus: The National Centre for Research in Vocational, Ohio State University)
[14] Kulik J A 1994 *Meta-analytic studies of findings on computer-based instruction Technology Assesment in Education and Training, ed Baker E L and O'neil H F* (New Jersey: Erlbaum)

A Model of Pantun Learning to Develop Elementary Student's Entrepreneurship Awareness

P V Asteria, B Yulianto
Department of Indonesian Language and Literature, Universitas Negeri Surabaya

primaasteria@unesa.ac.id

Abstract. The principle of entrepreneurship should be instilled from an early age. Pantun is one of Indonesian local culture taught for the elementary school student. Therefore, this study aims to describe (1) the ability to understand the content from pantun based entrepreneurship, (2) the ability to determine the messages from pantun based entrepreneurship, and (3) the ability to apply the messages from pantun based entrepreneurship into their real life. The research type is descriptive quantitative and conducted on the subject of 35 students from SDN Margomulyo I Ngawi. The data is the student's skill in reading comprehension from the pantun based entrepreneurship with the writing test items instrument. The results showed that the student's ability to determine the content from pantun based entrepreneurship shows achievement by an average of 9.3 (include an excellent category) and the ability of students to determine the entrepreneurship messages by an average of 7.8 (good). Meanwhile, the average student's ability to apply the entrepreneurship messages in pantun into their real life by an average of 8.7 (excellent). Overall, the student's skill in reading comprehension from the pantun based entrepreneurship as a whole reached an average of 86.4 (excellent category).

1. Introduction

Entrepreneurship education based on local wisdom is needed to counter the impact of global competition so that the country becomes independent and does not depend on other countries. Meanwhile, the number of Indonesian entrepreneurs today is still less than 2 percent or as many as 700 thousand people and still needed at least 4 million new entrepreneurs [1,2]. For comparison, in the United States reached 11 percent of the population, Singapore 7 percent, and Malaysia 5 percent.

In the entrepreneurship, human resources (HR) as a major aspect must have superior quality in order to create careful opportunities, design and plan intelligently, then it followed by a measured management for successful business. One of the HR development efforts is education. Formal education is implemented through learning activities from educational institutions so that it becomes the basis of empowerment and human resource development. Ideally, entrepreneurial concepts and values are integrated into the early learning curriculum, i.e. at the level of basic education [1,3].

There are various methods that can be done as an effort to cultivate entrepreneurship spirit in elementary school students [4,5]. This study subject is Indonesian Language and Literature with the topic of pantun in the elementary school. This choice based on thinking that the pantun is the indigenous cultural content of Indonesia so as to foster the local wisdom of the students to form a proud character of the country's own cultural richness. Pantun as one of the nation's cultural heritage must be preserved. But in fact, it is now rare that younger generations use or like pantun. This is very ironic because many cultural values of Indonesia are contained in the pantun. Therefore, researchers are interested to apply entrepreneurship-based pantun learning to elementary school students.

IOP Conf. Series: Materials Science and Engineering **296** (2018) 012050 doi:10.1088/1757-899X/296/1/012050

Research on pantun learning based on entrepreneurship has never been done. Thus, a study entitled A Model of Pantun Learning to Develop Elementary Students' Entrepreneurship Awareness is the first study that integrates entrepreneurship in pantun learning [6,7]. The general purpose of this research is to describe the application of learning pantun based entrepreneurship in elementary students. While the specific objectives include (1) describing the ability of elementary students in understanding the contents of pantun based on entrepreneurship, (2) describing the ability of elementary students in determining entrepreneurial messages in pantun based on entrepreneurship, and (3) describing the ability of elementary students in determining how to apply entrepreneurial messages in pantun into their lives.

2. Research methods

This research includes the type of descriptive quantitative research. This study took the population of SDN Margomulyo I Ngawi Class V students. SDN Margomulyo 1 Ngawi was chosen as a research site because the school is a developed state school in which there are students who have a diversity of backgrounds (religious, cultural, and socio-economic) [8]. Thus, the school is well suited to apply this research. The reason for taking the subject of the study of class V students is (1) the students of grade V have received the pantun lesson material, (2) the grade V students are mentally or psychologically capable of understanding the message of a literary work, and (3) the grade V students are still in the mid-semester and do not experience a lot of busyness, such as national exams or practice exams that are usually done by class VI when determining school graduation. The students of SDN Margomulyo I Ngawi Class V was taken one class from three classes. One class, the VA class, as a sample of the study amounted to 35 students. The sample of this research is students of SD Negeri Margomulyo I Ngawi class VA that consists of 35 students.

In this research, the validity and reliability test of the instrument is done by testing the pantun-based entrepreneurship reading test to the students of different classes, namely the VB class. If the data obtained from the test results of the instrument has been fulfilled and in accordance with that it should, means that the instrument developed by researchers is good and can be used to retrieve the actual data in research. In addition, instrument test results were also analyzed by involving several parties, namely peer examination and discussion with teachers as research partners. The collected data are scored using the specified rating and benchmark that have been determined. The assessment and benchmark rubric or formula used in determining the percentage of competency-based entrepreneurship skills as below [9,10].

Table 1. Table of assessment rubric

Question number	Aspect	Description	Maximum score
A	Understand the content of entrepreneurial-based pantun	Able to mention the contents of entrepreneurship-based pantun with complete and exact	1
B	Determining the values of entrepreneurship in pantun	Able to determine entrepreneurial values in pantun with complete and exact	1
C	Determine how to apply the value of entrepreneurship in pantun into real life	Able to determine how to apply entrepreneurship value in pantun into real life with complete and exact	1
Maximum Score per number			3
Maximum score of all questions = 3 x 10 numbers			30

IOP Conf. Series: Materials Science and Engineering **296** (2018) 012050 doi:10.1088/1757-899X/296/1/012050

Benchmark rating:

- $$\frac{\sum y}{\sum x} X\,100\% = P \qquad (1)$$

Information:

- P = The value of the ability to understand pantun based entrepreneurship
- $\sum y$ = Number of scores obtained by the students
- $\sum x$ = Maximum score
- Student scoring criteria are classified according to the following standards.

Score	Category
85 – 100	Very Good (A)
70 – 84	Good (B)
55 – 69	Enough (C)
40 – 54	Less (K)
< 40	Very Less (KS)

3. Result and discussion

3.1. Ability to determine the content of pantun based on entrepreneurship

From the data table below, the test results determining the contents of pantun based entrepreneurship can be classified into several levels of ability. Classification based on the scoring of students by counting using a predetermined valuation formula. Below is a percentage of the ability to determine the content of the entrepreneurial based pantun based on the scoring guidelines or benchmarks presented.

Table 2. Table of ability percentage of determining the pantun content based on entrepreneurship

Number	Score	Frequency	Percentage	Category
1	8.5 – 10	28	80%	Very Good (A)
2	7.0 – 8.4	6	17,1%	Good (B)
3	5.5 – 6.9	1	2,9%	Enough (C)
4	4.0 – 5.4	0	0%	Less (K)
5	< 4.0	0	0%	Very Less (KS)
Sum		35	100%	-

From the data table above, it can be described that the ability to determine the contents of pantun-based entrepreneurship shows the diversity of achievement of results. Students who score between 8.5-10.0 account for 80% of the total sample. The details are students who scored 8.5-10.0 as many as 28 students from 35 samples. Meanwhile, students who scored between 7.0-8.4 of 17.1% of the total number of samples, i.e a number of 6 students from 35 samples. While the score of 5.5-6.9% as much as 1 student. In the score criteria below 5.5 in the test results determine the content of the entrepreneurial based pantun is 0% or nil.

In this study, according to the results of data analysis shows that students, in general, have been able to determine the contents of pantun based entrepreneurship. This is indicated by almost all students getting a very good score on this aspect. The choice of words used to give answers generally is good enough. As in a fragment of student answers below.

" To work we need confident capital (A1)"

" Being a person must be independent (A21)"

According to the results of the study showed that the ability to determine the contents of pantun based on entrepreneurship varies greatly. The scope of the scores achieved students, namely between 6-10 scores. With the acquisition can be said that the ability to determine the content of pantun-based entrepreneurship at a very good level. There is one student who scores below the standard, i.e. students with code A5. With the existence of pantun-based learning entrepreneurship, indirectly assist students in finding the role of entrepreneurial values in their lives. In addition, learning pantun based entrepreneurship is also a good tool in the effort to internalize the value of entrepreneurship in themselves. Students will know the importance of entrepreneurial values without feeling directly patronized.

3.2. Ability to determine the entrepreneurial messages in pantun

After a test to the sample in terms of ability to determine entrepreneurial messages in pantun, indicates the existence of variation or heterogeneity of student competence. This is indicated by the scoring that varies from one student to another in a class. Clearly and comprehensively the results of the ability test determine the content of the entrepreneurship-based pantun are contained in the table below.

From the data table below about the test results determine the entrepreneurial messages in the rhyme can be classified into several levels of ability. Classification based on the scoring of students by counting using a predetermined valuation formula. Below is a percentage of the ability to determine entrepreneurial messages in pantun based on the scoring guidelines or benchmarks presented.

Table 3. Table of ability percentage of determining the entrepreneurial messages in pantun

Number	Score	Frequency	Percentage	Category
1	8.5 – 10	15	42,8%	Very Good (A)
2	7.0 – 8.4	13	37,1%	Good (B)
3	5.5 – 6.9	3	8,6%	Enough (C)
4	4.0 – 5.4	3	8,6%	Less (K)
5	< 4.0	1	2,9%	Very Less (KS)
Sum		35	100%	-

From the data table above, it can be described that the ability to determine the entrepreneurial messages in the pantun shows the variation in the achievement of results. Students who scored between 8.5-10.0 amounted to 42.8% of the total sample. The details are students who scored 8.5-10.0 as many as 15 students from 35 samples. Meanwhile, students who scored between 7.0-8.4 of 37.1% of the total number of samples, i.e a number of 13 students from 35 samples. While the scores between 5.5-6.9% and 4.0-5.4% each of 3 students. On score criteria below 4.0 in the test results determine entrepreneurial messages in the entrepreneurial based pantun of 2.9% or one student.

In this study, according to the results of data analysis shows that students, in general, have been able to determine the entrepreneurial messages in pantun. This is shown by almost all students getting good scores on this aspect. The choice of words used to give answers generally is good enough. However, there is a problem that is commonly found in student responses, i.e. messages mentioned more than one. That is, students are still not fully convinced when determining what messages are contained in the pantun. As in a reply the following student answers.

" Independent, diligent, and hard work (A9)"

" Responsibility and honest (A22)"

Student answers like the above snippet are not entirely wrong. This is because the rhyme in the matter does indeed include two messages explicitly. This, of course, will make students confused. Therefore, students write two messages that confuse it to minimize the possibility of wrong answers that have been written.

According to the results of research indicates that the ability to determine the entrepreneurial messages in the pantun varies greatly. The scope of the scores achieved students, namely between the

IOP Conf. Series: Materials Science and Engineering **296** (2018) 012050 doi:10.1088/1757-899X/296/1/012050

score of 3.0-10. With the acquisition can be said that the ability to determine entrepreneurial messages in pantun on a good level. There are seven students who score below the standard.

3.3. Ability to determine the implementation of entrepreneurship message in pantun into everyday life

The scoring criterion for the third aspect of entrepreneurship-based pantun learning is the ability to determine how to apply entrepreneurial messages in pantun into everyday life. The aspect which is judged in determining how the application of entrepreneurial messages in pantun into everyday life is the precision of providing. For example, how to apply entrepreneurial messages in pantun to everyday life. In addition, examples of how the application of entrepreneurial messages in pantun to everyday life must be in accordance with the circumstances they are experiencing relating to the value of certain entrepreneurs.

After a test to the sample in terms of ability to determine how the application of entrepreneurial messages in pantun in everyday life, indicates the existence of variation or heterogeneity of student competence. This is indicated by the scoring that varies from one student to another in a class. Clearly and comprehensively the results of the ability test determine how the application of entrepreneurial messages in pantun to everyday life is contained in the table below.

From Table below, data on test results determining how the application of entrepreneurial messages in pantun to everyday life can be classified into several levels of ability. Classification based on the scoring of students by counting using a predetermined valuation formula. The table below is a percentage of the ability to determine how the application of entrepreneurial messages in pantun into an everyday life based on the scoring guidelines or benchmarks presented.

Table 4. Table of ability percentage of determining the implementation of entrepreneurship message in pantun into everyday life

Number	Score	Frequency	Percentage	Category
1	8.5 – 10.0	24	68,6%	Very Good (A)
2	7.0 – 8.4	8	22,8%	Good (B)
3	5.5 – 6.9	3	8,6%	Enough (C)
4	4.0 – 5.4	0	0%	Less (K)
5	< 4.0	0	0%	Very Less (KS)
Sum		35	100%	-

From the data table above, it can be described that the ability to determine how the application of entrepreneurial messages in pantun into everyday life shows the diversity of achievement of results. Students who scored between 8.5-10.0 were 68.6% of the total sample. The details are students who scored 8.5-10.0 as many as 24 students from 35 samples. Meanwhile, students who scored between 7.0-8.4 of 22.8% of the total number of samples, i.e. a number of 8 students from 35 samples. While the score of 5.5-6.9 as much as 8.6% that is as many as three students. In the score criteria below 5.5 in the test results determine the content of the entrepreneurial based pantun is 0% or nil.

In this study, according to the results of data analysis shows that students almost entirely have been able to determine how the application of entrepreneurial messages in pantun on daily life. This is indicated by almost all students getting a very good score on this aspect. The choice of words used to give answers generally is good enough. As in a fragment of this student's answer.

" Trying to make the work without help (independent) and dare to show to others (A1)"

" Do not buy pencil box but own work (A15)"

According to research results show that the ability to determine how the application of entrepreneurial messages in pantun vary widely. The scope of the scores achieved students, namely between the score 6.0-10. With the acquisition can be said that the ability to determine the content of pantun-based entrepreneurship at a very good level. There are three students who score below the standard. With the existence of pantun-based learning entrepreneurship, indirectly assist students in finding the role of entrepreneurial values in their lives. In addition, learning pantun based entrepreneurship is also a good tool in the effort to internalize the value of entrepreneurship in themselves. Students will know the importance of entrepreneurial values without feeling directly patronized.

4. Conclusion

Based on the analysis of the results of the research can be concluded that pantun learning can be used to develop entrepreneurial awareness in elementary school students. The ability to read entrenched pantun-based entrepreneurship (the ability to determine the content of rhyme-based entrepreneurship, the determination of determining the entrepreneurial messages in pantun, and the accuracy of determining the application of entrepreneurial messages in pantun into everyday life) showed excellent results of 86, 4. Therefore, it is highly recommended that language teachers implement the model of pantun learning to develop elementary students' entrepreneurship awareness so that entrepreneurial values can be developed from early childhood. Thus, children will grow into individuals who have a high entrepreneurial spirit and have competitiveness to face the global era.

References

[1] Irawadi E P 2013 Minim Jiwa Kewirausahaan di Indonesia *Media Tempo Online*
[2] Khusniyati L T 2012 Kewirausahaan
[3] Drucker P F 1991 *Inovasi dan Kewiraswastaan (Innovation and Entrepreneurship)* (Jakarta: Erlangga)
[4] Yusuf N 2006 *Wirausaha dan Usaha Kecil* (Jakarta: PTKPNF Depdiknas)
[5] Riyanti B P D 2003 *Kewirausahaan dari Sudut Pandang Psikologi Kepribadian* (Jakarta: PT. Grasindo)
[6] Depdiknas 2008 *Kamus Besar Bahasa Indonesia* (Jakarta: Gramedia Pustaka Utama)
[7] Rahma L M 2010 Pengertian Pantun
[8] Kemendikbud 2013 *Kurikulum 2013* (Jakarta: Kementrian Pendidikan dan Kebudayaan)
[9] Suharto 1988 *Metodologi Penelitian dalam Pendidikan Bahasa, Suatu Pengantar* (Jakarta: Depdikbud)
[10] Arikunto S 2006 *Prosedur Penelitian Suatu Pendekatan Praktek* (Jakarta: PT. Rineka Cipta)

IOP Conf. Series: Materials Science and Engineering **296** (2018) 012051 doi:10.1088/1757-899X/296/1/012051

Parental involvement in elementary school-aged child's creativity

Suparmi[1], S P Suardiman[2], and A Kumara[3]

[1)] Student of Post Graduate Program, Yogyakarta State University
[2)] Yogyakarta State University [3)] Universitas Gadjah Mada

suparmi@gmail.com

Abstract. This study aims at describing the parental involvement in cultivating elementary school-aged child's creativity. The qualitative research was designed with multidisciplinary study approach. Eight students and some parents from public elementary schools of Ngawen 4th of Gunungkidul, Yogyakarta, and 4 students from an elementary school in Sleman, Yogyakarta were involved in the process of collecting the data. In-depth interview, observation, and documentation were used simultaneously to collect the data. The results showed that: 1) the subject had a level of intelligence quotient; the intelligence of verbal creativity above the average level, and creative behaviour on average, 2) interaction of parents and child-related discussions, experiences, and plans, academic problems in school were needed to boost the students' creativity, 3) interactions of parents and school-related participations in school were also encouraged to implant students' social awareness, 4) interaction among parents communicated each other to have a better result of academic awareness, and 5) Parents should install family norms to cultivate children's intelligence quotient.

1. Introduction

Nowadays, it is necessary for the next generation to be more creative for the development of the nation. The developing countries are expected not to underestimate creativity if they want to compete with the developed economic countries [1]. The study of Bierly III, Kolodinsky, & Charette [2,3] used a sample of 899 students from undergraduate programs consisting 51% of men and 49% of women with diverse backgrounds of educational, parental income, skin color, and ethnicity. The study inspired the present study to be undertaken.

One of the important factors that support the success of an educational process is the family. Related to the importance of family's involvement in cultivating children's creativity, the authors conducted a survey on the use of social media on 233 students at Ungaran Elementary school in grade 4th and 5th at Yogyakarta during 2013. The result showed that 76% (178 students) had their Facebook and Twitter accounts, and 24% (55 students) have no any account. Some of the aspects that influenced students in creating the account came from outside the school environment: 33% by their parents, 42% by their brothers, 8% by their friends, 8% by their siblings, and 9% on their self. Children who do not have accounts are allowed to search the internet with the assistance of parents.

2. Some Related Literature

According to Whitehead [4], Creativity is the principle of novelty. According to Renzulli [5], there are three points which become the criteria of giftedness called three-ring conception that is above average ability, above average creativity and task commitment. Furthermore, Monks [6] elaborated Rezunlli's model by adding the environmental characteristics of family, school, and peers. Cooperation between the six factors can lead to giftedness. Maurer Sternberg 1985; Sternberg & Lubart, 1995; Sternberg & Williams, 1996 [7]. Creative work requires applying and balancing three abilities that can all be developed. These capabilities are Analytic, Synthetic, and Practical Ability. Urban (1995) in Seo, Lee, and Kim [8], classified creativity into three components: the cognitive aspects, personality, and environmental conditions. Amabile [9] there are three components in individual creativity: "expertise, creative thinking skills, and intrinsic task motivation", each of which is essential for creativity and the three components interact with each other.

Based on the description above, a subject who has good creativity can be identified by conducting the cognitive test, verbal creativity test (VCT) and creative behavior scale as well as subject achievements and preferences based on school and family documents. Anwar & Sagala [10], asserted that parents are responsible in the field of education. Ki Soeratman [11], reminded that the atmosphere of family life is the best place to conduct individual education and social education. Yan [12] has developed a variable family social capital that involves four aspects: 1) parent-child interactions, 2) parent-school interactions, 3) interactions with other parents, 4) family norms.

Taylor (2014) said that parents support to enhance child's creativity by Four Creativity Essentials is 1) stress can killing creativity, 2) failure, 3) play is one way to cultivate creativity, 4) opportunity to create and make. Garcia [2] suggested to parents to encourage children's creative talent: 1) that of vigilance, 2) your child's interests, 3) inspirational offerings, 4) giving encouragement, 5) get creative. Tartakovsky [13] suggested nine ways in which parents can encourage the creativity of their children: 1) Give space to create, 2) keep games and simple activities3) spend time, 4) expose children to the world, 5) creativity discuss, 6) cultivate creative thinking, 7) give child freedom to be creative, 8) help facilities, 9) spend time for your own creativity. This research will use the variables of Yan [12] to identify parent involvement including parent and child interaction, parent and school interaction, interaction among child's parents, and family norms cultivation.

3. Methods

Qualitative data analysis was conducted inductively, and the researchers did not treat the subjects in the environment. The research setting was carried out in the Yogyakarta by taking the setting at Public Elementary School IV Ngawen, in Dukuh Candi, Gunung Kidul, and Sleman Public Elementary School Model, Wedomartani Village, Sleman, in Yogyakarta. Both elementary school models have a personal character of creative students who are able to actualize themselves up to the provincial level and Ngawen IV elementary school to the national level. Analysis begins by collecting data. The data obtained from the documentation, archive records, observation, deep interviews, and physical devices of the research subjects.

4. Discussion

Subjects in this study were two students and parents from Ngawen IV elementary school, four students, and parent from Sleman model of elementary school, Yogyakarta. Data collection was conducted in May to August 2015. Parental involvement focuses on 1) parent-child interaction, 2) parent-school interaction, 3) interaction between child parents, 4) and family norms. The source of parental involvement comes from the parents of the subject. The results of data collection process n the level of intelligence and child creativity intelligence presented in table 1 below:

Table 1. IQ and VCT Subject with Wechsler Norm

Subject	Age	IQ Score	Category	VCT Score	Category
D	12 year	147	Very superior	137	Very superior
I	11 year	128	Superior	111	High average
K	10 year	136	Very superior	122	Superior
R	11 year	136	Very superior	115	Superior
A	10 year	143	Very superior	120	Superior
H	11 year	144	Very superior	124	Superior

Based on Table 1, the subject has the level of Intelligence Quotient (IQ) and Verbal Creativity Test (VCT) above average with Wechsler norm. The subjects are still elementary school students. The data of children's creative behavior was obtained based on the scale of the Munandar's creative behavior. The results served on Table 2 below:

Table 2. Creative behavior

Category	Formula	Range	Freq	s
High	$(\pi + 1\ SD) \leq X$	$\leq 21,33$	- 6	-
Medium	$(\pi - 1\ SD) \leq X < (\pi + 1\ SD)$	$10,67 \leq X < 21,33$	-	100%
Low	$X < (\pi - 1\ SD)$	$<10,67$		-

Keterangan:

π = average
SD = standard deviation
X = score factual

Based on the calculation of the data on Table 2, the maximum score of the creative behavior scale is 1x 32 = 32 and the minimum score is 0 x 32 = 0, mean: 16, standard deviation: 32-0: 6 = 5.33. Thus, the subjects had moderate behavior.

The results of child's achievement and activity data were delivered on Table 3 below:

Table 3. Achievement subject

Subject	Achievement and subject preference
D	National Math Olympics and help make handicrafts brooches
I	Soccer and make toys that are used to dispel the birds in the fields "long-longan"
	Storytelling at Sleman district level, making robotics, and "dolanan anak"
K	Tennis at the district level and "dolanan anak"
R	Child and child artist and movie artist, drawing comics and "dolanan anak"
A	Little doctors at the district level and "dolanan anak" and also write stories
H	

Based on table 3 above, the subject had the diversity of presentation and preferences:

- Case study A
  The subject's parent is not a biological parent. The subject's biological mother died while giving birth to her second sister. The subject is originated from Lampung. Since the kindergarten, the subject lives together with the elder brother's siblings who later referred to as parents in the village of Kampung, Ngawen, Gunungkidul. In developing children's sustainable creativity, parents seek: to actively interact with the child to discuss any learning problems, the expected

school in future, and personal problems. The extracurricular selection also was consulted to his parents, especially on his mother. Parents often interact with schools in order to experience children's schooling and to cooperate on preparing all learning materials and the Olympics matter. With high discipline, parents can deliver the subject to be a national math Olympics participant.

- Case study B

The subject's parent is not a biological parent because the subject's biological mother has died when the subject is a first-grade elementary school in Klaten district. When the subject was in Second grade, he moved and lived with his foster father in the village of Kampung, Ngawen, Gunungkidul. The subject's father works as a farmer, and his adoptive mother is a housewife who manages a grocery stall at home.

Parents interact with schools to the extent if the school took the initiative to convey the development of his sons and daughters. In helping household work, the subject was given the obligation to sweep the yard in the morning. In the afternoon after the break, the subject helped parents in the fields. With these activities, the subject could work to make a toy that can dispel birds that are named "long-longan". Case study C

Mr. FW is a K's father who works as a motivator, and the wife is a housewife, a mother who is willing to give up her job to accompany her children. The subject is the third son of three brothers and sisters. Parents were conducted cultivating and practicing child's creativity sustainability by discussing with their child about learning difficulties, personal problems and getting a less satisfactory grade. The storytelling course was accomplished outside the school. Parental interaction with the school was arranged in certain events (due to the eruption of Kelud Mountain). Parents offered themselves to volunteer for cooperation on cleaning up the classroom environment. Habituation did with mentoring of watching TV, playing games both offline and online. Before the fourth-grade, parents prepared all the school stuff but since the fourth grade, the subject prepared their own school's equipment. The habit of reciting Qoran was also done every day after maghrib until isya.

- Case study D

Mr. IY is R's father who works to be Widyaiswara in government agencies, and his mother is a makeup artist. The subject of two brothers with sister. Creativity is a way for the child to develop their minds and not make childless socially appropriate with the age of child development. Parents, in fostering sustainable creativity, attempted to discuss with the child about the problem of learning difficulties. Parental interaction with the school was arranged in certain events (due to the eruption of Kelud Mountain). Parents offered themselves to volunteer for cooperation on cleaning up the classroom environment. Mother's subjects also often volunteer to assist children's activities at school, such as makeup artists on the annual stage event ahead of the classroom hikes, featuring extracurricular art activities at school.

- Case study E

Mr. SPY is the father of A who works as a lawyer, and the mother is a housewife. The subject also asked parents to get a source of learning outside the school sources, like the internet sources. Parents gave freedom and facilitated to work on the desired projects. Various courses were conducted by subject ranging from the first-grade elementary school, such as painting, piano, vocals, swimming, and chess. Parental interaction with the school was conducted in certain events (due to the eruption of Kelud mountain). Parents offered themselves to volunteer for cooperation on cleaning up the classroom environment. Interaction with schools related to children's activities was often arranged because the subject had solid activities such as filming. Parents also work with schools to prepare subjects if they are going to compete in schools such as "dolanan anak". Active parents are also in school committee organizations. Habituation did with mentoring watching TV, playing games both offline and online. The habit of praying with other families, especially the prayers of Maghrib and Isya.

- Case study F
  Mr. AS is H's parents who work as village apparatus and mother works as a kindergarten teacher. The subject of two brothers with a younger brother. The subject also lives with his grandmother. Parents of the subject who interpreted creativity is very important as an effort how children are able to find answers to the problems faced. Parental interaction with the school is done in certain events (due to the eruption of Kelud mountain) parents offer themselves to volunteer for "*gotong-royong*" clean up the classroom environment. Parents also work with schools to prepare subjects if they are going to compete in small doctor competitions, scientific writing and "dolanan anak". Habituation did with mentoring watching TV, play games both offline. Before the fourth-grade parents prepare all the school equipment but since the fourth grade the subject of preparing the school's own equipment. The habit of praying with other families, especially the prayers of subuh and isya. The habit of reciting Qoran is also done every afternoon in the month of Ramadhan in the nearest mosque.

5. Discussion

Subjects meet creativity requirements of creative talent in the level of intelligence above average, commitment to the task, and creativity [5] and to children's environment [6]. Urban (1995) in Seo, Lee, and Kim [8], the meaning of creativity is also interpreted by various parents. Other parents consider a child who has a creativity as a child who is clever, independent and whatever ideas and actions make money. It is compatible with analytical abilities, synthetic abilities, and practical abilities [7]. Other parents argue that creativity is the intelligence, skill, and sense of the children that will be useful in their life [9].

Creativity is defined as self-deprecation of a child that is able to solve problems. This opinion is in accordance with basic needs instinct from Maslow, 1970 about self-actualization [14]. The parents consider creativity as a process and the result which is important for their children in accordance with the process of creativity [15]. The subject's parents come from the various job, economic backgrounds, and residences but have a common opinion that creativity is important for their children.

Various activities can represent the process of children's creativity, include discussion, trusting, guiding, supervising, giving examples, appreciating, giving opportunities, facilitating, and relating positively [1,3,11,16–19]. Parents also have similarities in adjusting positive habits of family values. However, the parents of the Elementary School Model of Sleman set up their own community to advise schools and to receive information from the school. Parents are also looking for sources outside the school so that their children can actualize themselves more widely [12,20].

6. Conclusion

Based on data analysis and discussion of the findings, this study concludes that creativity is a very important competence so that it is required to be explored and trained continuously. The source of the subject's aspiration to be creative comes from Television, hearing stories, courses, and following the competition. Parents of both the Elementary School of Ngawen 4 and the Elementary School Model adapt the family values with positive habits. For parents of children who come from SDN Ngawen 4 Gunung Kidul, parents' involvement with children limited to discuss the learning problems in school and personal problems. The Elementary School of Ngawen 4 holds additional learning if the students are in sixth grade to face the final exam. The Elementary School Model located near the urban areas of part of the Subject involves activities from outside the school venues such as tennis and theater. Parents provide freedom and facilities to their children. This research provides important information to parents: 1). optimizing child's creativity on an ongoing basis, parents need to take time to engage directly with the child in creative activities, 2). parents need to provide simple facilities so that children can try to answer their curiosity, and 3). parents need to seek information outside the school so that child can express themselves in a wider area.

References

[1] Shaheen R 2010 Creativity and education *Sci. Res.* **1** 166–9
[2] Garcia L 2014 Creativity Is King: What Art Can Do for Your Kids
[3] Munandar S C U 2009 *Pengembangan kreativitas anak berbakat* (Jakarta: Rineka Cipta) [4] Whitehead A N 2009 *Filsafat Proses, Proses dan Realitas Dalam Kajian Kosmologi* (Yogyakarta: Kreasi Wacana)
[5] Renzulli J S 1998 Three-ring conception of giftedness. In Baum, S. M., Reis, S. M., & Maxfield, L. R. (Eds.). Nurturing the gifts and talents of primary grade students *Mansf. Center, CT Creat. Learn. Press*
[6] Monks F J, Knoers A M P and Siti R H 2006 *Psikologi perkembangan pengantar dalam berbagai bagiannya (Ontwikkelings Psychologie Inleiding Tot De Vesschillende Deelgebieden)* (Yogyakarta: Gajah Mada University Press)
[7] Steinberg R J and Williams W M 2003 Teaching For Creativity: Two Dozen Tips
[8] Seo H-A, Lee E A and Kim K H 2005 Korean Science Teachers' Understanding of Creativity in Gifted Education *J. Adv. Acad. ProQuest Educ. Journals* **XVI** 98 – 105
[9] Amabile T M 1997 Motivating Creativity in Organizations: On Doing What You Love and Loving what you do *Calif. Manage. Rev.* **40**
[10] Anwar Q and Sagala S 2004 *Profesi jabatan kependidikan bagi guru sebagai upaya menjamin kualitas pembelajaran* (Jakarta: Uhamka Press)
[11] Soeratman K 1977 *Karya Ki Hajar Dewantara, bagian pertama Pendidikan* (Yogyakarta: Majelis Luhur Persatuan Taman Siswa)
[12] Yan W 1999 Successful African American Students: the role of parental Involvement *J. Negro Educ. Howard Univ. Proquest Educ. Journals* **68** 5
[13] Tartakovsky M 2012 9 Ways to Support Your Child's Creativity
[14] Feist J and Feist G J 2008 *Theories of personality Edisi Keenam translated by Yudi Santoso* (Yogyakarta: Pustaka Pelajar)
[15] Sudarminta 1991 *Filsafat Proses, Sebuah Pengantar Sistematik Filsafat Alfred North Whitehead.* (Yogyakarta: Kanisius)
[16] Epstein J L and Al. E 2009 *School, family, and community partnerships your handbook for action* (California: Corwin Press A Sage Company)
[17] Peppler K A and Solomou M 2011 Building creativity: collaborative learning and creativity in social media environments *Q Emerald Gr. Publ. Ltd.* **19** 13–23
[18] Davis G A 2004 Creativity, self-actualization, and you dalam creativity is forever
[19] Supriyadi D 1994 *Kreativitas, kebudayaan, dan perkembangan iptek* (Bandung: Alfabeta) [20] Pishghadam R and Zabihi R 2011 Social and cultural capital in creativity *Can. Soc. Sci.* **7** 32–8

IOP Conf. Series: Materials Science and Engineering **296** (2018) 012052 doi:10.1088/1757-899X/296/1/012052

Learning outcomes through the cooperative learning team assisted individualization on research methodology' course

N F D B Pakpahan

Faculty of Engineering, Universitas Negeri Surabaya, Indonesia

nurmipakpahan@unesa.ac.id

Abstract. All articles *must* contain an abstract. The research methodology is a subject in which the materials must be understood by the students who will take the thesis. Implementation of learning should create the conditions for active learning, interactive and effective are called Team Assisted Individualization (TAI) cooperative learning. The purpose of this study: 1) improving student learning outcomes at the course research methodology on TAI cooperative learning. 2) improvement of teaching activities. 3) improvement of learning activities. This study is a classroom action research conducted at the Department of Civil Engineering Universitas Negeri Surabaya. The research subjects were 30 students and lecturer of courses. Student results are complete in the first cycle by 20 students (67%) and did not complete 10 students (33%). In the second cycle students who complete being 26 students (87%) and did not complete 4 students (13%). There is an increase in learning outcomes by 20%. Results of teaching activities in the first cycle obtained the value of 3.15 with the criteria enough well. In the second cycle obtained the value of 4.22 with good criterion. The results of learning activities in the first cycle obtained the value of 3.05 with enough criterion. In the second cycle was obtained 3.95 with good criterion.

1. Introduction

Research methodology is a lecture that must be taken by students before making thesis or scientific research. Material research methodology constitutes the materials supporters that are to be understood by students to finish the thesis. The thesis is the duty of the end of scientific research written by a student to meet the requirements got a degree. This indicates of understanding of material research methodology a decisive factor in seeking success students finish the thesis [1].

Based on observation and experience researchers, almost 28 % students academic year 2010 in the department of civil engineering still has yet to complete a thesis, and have reached ending time of his studies. One of a contributing factor is a student do not have an understanding that supports its ability to develop a proposal thesis (plan research) and implement research. It means, students have never had the capacity who capable in making proposal research or proposition thesis, because they have not yet had a good understanding in preparing the problem who want to study, moreover about procedure research . Learning that is teacher-centered consisting of the as a source of information and learning dominated by teachers, less member opportunity to school tuition to develop the ability of diverse. As a result school tuition less play an active role in learning, and created atmosphere active and democratic.

Problems now of how to implement learning that can be pushed students to hurry to understand matter research methodology as a basis supporting the completion of thesis students. The application of learning that would improve liveliness students and expected to improve study results is learning model cooperative type Team Assisted Individualization (TAI) [2]. Learning model aims to address a

IOP Conf. Series: Materials Science and Engineering **296** (2018) 012052 doi:10.1088/1757-899X/296/1/012052

learning disability students individually. Learning activities more used to problem-solving, typical of type TAI this is every student individually learn matter learning already prepared by lecturers. Learning outcomes individual brought to groups to discuss and mutual discussed by members of the group, and all members of the group liable to all answers as the responsibility of us all. After going through a learning process it is hoped that students get the result of learning which satisfactory. According to Slavin [2], a method of TAI is a program pedagogical who are trying to adopt learning by taking into account individual differences student all at academic. The purpose of TAI to minimise individual teaching is ineffective, and an increase of knowledge, skill, and motivation the student to study in a gregarious manner.

Based on background above it can be formulated the following problems: 1) How to increase the learning outcomes of students after implementation of cooperative learning with *TAI* on research methodology' course in the Department of Civil Engineering Universitas Negeri Surabaya?; 2) How to increase the teaching activities in class after implementation of cooperative learning with *TAI* on research methodology' course in the Department of Civil Engineering Universitas Negeri Surabaya?

2. Basic Theory

2.1. Cooperative learning type Team Assisted Individualization (TAI)

Cooperative learning-type Team Assisted Individualization (TAI) according to Helda [3] is a combination of learning, collaborative learning with individual learning. Gradually, each member of the group given the problems they have to do it yourself first. After that assessment is carried out together in groups. If reserved the first stage had been completed properly, then every student do some problems in the next stage.

According to Slavin in a Huda [2,4], TAI is a program pedagogical trying to adopt learning by fine distinctions individual student all at academic. TAI development can support the practices of the classroom, as grouping students, grouping ability in the classroom, teaching program, and learning based on computer. The purpose of Team Assisted Individualization (TAI) to minimise teaching individual that has proved less effective, in addition, to increase knowledge, ability, and motivation of students in a learning group.

Excess of TAI in Slavin: 1) improve learning outcomes students; 2) increase the motivation to study for on the self-student; 3) decrease treatment that annoys; 4) program or this model can get help students who are weak.The feebleness of TAI according to Slavin among other: 1) it takes a long time to make and develop a device of learning; 2) as the number of students that is great in a class, the teacher will have any difficulty in provides guidance to the students; 3) not all matter be able to apply the TAI use; 4) cause dependency students where students who less clever indirectly would depend on the students well [2].

2.2. Learning outcomes

According to Gagne [5], learning activity was complex. Learning outcomes in the form of capabilities. After having learned people skills, knowledge, attitude, and value. The emergence of the capabilities of stimulation is derived from the environment and the cognitive process conducted by learners.Thus learning is a set of the cognitive process that changes the nature of environmental stimulation, passing of information processing, become capital new. Piaget next contends that the knowledge formed by the individual. For individual conduct continuous interaction with the environment. That environment undergoing change. With the existence of interaction with environment intellect hence function keeps growing.

According to Bloom [2,4,6], learning outcomes includes: 1) cognitive domain, covering knowledge (knowledge, memory), comprehension (understanding, explains, summarizing, example) , application (applying), analysis (outlines , ascertaining the relations), synthesis (organize, plotting, forming new buildings), and evaluation (assess); 2) domain affective, covering receiving (attitudes receiving), responding (give a response), valuing (value), organization (organization), characterization; 3) psychomotor domain, covering initiatory, pre-routine, and routinized. Psychomotor also includes skills, productive, technique, physical, social, managerial, and intellectual.

IOP Conf. Series: Materials Science and Engineering **296** (2018) 012052 doi:10.1088/1757-899X/296/1/012052

Domain cognitive covering [4,6]: 1) memory or knowledge (knowledge), namely the ability remember a material that has been learned; 2) understanding (comprehension), namely the ability catch understanding, translate and interpreting; 3) the application of (application), namely the ability to use a material that has been learned in a new situation and real; 4) analysis (analysis), namely the ability identify and connecting among parts to build a whole; 5) synthesis (synthesis), namely the ability concluded and united a separate piece to build a whole; 6) an evaluation, namely the ability study value or the price of something, as a statement or research reports based a criteria [7].

Based on the definition of above, can be taken the conclusion that study results are a result of learning activities that transformed the student competency on the knowledge.

3. Methodology of research

3.1. Location and time research

This research was conducted at the Department of Civil Engineering, Faculty of Engineering Universitas Negeri Surabaya in the odd semester of academic year 2013/2014. Data collection was conducted in September 2016 until November 2016.

3.2. Subject and object research

This study uses the subject of the action (principal subject) that 30 students of Building Technical Education, which is the programmed course of research methodology. The subject broadcaster is lecturer research methodology. Whereas the object of this research is the application of Team Assisted Individualization (TAI) cooperative learning.

3.3. Research methods

References used in the Classroom Action Research is a model Kemmis and MC Taggart [7]. Steps models such as the following picture.

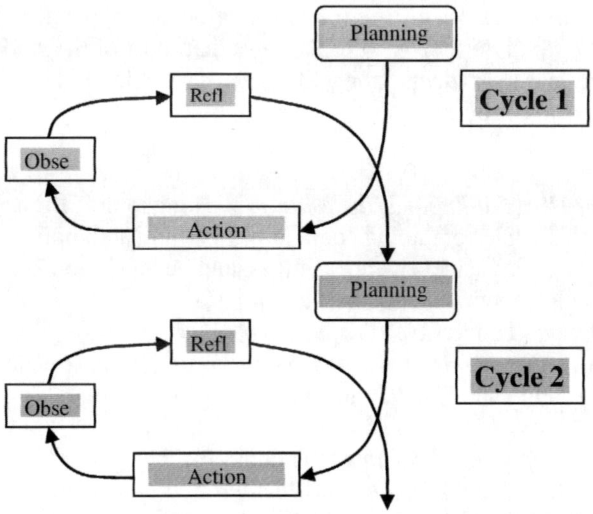

Figure 1. Step classroom action research, a model Kemmis dan M.C Taggart [7,8]

3.4. Methods of data collection

This study uses the instrument: (1) an instrument learning outcomes tests, (2) observation sheet activities of lecturers presenting cooperative learning type TAI on course research methodology [9].

Learning outcomes tests were analyzed using the criteria required that 75% of the students master the material tests or obtain a minimum score of 75. Learning outcomes tests conducted in each cycle, if the student's understanding achieved as required, then the cycle will be completed action.

4. Result and discussion

4.1. Results of research

*4.1.1. Learning Outcomes.*The student learning outcomes related to intellectual indicated by the value obtained after completing the test. Cognitive competence of students after applied Problem Based Instruction [10,11] can be seen in the following table.

Table 1. Frequency distribution of student learning outcomes -test scores on application of cooperative learning type *TAI* cycle to cycle

Score Interval	Cycle 1			Cycle 2		
	Mean	Frequency	%	Mean	Frequency	%
0 - 30	-	0	0	-	0	0
31 - 54	50	1	3.34	-	0	0
55 - 74	59.44	9	30.00	67.25	4	13.34
75 - 89	75.23	13	43.33	81.06	16	53.33
90 - 100	85.09	7	23.33	87.81	10	33.33
Number	73.27	30	100	82.63	30	100

Table 2. Recapitulation of learning outcomes on application of cooperative learning type Team Assisted Individualization cycle to cycle

Exhaustiveness	Cycle 1			Cycle 2		
	Mean	Frequency	%	Mean	Frequency	%
Non	58.50	10	33.33	67.25	4	13.33
Complete	82.72	20	66.67	85.00	26	86.67
Number	73.27	30	100	82.63	30	100

Criteria for the achievement of learning outcomes is a score ≥ 75. That is, the students are said to have competence when it got a score of at least 75. In the table indicated that the percentage of learning outcomes in cycle 2 with the implementation of cooperative learning Type *Team Assisted Individualization to* reach 86.67% $\geq$ 75%. It shows that most of the students have managed to achieve complete, as many as 26 students from 30 students. While students are not complete in learning outcomes are 4 students or by 13.33%.

References

[1] Jackson N 1978 *Civil Engineering Materials* (Hongkong: ELB & MacMillan)
[2] Slavin R i cols 1984 Combining cooperative learning and individualized instruction: Effects on student mathematics achievement, attitudes, and behaviours *Elem. Sch. J.* **84** 409–22
[3] Helda T and Elizabeth N *Steeps in the Action Research in the Action Research Reeder* (Victoria Australia: Deakin University)
[4] Huda M 2015 *Cooperative Learning* (Yogyakarta: Pustaka Pelajar)
[5] Gagne R M and Leslie J B 1979 *Principles of Instructional Design* (New York: Prentice Hall Inc.)
[6] Isjoni 2011 *Cooperative Learning* (Bandung: Alfabeta)
[7] Kember and David 2000 *Action Learning and Action Research* (USA: Kogan Page Limited)
[8] Rukmini E 2016 Deskripsi Singkat Revisi Taksonomi Bloom
[9] Khasinah S 2013 Classroom Action Research *J. Pionir* **1** 107–204
[10] Nur M 2004 *Learning Strategies* (Surabaya: Center for Science & Mathematics School)
[11] Nur M 2000 *Problem Based Teaching* (Surabaya: University Press: Universitas Negeri Surabaya)

IOP Conf. Series: Materials Science and Engineering **296** (2018) 012053 doi:10.1088/1757-899X/296/1/012053

Engaging Pre-Service Teachers to Teach Science Contextually with Scientific Approach Instructional Video

E Susantini [1], I Kurniasari[2], A N M Fauziah[3], T Prastowo[4], A Kholiq[4], L Rosdiana[3]

[1]Department of Biology, Universitas Negeri Surabaya, Surabaya, Indonesia
[2]Department of Mathematics, Universitas Negeri Surabaya, Surabaya, Indonesia
[3]Department of Natural Sciences, Universitas Negeri Surabaya, Surabaya, Indonesia
[4]Department of Physics, Universitas Negeri Surabaya, Surabaya, Indonesia
Jl. Ketintang, Surabaya 60231, Indonesia

endangsusantini@unesa.ac.id

Abstract. Contextual teaching and learning/CTL presents new concepts in real-life experiences and situations where students can find out the meaningful relationship between abstract ideas and practical applications. Implementing contextual teaching by using scientific approach will foster teachers to find the constructive ways of delivering and organizing science content. This research developed an instructional video that represented a modeling of using a scientific approach in CTL. The aim of this research are to engage pre-service teachers in learning how to teach CTL and to show how pre-service teachers' responses about learning how to teach CTL using an instructional video. The subjects of this research were ten pre-service teachers in Department of Natural Sciences, Universitas Negeri Surabaya, Indonesia. All subjects observed the instructional video which demonstrated contextual teaching and learning combined with the scientific approach as they completed a worksheet to analyze the video content. The results showed that pre-service teachers could learn to teach contextually as well as applying the scientific approach in science classroom through a modeling in the instructional video. They also responded that the instructional video could help them to learn to teach each component contextual teaching as well as scientific approach.

1. Introduction

Science education has been reformed to encourage teachers to contextualize materials taught, providing an opportunity for students to construct how science is developed and how one concept relates to another [1]. Contextual teaching presents new concepts in real-life experiences and situations where students can discover the meaningful relationship between abstract ideas and practical applications in the context of their use. In contextual science teaching, it is also important for teachers to strengthen the use of scientific approach in which students actively find the knowledge to solve the problem using the scientific method and science process skills. Scientific approach makes science taught the way it found through objective data rather than conventional lectures [2,3,4]. Hence, implementing contextual teaching together with scientific approach will foster teachers to find the constructive ways of delivering and organizing science content. In this paper, we developed an instructional video to the model adequate example of contextual teaching integrated with a scientific approach. We propose that the use of this instructional video can engage pre-service teachers to learn how to teach science contextually as well as corroborate scientific approach along with material taught. In investigating how the instructional video affects learning process of pre-service teachers in applying contextual teaching along with scientific approach, this study addressed these research questions:

- Could the instructional video engage pre-service teachers in learning how to teach CTL with the scientific approach?
- What were pre-service teachers' responses about learning how to teach CTL using instructional video?

1.1. Contextual teaching in science

Contextual teaching is supported by cognitive psychology theory, stating that "cognition is contextually dependent and must be described in that context before the material is understood at all." There are two complementary usages of word context [5]. First, context used to denote domain specificity which relates to the disciplinary knowledge that the learners wish to acquire, while the later used to denote signal task that contains true-to-life problems for the learners in the process of acquiring or applying the knowledge [6, 7]. Contextual learning occurs when students process new information or knowledge in a way that it makes sense to them in their own frames or reference (their own memory, experience, and response) [8]. In order to do so, teachers need to design learning environment with appropriate and desirable contexts into which to fit students' new knowledge. Five strategies called Relating-Experiencing-Applying-Cooperating-Transferring (REACT) to apply contextual teaching. Relating refers to learning within context of learners or learners' prior knowledge. Experiencing refers to learning by doing, including exploration, investigation, discovery, and invention. Applying refers to the activities in which the concepts are used or implemented. Cooperating refers involvement of other learners during the learning process, including sharing, responding, communicating. Transferring refers to the use of classroom-acquired knowledge in new context or new situation. There are seven components of contextual teaching and learning: 1) constructivism; 2) modelling; 3) questioning; 4) learning community; 5) inquiry; 6) authentic assessment; and 7) reflection. In constructivism, students build their own knowledge by testing ideas based on prior knowledge or experience, applying these ideas to a new situation, and integrating the new knowledge gained with pre-existing knowledge [9]. In constructivist classroom, students are actively engaged in hands-on activities and encouraged to gain knowledge through exploration [8]. Inquiry has its origins in the practices of scientific inquiry, emphasizing students to pose questions, gather and analyse data, and construct problem-solving or conclusion based on evidence [10]. Modelling in contextual teaching can be performed by teacher himself/herself or even involving students to give example to other students relating to the material taught. Questioning aims to stimulate and foster students thinking skills; this can be conducted by encouraging students to ask questions to be answered by other students or whole classroom or simply asking the students about what they have not understood or they want to know. Learning community refers to increase the use of group work and help low-achieving students to learn with the assistance of high-achieving students. Authentic assessment utilized tasks that are real example of extended criteria performances of actual learning goals [11]. This kind of assessment also used rubrics and other criteria checklists as standards to improve learning and teaching, providing multiple opportunities for students to learn and practice the desired outcomes as well receiving feedback and reflection. Last component, reflection, engages students and teachers to review, think, and evaluate their learning process and what they have just learned.

In science, contextual teaching can be conducted by illuminating theoretical practices and providing an opportunity for hands-on investigation. The contextual hands-on investigation should increase the degree of openness, for instance by posing the problem, but methods and answers are left open; offering students to discover relations he doesn't already know; and confronting students with raw phenomena with open problems, answers, and methods [12]. Contextualized science should be taught like scientist' science or open-ended research in which 1) the task is simplified, 2) students' motivation and engagement are enhanced by the perception that their practical work is authentic. Essentially, contextual teaching in science should also present concepts in familiar contexts and tangible examples or experiences compared to abstract conceptual models [1].

1.2. Teaching Science with Scientific Approach

As an attempt to increase the quality of education, one of the main characteristics of the new curriculum in Indonesia, known as Curriculum 2013, is the implementation of scientific approach throughout all of the subjects, including sciences (biology, chemistry, and physics). Curriculum 2013 describes five steps of scientific approach which are commonly known as 5M:

- Observing; involving the ability to use five senses to identify existing problems.
- Questioning; at which teacher encourage students to generate testable question; such as how a phenomenon happens or why the problem occurs. Scientific questions eventually lead the students to generate a hypothesis of the problem.
- Experimenting; involving any efforts collecting data to reject or accept hypothesis; such as experimentation, investigation, surveys, or interviews with the expert
- Associating; at which the collected data and information are associated one to another in order to avoid incorrect interpretation and affirm consequence
- Communicating result; involving activities used by students to share their discovery to classroom scale. This stage provides an opportunity for students to have comments, suggestions, or critics to their discovery.

According to this preceding discussion, scientific approach is strongly related (and can be indistinguishable in practice) to the inquiry within contextual teaching framework in which students generate problems or testable questions, observe phenomena, analyze data, communicate result, and evaluate what has been learned. This emphasizes the importance of combining contextual teaching with the scientific approach as well as training pre-service teachers to apply these pedagogy strategies to support educational reform efforts in Indonesia.

1.3. The use of instructional video in pre-service teacher education

Bandura's Social Learning Theories used as the theoretical framework for using instructional video in pre-service teacher education. Instructional video integrates modeling and video as visual cues in the expectation that the teachers will engage themselves in specific behavior which is planned to teach [13]. Video with classroom instruction also builds bridges between learning theory and teaching practice in the actual classroom. There are extensive research reports that observing video and discussing the whole content or edited segments/clips of video can raise insightful issues relating to good teaching practice [14, 15, 16]. There was also study confirmed that using video in teacher education, especially by embedding exemplary strategy in targeted-performance, likely to increase the ability to plan a lesson and observe authentic classroom sequences [17].

In this paper, our main purpose in providing instructional video to pre-service teachers is constructing knowledge about 'what to do' in the classroom. When they observe instructional video, they receive source model to apply pedagogical strategies as well as a reference regarding their need to modify these strategies in their future beginning career. However, at this point, pre-service teachers are not only become a 'viewer' but also an observer at the same time to reflect their own beliefs and practices. Therefore, our secondary purpose in providing the instructional video is encouraging pre-service teachers to critically analyze and evaluate teaching actions of a model teacher then consider how they will act to handle the problem in the similar condition [18]. In conclusion, the instructional video takes three important roles in pre-service teacher's education: (1) improving quality of teaching; (2) potentially enhancing drive to learn, memorize, and conduct specified teaching skills; and (3) helping to solve a specific problem that may be raised from actual classroom [19,20,21].

IOP Conf. Series: Materials Science and Engineering **296** (2018) 012053 doi:10.1088/1757-899X/296/1/012053

2. Research methods

2.1. Participants

Participants of this study were ten (10) pre-service teachers in Department of Natural Sciences, Universitas Negeri Surabaya, Indonesia. These pre-service teachers had been taught about a general overview of contextual teaching and learning as well as scientific approach and hadn't experience to teach in the actual classroom. All participants' names used in this study were pseudonyms.

2.2. Classroom procedure

All participants observed the instructional video which demonstrated contextual teaching and learning combined with the scientific approach as they completed a worksheet to analyze the video content [22]. During this activity, they were supervised by a lecturer (one of the authors). The worksheet asked pre-service teachers replayed the video on their own to discuss: (1) which scene showed characteristics of contextual teaching and learning, (2) which scientific approach steps appeared at each phase of contextual teaching and learning, (3) what social skills could be gained from learning activities and (5) what strength and weakness of learning activity could be noticed. Through this discussion, the lecturer explored how pre-service teachers perceive about how to teach any materials using contextual teaching and learning which was combined with a scientific approach. After observing the video, they were asked to teach other science topics based on the video modeling in the microteaching sessions. They only had 30 minutes to perform with the performance was randomly assigned. Their performances were observed and evaluated individually. The observation was conducted by the lecturer and two independent observers, while the evaluation was only prompted by the pre-service teachers by the lecturer. Observation and evaluation covered their teaching skills in applying contextual teaching and learning combined with a scientific approach. The end of the session, they were given a questionnaire to get their response to the video modeling and whole learning process.

2.3. Instructional video

Instructional video used in this study was developed to provide adequate modeling of contextual teaching and learning in the science classroom with a scientific approach. This instructional video was validated by three independent educational experts. A lesson in the video was taught by one of the authors on the subject of 'The Motion of Human Body' to a class of 13-14 year-old-students. The purpose of this lesson was to provide an opportunity to students to construct their own knowledge how simple machines work through contextual teaching and learning combined with a scientific approach. The instructional video was also featured with captions, showing which characteristic of contextual teaching was demonstrated at specific scenes and what element of scientific approach inserted along. Another feature, lecture notes, emerged before related scenes to explain the meaning of characteristics of contextual teaching.

In the video, firstly, the model teacher showed a video about how to smash correctly in a badminton game. Model teacher and students then discussed that when a server served, the shuttlecock should pass over the service line on the opponent's court or it would count as a fault. Both teachers and students agreed that a good serving should reach far enough distance at opposite court. Model teachers then asked some students to demonstrate a serving like the player did in the video. Some students served in the great distance but some students did not. The model teacher then encouraged students to ask how this difference could happen; what made one serving could reach great distance and what made another one could not. Students have then brainstormed a hypothesis that hand position might affect the serving distance. To investigate this hypothesis, teachers divided students into groups and asked them to conduct an investigation and write the result on the worksheet given. Model teacher distributed three objects which formed motion in the human body: high-heeled shoes, volleyball that should be supported with out-stretched arm and badminton racquets that should be held in the grip. A group could only work with one object. Within their groups, each group investigated how their object produced different force as they applied the different position of pivot point called fulcrum. The result of this investigation was written in the portfolio and shared with other groups through a classical presentation.

Groups with racquet and shuttlecock held a racquet (load) toward the wrist and curled it around with fingers (force), demonstrating first-class motion. First-class motion, they said, has fulcrum between force and load, thereby very useful to increase force. Groups with high-heeled shoes explained that wearing high-heeled shoes demonstrated second-class motion. The groups said that second-class motion locates load between force and the fulcrum; therefore resistance (weight of the body) is always greater than the effort applied to the heel. Groups with volleyball explained that supporting volleyball by stretching out the arm was third-class motion at which force was located between the load and the fulcrum. The fulcrum was the upper hand (at the elbow joint, precisely), and the force was applied by shoulder joint of lower hand, while the load or resistance was applied at the bottom end of upper hand (fingers). Hence, third-class motion increases the distance over which the force is applied. Comparing and communicating data of investigation result among groups hereby bridged model teacher and students to conclude that positioning hand like third-class motion (which is locating applied force between the load and the fulcrum) which will likely increase the distance of a serving, helping a badminton player to score point easily. They also constructed conclusion that bones and joint forms fulcrum and work exactly the same like the motion, one type of simple machine. Hence, as the muscle crosses the joint applying force (effort or input force/applied force), the body can move a load (resistance or output force). The class then ended as the model teacher reminded students to learn the next topic.

2.4. Data collection and analysis

In examining how pre-service teachers learned to teach with the contextual and scientific approach, we monitored how they planned their learning activities through their designed lesson plan and observed their teaching skills throughout microteaching using criteria and rubrics performances. Observers assessed pre-service teachers' teaching skills using four-point Likert scale: 1 = 'poor', scale 2 = 'fair', 3 = 'good', and 4 = 'advanced'). These teaching skills involved two criteria: (1) performing contextual teaching and learning and (2) integrating scientific approach into science topics. Score 3.0 or above indicated adequate teaching skills. We also examined the video-analysis worksheet (Appendix) to align the data how instructional video affected the teaching skills. Data about participants' responses were collected using questionnaire with open-ended questions asking whether the modeling through instructional video could help them to teach science contextually with a scientific approach. Prior to this questionnaire submission, the lecturer told the pre-service teachers that their answers or responses to the questions would not affect their grade. Data about pre-service teachers' views were analyzed descriptively.

3. Results and discussion

3.1. Could pre-service teachers teach contextually and scientifically with instructional video?

According to their average score (Table 1), pre-service teachers showed adequate skills in teaching science contextually and incorporated the learning activities to the scientific approach. They also taught with systematical fashion and selected relevant materials. We examined their teaching performance based on essential characteristics of contextual teaching learning and scientific approach elements.

Table 1. Pre-service teachers' performance in conducting contextual teaching with scientific approach

Performance aspect	Average score
Teaching science contextually	
a. Learning activities were relevant with seven components of contextual teaching and learning:	
1. Constructivism	3.3
2. Modeling	3.1
3. Questioning	3.4

4. Learning community	3.8
5. Inquiry	3.6
6. Authentic assessment	3.2
7. Reflection	3.4
b. Characteristics of contextual teaching and learning were conducted systematically	3.7
c. The material was relevant to be taught with contextual teaching and learning	3.5
Learning activities with scientific approach	
a. Learning activities were focused on teaching scientific approach	3.8
b. Developing scientific approach, including:	
Observing	3.8
Questioning	3.7
Experimenting (collecting data)	3.9
Associating	3.6
Communicating data	3.9
Creating (if any)	3.5
c. The applied scientific approach was relevant to be used with contextual teaching and learning	3.6

The performed pre-service science teacher number 1, taught heat and temperature by asking her students to make their own temperature scale in thermometer. Students were encouraged to generate questions and hypothesis then conducted an experiment which asked them to compare the temperature of freezing and boiling temperature of water between standard alcohol thermometer and non-scaled thermometer. She introduced her lesson with a brainstorm asking why people used a thermometer instead of simply declaring that something was hot and cold. Her constructivism was started well but she did not explore further this constructivism throughout her classroom. She also used scientific approach well, leading students observing that heat transfer happened as a hand touched warmer object to cooler object, generating testable questions, conducting an experiment, collecting and analyzing data, and having them to compare and communicate temperature reading using their scaled thermometer and commercial/standard thermometer. However, pre-service science teacher number 1 was still weak in utilizing authentic assessment in her classroom. On her video-analysis worksheet, pre-service science teacher number 1 wrote, "Though the caption said so, I barely noticed that the model teacher in the video conducted an assessment of students' practical work."

Pre-service science teacher number 2 wanted to teach the concept of a sound wave with its application in animals' ultrasounds. In order to so, she divided the material into one of the constituents at which she encouraged her students to discover the influence of pendulum length to the pendulum period. Similar to pre-service science teacher number 1, pre-service science teacher number 2 was so good at designing practical work as well as teaching with a scientific approach. Unfortunately, the way she contextualized and constructed the material into students' real-world situations was not relevant with students' practical work. She asked her students to sing together to model that the way human sing and talk were actually vibrations of vocal cords whilst the experiment utilized a pendulum which would be swung with different amplitudes and length. Compared to pre-service science teacher number 1, pre-service science teacher number 2 showed better skills in assessing authentically and encouraging reflection at the end of the lesson. However, at this point, both of them were still weak at bringing materials into real-world situations.

Pre-service science teacher number 3 taught her students to investigate the relationship between electrical charge and type of conductor materials. She initiated her lesson in real-world phenomena that houses and skyscraper buildings are essentially protected from direct strikes of lightning by installing proper lightning protection called lightning rod. She and her students brainstormed how lightning rod worked against lightning surges. To study how a lightning rod works, pre-service science teacher number 3 showed electroscope, saying that they would learn the working principle of the lightning rod by conducting an investigation with an electroscope. Teacher modeled how electroscope worked by rubbing sheet mica to hair, bringing the ruler near the metal plate of an electroscope, and showing how foil leaves hanging from the bottom of the electroscope spread apart. The investigation continued with groups of students tried various materials to be detected with an electroscope, ranging from strong conductor materials to strong isolator materials, such as glass, wool, aluminum ruler, and sheet of mica. As they collected and analyzed data, students communicated investigation result and found out that the greater the charge, or the stronger the materials as a conductor, the farther the foil leaves of electroscope spread. They then discussed and reflected how the way electroscope work could explain the mechanism of a lightning rod in protecting buildings from a lightning strike. There was no further elaboration how investigation conducted by the students related to the initial learning goal of finding correct lightning rod materials. Pre-service science teacher number 3's contextualizing effort was interesting except she totally forgot that electroscope used detected static electricity by using thin metals which would be separated when it was charged. However, it was good that pre-service science teacher number 3 had the ability to notice that electroscope can detect good conductor, whilst the conductor was the perfect material to allow the electrical current to flow the ground. At this point, pre-service science teacher number 3 was the first pre-service teachers who almost successfully tried contextualizing her material. The authentic assessment was hardly observed in pre-service science teacher number 3's teaching.

Pre-service science teacher number 4 who almost successfully contextualized her material with the scientific approach as she encouraged her students to investigate why people could be suffered from hypertension. At first, she explored students' prior knowledge by asking them to have her blood pressure measured with automatic blood pressure monitor. They found out that blood pressure of one and others was quite different, could be really high and very low, and related the phenomena with an increased number of hypertension sufferers. They generated questions how could this happen and constructed a hypothesis that hypertension should be a related area of blood vessels. In groups, students conducted an experiment with two different set of apparatus. One apparatus was a bottle containing water connected to beaker glass with a wide plastic pipe, while another one was the same apparatus but with a narrow plastic pipe. They measured the diameter of the pipe with a ruler and pressed the bottle, thus the water was flushed out in measurable volume. Students analyzed then presented that the pressure in a small area was very high, resulting high force to pump the water out, as Pascal suggested. At presentation, pre-service science teacher number 4 asked the students to assess their friends' performance in communicating investigation. Pre-service science teacher number 4 then engaged her students to reflect their learning process, concluding that hypertension sufferer had narrowed blood vessels, thus their heart worked so hard in pumping the blood. Her general contextual teaching strategy was good, but she still needed to improve her ability utilizing teaching aids. It would be better if she learned how to measure blood pressure correctly and provided vernier caliper, thus her students could precisely measure the pipe diameter.

Pre-service science teacher number 5 planned to teach her students about a simple test for color additives in foods. She started the lesson by asking the students to come forward and drinking two different soft drinks. One student tried the red soft drink and her tongue turned red accordingly. Another student who drank colorless soft drink did not show color change on her tongue. She further explained that color additives were commonly found in foods, beverages, and cosmetics to improve their look. Some color additives were harmless but some other color additives could pose health hazards, thus retailers and consumers should know how to identify color additives as well as to differenciate natural color additive and artificial color additive. In groups, pre-service science teacher number 5 subsequently facilitated her students to conduct a simple test for color additives. Following an open worksheet, they extracted curcumin, an example of natural color additive, with ethanol solvent. This extract was then filtered and added into sodium hydroxide (NaOH). Curcumin extract remained yellow even though it mixed with NaOH. They were agreed that this was used as control representing natural color additive.

Pre-service science teacher number 5 asked her students to add five drops of NaOH on two different tested foods: cheese pasta and candy. Students surprisingly discovered that mixture of cheese pasta and NaOH immediately turned from yellow into red. Same color changes also occurred in the tested candy. These data draw a conclusion to pre-service science teacher number 5 and her students that there was dye or colour additives added into their food, the colour changed immediately to red. On the contrary, if the food was pure or used natural additive, there would not be any colour change occurred. Pre-service science teacher number 5 also conducted an assessment of social skills during this presentation. Constructivism and contextual approach conducted by pre-service science teacher number 5 was well developed although she still required teaching her students how to use pipette properly.

Pre-service science teacher number 6 and pre-service science teacher number 7 wanted to teach her students the working principle of mixture separation. Pre-service science teacher number 6 taught her students about chromatography while pre-service science teacher number 7 taught about filtration. As the lesson started, pre-service science teacher number 6 demonstrated a mixture of limestone and water which then separated with filtration. She encouraged her students to ask whether all kinds of mixture could be separated by filtration. Her video about mixture separation led students to ask and assume that mixture separation was determined by the characteristics of mixture itself. pre-service science teacher number 6 continued her lesson by engaging her students to investigate ink separation using chromatography. Students formed groups and paint a dot on a piece of filter paper. This filter paper was then attached to paper clip and hang down inside a glass containing water, thus the tip of the paper touched the water (but the dot did not). Students observed that the filter paper absorbed the water, dispersed the painted dot into different colours. In group presentation, pre-service science teacher number 6 and her students reflected that chromatography could separate the ink and brainstormed that other type of mixture, such as water, passed different mixture separation called distillation, producing drinking water. In other hand, pre-service science teacher number 7 started the lesson with news from Tasikmalaya involving phenomena of water scarcity during drought season. She contextualized and constructed her material well as she read the headline that Tasikmalaya inhabitants had no access to clean water in the drought season. The inhabitants decided to use water from a muddy pond nearby since their well was dry and governments still had not supply Tasikmalaya with clean water, exposing themselves to health risk causing by drinking poor water. Pre-service science teacher number 7 then encouraged her students to redefine that muddy water was actually kind of mixture, composing of water with soil precipitates. She also tried to give hints that a mixture could be separated in many different ways. They brainstormed problem solving to make the inhabitants could obtain clean water and finally came to an idea that water filtration might be helpful. Pre-service science teacher number 7 then divided her students into groups and asked them to design a water filter from available materials, such as pebble, sands, coconut fibres, 600 ml plastic bottle, cotton, and active carbon. Students shared their water filter design and communicated the reason why they selected one material over another to form certain water filter structure. Other groups had responsibility to assess group who voluntarily presented their works. Pre-service science teacher number 7 and her students then reflected on their learning activity, declaring that filtration was one of techniques that could separate liquid materials from the solid one and would be promising to be utilized in Tasikmalaya.

Pre-service science teacher number 8 planned to teach her students about environmental pollution. Interestingly, she came to the class and asked her students to look around the classroom. Pre-service science teacher number 8 asked the students to take some snack and candy wrappers in the floor to be thrown away in the dustbin. She emphasized that polluted environment, with tons of garbage and hazardous chemicals, could affect living organisms. Pre-service science teacher number 8 continued her explanation to a footage showing environmental damage and asked her students to mention characteristics of pollution. She also asked her students to generate testable question related to the footage. They agreed to investigate the influence of pollutants on the survival of aquatic organisms. Within their groups, students conducted an experiment in which they observed how a fish could survive in two different environments. The first environment was a container with clean, colorless water, while another environment was a container with water from washing laundry. Pre-service science teacher number 8's students discovered that the fish immediately died after exposure to washing laundry water. Fish operculum flapped over so fast at first minutes, then slowed down to death. On the contrary, fish's operculum in clean water tank moved stably. Groups compared and contrasted the data, delivering the result to their friends. Pre-service science teacher number 8 directed her students to relate investigation result with a hypothesis, reflecting that pollutants might harm survival of aquatic organisms.

Pre-service science teacher number 9 decided to teach her students to conduct standard measurements with a vernier caliper. She took two measuring instruments, ruler and vernier caliper to the classroom. Pre-service science teacher number 9's students, who had never seen vernier caliper before, was eager to ask why someone measured things with those tools. Later, pre-service science teacher number 9 pointed two students to measure the length of her desk with their hands. One student said that pre-service science teacher number 9's desk was five hands long, while the other one insisted that pre-service science teacher number 9's desk was only four hands in length. Pre-service science teacher number 9 ignited the discussion by asking how those two students measured one same thing but with different results. Afterwards, she asked other two students to measure the same table with a ruler. Compared to prior measurement, these two students reported the same length. This modeling led the students to a discussion about the importance of measuring things with standard instruments. Pre-service science teacher number 9 also demonstrated how to use a vernier caliper correctly. She grouped her students and encouraged them to try measuring two objects (a small wooden block and a cap of a plastic bottle) with a vernier caliper. Prior to this activity, pre-service science teacher number 9 already brainstormed what standard measurement was. She also looked around and helped their students to use vernier caliper. Each group presented their measurement data and discussed the result. Lastly, pre-service science teacher number 9 concluded the lesson and asked the students to write a report about their measuring activity that day.

As part of explaining the human excretory system, pre-service science teacher number 10, the last performed pre-service teacher, taught her students to prove that exhaled air from lungs contained carbon dioxide. She played a commercial break showing a marathon runner. The runner was covered in sweats. Pre-service science teacher number 10 modeled herself cleaning her forehead with a facial tissue. The facial tissue became wet, leading pre-service science teacher number 10 to engage her students to discover that skin could excrete water. They then discussed what another organ in excretory system, in particular, lungs, would excrete in response to metabolism excess. To investigate this question, pre-service science teacher number 10 asked her students to make groups then prove that one of the excretory products was carbon dioxide. Following a worksheet, the students mixed limestone with water and let them for few minutes to form precipitates. Students then filtered the upper layer of this solution and transferred them into new glass. Later, students blown out some air through a straw for 15 seconds, continued for prolonged time, then observed the changes occurred in limestone water. They discovered that the limestone water turned cloudy and formed other precipitates on the bottom. Students collected data and successfully analyzed that if limestone reacted with carbon dioxide, it would form calcium carbonate precipitates. As the groups shared their findings, this investigation cleared out that excretory system, specifically lungs, produced carbon dioxide. pre-service science teacher number 10 and her students reflected what they learned that day and agreed to make a mind map about the excretory system in the next meeting.

Inconsistency with their well-contextualized teaching, pre-service science teacher number 5, 8, 7, 9 and number 6 correctly identified all of model teacher's behaviors representing characteristics of contextual teaching and scientific approach elements. As pre-service science teacher number 7 asserted:

> "Model teacher started *constructivism* approach as she played a show of badminton match. Later, as part of *modeling*, she asked several students demonstrated how a badminton player made a serve. At this early stage, we knew that the students *observed* both of badminton show and live-show of serving from their friends. As students became interested and *questioned* how badminton techniques related to the human locomotion, model teacher constructed the knowledge further by asking them to *make groups* (learning community) and

investigated the problem (inquiry) through an open-ended worksheet and portfolios. Within investigation, students *collected data* and *made the association* to their prior knowledge about the motion. The model teacher then directed the students *to communicate* their investigation portfolios. She said that she would *assess* students' involvement in the presentation. Accordingly, she complimented on groups who contributed more to discussion but still encouraged passive groups to do so. Last, model teacher *reflected* on learning process with her students, reviewing how motion system could work in the human body and how this principle can be a very useful trick to score a serving point."

Interestingly, although some pre-service teachers, such as pre-service science teacher number 4, 10, 3 and number 2 misidentified inquiry and authentic assessment during video-analysis, our preceding description about their microteaching indicated that they showed adequate skills within those criteria. For instance, this misidentification could be seen as pre-service science teacher number 2 asserted, "Inquiry can be seen as students work on their worksheet. Authentic assessment occurs as students generate research questions, invents a hypothesis, and performs an investigation." In her microteaching, although pre-service science teacher number 2 was still poor at contextualizing her material, she conducted an inquiry and authentic assessment well. Pre-service science teacher number 4 even had a good idea of authentic assessment by involving her students to do peer-assessment toward their friends' presentation. Perhaps, evaluation and reflection conducted by the lecturer upon the video analysis provided great help in noticing correct scenes.

3.2. What were they perceive?

All pre-service teachers perceived that the instructional video helped them to identify as well as learned how to contextually teach science with the scientific approach (Table 2), including each characteristics representing contextual teaching and stages of the scientific approach. The lowest response showed the perception that the instructional video did not use proper Bahasa Indonesia (40%).

Table 2. Pre-service teachers' views about learning with instructional video

Response	Response was given (%)
The video helps me to:	
Identify seven essential characteristics of contextual teaching and learning	100
Learn how to teach *constructivism* within contextual teaching and learning	100
Learn how to teach *modeling* within contextual teaching and learning	100
Learn how to teach *questioning* within contextual teaching and learning	100
Learn how to teach *learning community* within contextual teaching and learning	100
Learn how to teach *inquiry* within contextual teaching and learning	100
Learn how to teach *authentic assessment* within contextual teaching and learning	100
Learn how to teach *reflection* within contextual teaching and learning	100
Learn how to teach *observing* within scientific approach	100
Learn how to teach *questioning* within scientific approach	100
Learn how to teach *experimenting* within scientific approach	100
Learn how to teach *associating* within scientific approach	100
Learn how to teach *communicating data* within scientific approach	100
The video was easy to understand	90

In line with Table 2, pre-service science teacher number 4 also gave positive comments, "I love the idea learning to teach contextually with the [instructional] video. It encourages me to apply this approach in my future classroom." pre-service science teacher number 7 added, "[This] video is very useful reference for teaching practice." pre-service science teacher number 1 suggested,

> "It is very convenient to learn with the video because the caption helps me to identify specific scenes which are related to characteristics of contextual teaching. However, it would be nice if the video can show the clearer stage of authentic assessment."

Moreover, most pre-service teachers suggested that the instructional video should use grammatically correct Bahasa Indonesia. They perceived that using some Javanese (one of the local languages in Indonesia) or modern slang Bahasa Indonesia was not appropriate in the classroom.

In general, this study showed that pre-service teachers could learn how to teach science contextually along with scientific approach to the instructional video that we had developed. They noticed the scenes in the instructional video as visual cues, which then attaching mental images in their system thinking; thereby helping pre-service teachers to construct more concrete knowledge and build greater interests in learning than verbal description does [19,13, 23]. Taking notes (on the worksheet) and discussing model teachers' teaching behavior, as performed by pre-service teachers' in this present study, also helps them to increase classroom awareness. Previous work reports that video lesson analysis improves pre-service teachers' ability to pay attention to noteworthy events classroom interactions [24,25]. Higher scores on the video analysis task are also associated with better instructional quality [26]. Furthermore, pre-service teachers who write their own observations at video clips show better classroom awareness [27]. A Recent study also indicated that peer dialogue during the use of video potentially heightens teacher's awareness about their own teaching practice [28].

The tendency of learning to teach contextually using instructional video was also affected by pre-service teachers' reflective behaviors. The video made them evaluate that model teacher should be more open to the students about authentic assessment and interact with proper classroom language. Teachers often evaluate model teacher's pedagogical approaches as viewed in the video, offering advice on what the teacher should have done differently [15]. The reflective behaviors engage pre-service teachers to learn in meaningful ways, improving their teaching practice based on their pedagogical knowledge and beliefs. For instance, pre-service teachers decided to apply peer-assessment during group presentation to strengthen authentic assessment and spoke in grammatically corrected Bahasa Indonesia.

We also find that these reflective behaviors may depend on the interaction between the facilitator and the pre-service teachers during the learning process. At the early microteaching session, when pre-service science teacher number 1 and pre-service science teacher number 2 taught with the poor contextual framework, the lecturer evaluated their techniques and strategies for teaching and asked the pre-service teachers to focus how their material would be learned within real-world situations which were familiar to students. Lecturer also raised issues on how constructivism and scientific approach built by reducing cookbook experiment in the instructional video. Pre-service teachers then improved ways of teaching. In the preceding description, we noticed that they used recent news on newspaper to contextualize their material and left the problem, method, and answers open within practical work to emphasize constructivism as well as scientific approach. Previous research also reveals that facilitator supports video-mediated teacher education by focusing issues to the learning goal, managing the noteworthy scene to be viewed and rewind, and explored understanding and interpretation upon viewing the video [15].

4. Conclusion

We concluded that pre-service teachers could learn to teach contextually as well as applying the scientific approach in science classroom through a modeling in the instructional video. Pre-service teachers' teaching skills also showed that they became reflective about their own teaching practice and beliefs upon viewing the video and receiving an evaluation from the facilitator. Even so, we still can find that pre-service teachers may be weak at some criteria, such as utilizing teaching aids incorrectly, selecting irrelevant teaching aids, and associating one teaching ideas with the mismatched real-world situation. At this point, it would be unrealistic to expect that pre-service teachers could teach contextually without single failure or weakness after one session learning with the video. Long periods of training are needed to sustain the way they learn to teach and improve their pedagogical content knowledge.

Moreover, all of the teachers responded that the instructional video could help them to learn to teach each component contextual teaching as well as scientific approach. They also saw that the video was easy to understand and valuable as a reference of teaching practice in their future career as a beginning teacher. It seems likely that the pre-service teachers perceived that classroom languages should be spoken in formal Bahasa Indonesia and avoid the use of local languages or slang words. We also consider that facilitator takes important parts to help pre-service teachers focus on essential characteristics of targeted-performance. In this study, we can observe that facilitator manages peer-dialogue during video analysis, prompt pre-service teachers to identify which classroom events represent components of targeted-performance, and helps pre-service teachers to evaluate their own teaching practice or notice missing elements of targeted-performance.

Acknowledgement

The study was funded by Ministry of Research and Technology (Kemenristek)-Directorate General of Higher Education (DIKTI) of Indonesia and Islamic Development Bank (IDB) as Excellent Research of Higher Education "The Development and Upgrading of Seven Universities in Improving the Quality and Relevance of Higher Education in Indonesia" that was given to Universitas Negeri Surabaya in 2016. The authors thanked (1) students who directly involved in the instructional video-filming, (2) all of the pre-service teachers who agreed to take part in the study.

References

[1] Klassen S 2006 A Theoretical Framework for Contextual Science Teaching. Interchange 37/1-2 31-62
[2] Wieman C 2007 Why not try a scientific approach to science education? *Change* 1-15
[3] Wieman C and Gilbert S 2015a Taking a Scientific Approach to Science Education, Part I – Research *Microbe,* **10** chapter 4 152-156.
[4] Wieman C and Gilbert S 2015b Taking a Scientific Approach to Science Education Part II – Changing Teaching *Microbe* **10** 203-207
[5] Tweney R D 1992 Serial and parallel processing in scientific discovery In R N Giere (Ed), Minnesota studies in the philosophy of science (*Cognitive models of science* vol. XV) (Minneapolis, MN: University of Minnesota Press) 77-88.
[6] Baker E, O'Neil H F Jr and Linn R L 1994 Policy and validity prospects for performance-based assessment *Journal for the Education of the Gifted* **17** 332-353
[7] Ebenezer J V and Gaskell P J 1995 Relational conceptual change in solution chemistry *Science Education* **79** 1-17
[8] Crawford M L 2001 *Teaching Contextually: Research, Rationale, and Techniques for Improving Student Motivation and Achievement in Mathematics and Science* (Waco: CCI Publishing)
[9] Berns R G and Erickson P M 2001 Contextual Teaching and Learning: Preparing Students for the New Economy (The Highlight Zone: Research @ Work).
[10] Hmelo-Silver C, Duncan R and Chinn C 2007 Scaffolding and Achievement in Problem-Based and Inquiry Learning.
[11] Sears S 2003 *Introduction to Contextual Teaching and Learning* (Bloomington: Phi Delta Kappa Educational Foundation).
[12] Roth W and Roychoudhury A 1993 The development of science process skills in authentic contexts *Journal of Research in Science Teaching* **30** 127-152.

[13] Bellini S, and Akullian J 2007 A meta-analysis of video modeling and video self-modeling intervention for children and adolescents with autism spectrum disorders *Exceptional Children* **73** 264-287

[14] Van Es E A.and Sherin M G 2006 How Different VideoClub Designs Support Teachers in "Learning to Notice".*Journal of Computing in Teacher Education* **22** number 4 pp 125-135.

[15] Van Es E A and Sherin M G 2008 Mathematics teachers' ''learning to notice in the context of a video club *Teaching and Teacher Education* **24** pp 244–276.

[16] Star J R, Lynch K and Perova N 2011 Using video to improve mathematics' teachers' abilities to attend classroom features: A replication study In Mathematics teachers' noticing: Seeing through teachers' eyes ed Miriam G, Sherin, Victoria R, Jacobs, and Randolph A, Philipp (New York: Routledge) pp 117-133.

[17] Seidel T, Blomberg G, and Renkl A 2013 Instructional strategies for using video in teacher education *Teaching and Teacher Education,* **34** pp 56-65

[18] Blomberg G, Renkl A, Sherin M G, Borko H and Seidel, T. (2013).Five research-based heuristics for using video in pre-service teacher education.*Journal for Educational Research Online* **5** number 1 pp 90-114

[19] Agommuoh P C and Nzewi U M 2003 Effects of Videotape Instruction on Secondary School Students Achievement in Physics *Journal of STAN* **38** pp 88-93.

[20] Gaudin C and Chaliès S 2015 Video viewing in teacher education and professional development *Educational Research Review* **16** pp 41-67.

[21] Lin P 2005 Using research-based video cases to help pre-service primary teachers *International Journal of Science and Mathematics Education* **l 3** 351-377.

[22] Susantini E, Kurniasari I, Fauziah A N M and Prastowo T 2016 *Proposal Penelitian Unggulan Perguruan Tinggi: Pengembangan Video Pembelajaran Pendekatan Saintifik untuk Mengembangkan Keterampilan Mengajar Calon Guru MIPA* [Research Proposal on Research for Higher Education: Developing Instructional Video with Scientific Approach for Improving Teaching Skills in Mathematics and Science Pre-service Teachers] Surabaya: Unesa University Press.

[23] Escalada L T and Zollman D A 1997 An Investigation on the Effects of Using Interactive Digital Video in a Physics Classroom on Student Learning and Attitudes *Journal of Research in Science Teaching* **34** chapter 5 467-489

[24] Alsawaie O N and Alghazo I M 2010 The effect of video-based approach on prospective teachers' ability to analyze mathematics teaching *Journal of Mathematics Teacher Education* **13** 223-241

[25] Kersting N B, Givvin K B, Sotelo F L and Stigler J W 2010 Teachers' Analyses of Classroom Video Predict Student Learning of Mathematics: Further Explorations of a Novel Measure of Teacher Knowledge *Journal of Teacher Education* **61** 172-181

[26] Kersting N B, Givvin K B, Thompson B J, Santagata R and Stigler J W 2012 Measuring Usable Knowledge: Teachers' Analyses of Mathematics Classroom Videos Predict Teaching Quality and Student Learning *American Educational Research Journal* **49** 568-589

[27] Fadde P and Sullivan 2013 Using interactive video to develop teachers' classroom awareness *Contemporary Issues in Technology and Teacher Education* **13** 156-174

[28] Kourieos S 2016 Video-Mediated Microteaching – A Stimulus for Reflection and Teacher Growth *Australian Journal of Teacher Education* **41** 65-80

IOP Conf. Series: Materials Science and Engineering **296** (2018) 012054 doi:10.1088/1757-899X/296/1/012054

Training of Trainers (ToT) Program in Team Teaching

Werry Febrianti[1*], Leo Hari Wiryanto[2]

[1]Study Program of Mathematics,
Sumatera Institute of Technology, Lampung Indonesia
[2]Faculty of Mathematics and Natural Sciences,
Bandung Institute of Technology, Jalan Ganesha 10 Bandung Indonesia

*Email: werry.febrianti@itera.ac.id

Abstract. The first year students in Sumatera Institute of Technology (ITERA) follow the first year program (TPB). They will learn about mathematics, physics, chemistry, and all of the basic subjects that they need for learning in ITERA. They will study in the big classrooms with different background department of their friends. This situation makes the lectures become more challenging in teaching their lessons. Besides the classrooms, the experience of the lecturers is still need to be improved because the lecturers are young and less of experience in teaching so that they need guidance from their senior lecturer. Because of that situation, *Training of Trainers* (ToT) program in team teaching is one of the solution that can increase the young lecturers's ability so that they can teach well in the massal conditions of the classrooms. ToT program in team teaching indicated the better result than regular teaching.

1. Introduction

Calculus I, in Sumatera Institute of Technology (ITERA), is called Elementary Mathematics I. Elementary Mathematics I in ITERA with larger students and because of lecturers who teach are still young and unexperience become a challange in teaching and learning process in the first year program (TPB). Davis[1] said that team teaching most often refers to two or more faculty members who jointly produce a course. Easterby-Smith and Olve[2] said that team teaching involves two or more teachers collaborating over the design or implementation of the same course. Then, Hathcher & Hinton[3] said that in team teaching, instructor will held weekly meeting for course preparation. Hecht, Roberts, and Schoon[4] has reported about team teaching on the teachers increasing their motivation and improves teacher affect.

It caused an idea to held *Training of Trainers* (ToT) program in team teaching for the young lecturers so that they can be ready to teach in the classrom and become more confident.ToT program in team teaching is given by the senior lecturer to the unexperience young lecturers. ToT program in team teaching is a weekly program so that the senior lecturer will teach the young lecturers to teach the lesson that will be teach a week later. ToT program in team teaching could be conclude as a preparation program to make the young lecturers more ready to teach in the classroom.

2. Method's Research

This research will compare the study result of Elementary Mathematics I in class D in odd semester 2015 with the study result of Elementary Mathematics I in class TPB 4 in odd semester 2016. There were some differentiation with the learning process in Elementary Mathematics I in class D in odd semester 2015 with the learning process in Elementary Mathematics I in class TPB 4 in odd semester 2016. Because of that differentiation, this research will look for the learning process that the lecturers did in Elementary Mathematics I in class D in odd semester 2015 with the learning process that the lecturers did in Elementary Mathematics I in class TPB 4 in odd semester 2016. Then, this research will see the tutorial process in Elementary Mathematics I in class D in odd semester 2015 with the tutorial process in Elementary Mathematics I in class TPB 4 in odd semester 2016. Futhermore, this research will also compare the study result of Elementary Mathematics I in class D in odd semester 2015 with the study result of Elementary Mathematics I in class TPB 4 in odd semester 2016.

3. The Result of the Research and Discussion

This research concerned about the learning process that lecturers's did in Elementary Mathematics I in class D in odd semester 2015 with the learning process that lecturers's did in the Elementary Mathematics I in class TPB 4 in odd semester 2016. Then, this research focused on the tutorial process in the Elementary Mathematics I in class D in odd semester 2015 and the tutorial process that lecturers's did in Elementary Mathematics I in class TPB 4 in odd semester 2016. This research also focused on the study result in learning process with ToT program in team teaching that was implemented in learning process in Elementary Mathematics I in class TPB 4 in odd semester 2016. This study result would compared with the study result in Elementary Mathematics I in class D in odd semester 2015.

The learning process in Elementary Mathematics I in class D in odd semester 2015 didn't implement the ToT program in team teaching. The learning process was given by the the young lecturers in ITERA without ToT program in team teaching from the senior lecturer. This research took the sample randomly so that the result point out the class D in odd semester 2015 as the sample class. Thus, the students in the class D in odd semester 2015 are 90 students. It means that the lesson is in the big classroom. Then, the tutorial process in class D in odd semester 2015 didn't taught by their lecturer but from their seniors who had choosen from the academic to teach them. It was different from the tutorial process in class TPB 4 in odd semester 2016. Then, tutorial in the class D in odd semester 2015 didn't divide into two small classroom so that it has made the tutorial become not effective and not efficient.

The learning process in Elementary Mathematics I in class TPB 4 in odd semester 2016 implement the ToT program in team teaching. Before teaching the lesson in the big classroom, the unexperience young lecturers has been taught by the senior lecturer. ToT program in team teaching increased the ability of the young lecturers in understanding the matter from the lesson that they will be teach so that this program maked them become more confident and ready to teach the lesson to their students. Futhermore, ToT program in team teaching also discuss about the problem that will be teach in tutorial class. The problems is given by the senior lecturer in the ToT program in team teaching so that it maked the uniform problem that was given to the students in TPB class in 2016. This research taked the sample randomly so that the result point out the class TPB 4 in odd semester 2016 as the sample class. Thus, the students in the class TPB 4 in odd semester 2016 are 87 students. It means that the lesson is in the big classroom. Then, the tutorial process in class TPB 4 in odd semester 2016 was taught by their own lecturers. Tutorial in the class TPB 4 in odd semester 2016 divided into two small classroom so that it has made the tutorial become more effective and more efficient. This tutorial usually discuss about the problem from the lesson that has been taught in the big classroom. The goal of this tutorial in discussing many problems to make the students more understanding about the lesson and also to make their motoric nerve become fully-fladged. The students also comfort with the system in tutorial in Elementary Mathematics I with their own lecturers because they don't need adaptation so

IOP Conf. Series: Materials Science and Engineering **296** (2018) 012054 doi:10.1088/1757-899X/296/1/012054

Table 1. The result of the learning process in Elementary Mathematics I in class D and class TPB 4

The Mark in Elementary Mathematics I	Class D	Class TPB 4
A	1	2
AB	2	4
B	3	9
BC	8	7
C	14	25
D	31	24
E	29	12

that it makes the situation really condusive in learning process. This tutorial also effective in small classroom because the students more active and more pay attention with the lesson and the problems that they learn and solve so that it gives effect to the result of the learning process.

The result of the learning process in Elementary Mathematics I in class D in odd semester 2015 without ToT program in team teaching and the result of the learning process in Elementary Mathematics I in class TPB 4 in odd semester 2016 with ToT program has been shown in Table 1.

According to the result of the mark in Elementary Mathematics I can be seen that the result of the learning process in class TPB 4 with ToT program in team teaching to their lecturers has better mark than the result of the learning process in class D without training of the ToT program in team teaching. It has been influenced by the ToT program in team teaching to the unexperience young lecturers so that they become more ready to teach in the big classroom. Then, this result also influenced by the sum of the students that they teach in tutorial class. Because of the students in tutorial class is smaller than regular class so it has made the learning process in tutorial class more effective than regular class. It can be seen from the students who become more active in understanding the lessons that they learned.

4. Conclusion

Based on the research about training of trainers (ToT) program in team teaching, it can be conclude that the necessary to hold on workshop in learning process before teaching process followed by focus on the matter or the lessons that will be teach. This goal of this program to increase the ability of the young lecturers.

Acknowledgment

This paper supported by P3MI Research Grant ITB.

References

[1] Anderson R S and Speck B W 1998 Oh what a difference a team makes : why team teaching makes a difference *Teaching and Teacher Education* **14** 671-686

[2] Carpenter D M, Crawford L and Walden R 2007 Testing the efficacy of team teaching *Learning Environments Research* **10** 53-65

[3] Easterby-Smith M and Olve N G 1984 Team teaching : Making management education more student centered *Management Education and Development* **15** 221-236

[4] Hatcher T and Hinton B 1996 Graduate students' perceptions of university team teaching *College Student Journal* **30** 367-377

[5] Hecht J B, Roberts N K, and Schoon P L 1996 Teacher teams and computer technology : Do combined strategies maximize student achievement ? *Journal of Research on Computing in Education* **28** 318-328

IOP Conf. Series: Materials Science and Engineering **296** (2018) 012055 doi:10.1088/1757-899X/296/1/012055

Talent identification model for sprinter using discriminant factor

N W Kusnanik[1], A Hariyanto[1], Y Herdyanto[1], A Satia[2]

[1]Faculty of Sports Science, Universitas Negeri Surabaya, Indonesia
[2]Faculty of Sports Science, Universitas Negeri Jogjakarta, Indonesia

niningwidyah@unesa.ac.id

Abstract. The main purpose of this study was to identify young talented sprinter using discriminant factor. The research was conducted in 3 steps including item pool, screening of item pool, and trial of instruments at the small and big size of samples. 315 male elementary school students participated in this study with mean age of 11-13 years old. Data were collected by measuring anthropometry (standing height, sitting height, body mass, and leg length); testing physical fitness (40m sprint for speed, shuttle run for agility, standing broad jump for power, multistage fitness test for endurance). Data were analyzed using discriminant factor.
The result of this study found that there were 5 items that selected as an instrument to identify young talented sprinter: sitting height, body mass, leg length, sprint 40m, and multistage fitness test. Model of Discriminant for talent identification in sprinter was D = -24,497 + (0,155 sitting height) + (0,080 body mass) + (0,148 leg length) + (-1,225 Sprint 40m) + (0,563 MFT). The conclusion of this study: instrument tests that have been selected and discriminant model that have been found can be applied to identify young talented as a sprinter.

1. Introduction

Science and technology in sports are the important things that can be used to develop sports achievement. One of the goals of doing sports is to achieve the excellent performance. Achievements can be obtained by conducting structured and systematic in sports coaching. Sports coaching should be based on scientific concepts that supported by sports science and technology. Through science and technology can carry out the training process, identifying, selecting, and predicting young talented athlete effectively and efficiently. The aim of talent identification is for identifying and selecting young talented athlete that valid and reliable. This is necessary to obtain accurate results in predicting young talented athlete for future. It is hoped that young talented athlete who is identified and selected can compete at the international level.

There are some talent identification models that have been developed by some countries including Germany, Japan, China, Scotland, and Australia [1–5]. Some literature report that there is some talent identification which specifies for sport such as volleyball [6–14]. Talent identification in Indonesia needs to be developed systematically and continuously, especially for a sprinter. Therefore, the purpose of this study is to identify young talented sprinter using discriminant factor in order to identify young talented sprinter effectively and efficiently.

IOP Conf. Series: Materials Science and Engineering **296** (2018) 012055 doi:10.1088/1757-899X/296/1/012055

2. Methods

This study was conducted at 315 elementary school students with an aged range of 11-13 years old. The research developed talent identification model based on Aussie Sport's talent search instruments. The product of this study was a model of talent identification using discriminant factor that can identify young talented sprinter. Data were collected by measuring anthropometric including standing height, sitting height, body mass, leg length; testing physical fitness (40m sprint for speed, shuttle run for agility, standing broad jump for power, multistage fitness test for endurance). Data were analyzed using discriminant factor.

3. Results and Discussion

The result of talent identification based on collecting the item pool, screening of item pool by consulting with the sprinter coaches and sports scientists and test measurements expert for sports. Then produced Selected Test Item Design (STID) as follows: anthropometric measurements include standing height, body weight, sitting height, and leg length; physical fitness tests include sprint 40m, vertical jump, standing broad jump, sit up, push up, flexibility and multistage fitness test.

The object identification was to determine whether the research variables can be used to test the differences of a group against other groups. The variables are standing height, sitting height, body mass, leg length, sit up, push up, flexibility, vertical jump, sprint 40 m, standing broad jump and Multi Fitness Test (MFT). Based on the identification using the test of Equality of Group Means shown in Table 1:

Table 1. *Tests of Equality of Group Means*

	Wilks' Lambda	F	df1	df2	Sig.
Standing Height	.704	131.810	1	313	.000
Sitting Height	.718	123.181	1	313	.000
Body Mass	.767	95.290	1	313	.000
Leg Length	.703	132.048	1	313	.000
Sit Up	.983	5.313	1	313	.022
Push Up	.992	2.532	1	313	.113
Flexibility	.988	3.765	1	313	.053
Vertical Jump	.799	78.762	1	313	.000
Sprint 40m	.725	118.843	1	313	.000
Standing Broad Jump	.893	37.670	1	313	.000
Multistage Fitness Test	.686	143.157	1	313	.000

As clearly shown in Table 1 that observed from F-test, therefore variable standing height, sitting height, body mass, leg length, sit up, vertical jump, sprint 40m, standing broad jump, and MFT has P-value (Sig) 0.000 <0, 05 level of significant so H1 accepted and Ho rejected. This means that the variable can be used to identify the talent category of the sprinter athlete. The push-up and flexibility arm has P-value (Sig) 0.113 <0.05 level of significant so H1 is accepted and Ho is rejected. This means both of these variables cannot be used to identify the talent category of sprinter athletes.

After knowing that the variables can be used as discriminant variables, it is necessary to know how big the differences of each discriminant variable in two categories can be seen from the equation of discriminant function. To know the discriminant function equation can be seen in Canonical Discriminant Function Coefficients as seen in Table 2:

Table 2. *Canonical Discriminant Function Coefficients*

	Function
	1
Sitting Height	.155
Body Mass	.080
Leg Length	.148
Sprint 40m	-1.225
Multistage Fitness Test	.563
(Constant)	-24.497
Unstandardized coefficients	

Another crucial finding as clearly depicted in Table 2 that the coefficients of each variable can be formed in a discriminant function, as the following discriminant functions: D = -24.497 + (0.155 Sitting Height) + (0.080 Body Mas) + (0.148 Leg Length) + (- 1.225 40m Sprint) + (0.555 Multistage Fitness Test). Based on the equation it is seen that the average value of 40 m sprint is the most dominant to predict the difference of talent category (Sprinter and non Sprinter) because it has the highest coefficient that is 1.225 and the lowest in predicting sprinter talent category is body mass with coefficient 0,080.

In Phase 1, a Selected Test Instrument Design (RITT) was prepared based on the results of some literature reviews and consultation with a team of sports experts. The selected instrument was found 11 items. For anthropometric measurements consisting of body height, body mass, sitting height, and leg length; for physical fitness tests such as 40 m sprints, vertical jumps, standing broad jumps, arm muscle strength (push up), abdominal muscle strength (sit up), flexibility and multistage fitness tests (MFT).

The selection of the RITT refers to a sports search model that has been modified by [11]. The number of instruments based of Aussie Sport [5] model was 10 items instrument consisting of 4 anthropometric measurements of body height, sitting height, body mass, spam arm, and 5 physical fitness tests including 40m sprint, vertical jump, shuttle run 5m, basketball throw, throw and catch tennis ball, and multistage fitness test. The instrument of the modified sports search model has been tested for its validity and reliability [15].

Based on the trial results and discriminant analysis showed that there were 5 items of measurement tests consisting of anthropometric measurements (body mass, leg length, and sitting height); and physical tests (40m sprints) and Multistage Fitness Test (MFT), later referred to as Selected Instruments Test (SIT). Based on the discriminant equation it seems that 40m sprint was the most dominant value in predicting young talented for sprinters. 40m sprint is one form of test to determine the level of one's physical fitness in the component speed. Sprinters require components of physical fitness, especially acceleration and speed in supporting their performance. Furthermore, sprinter also requires cardiovascular endurance as one component of physical fitness that must be owned. This is because the endurance capability is needed by a sprinter in doing various exercise activities. The Sprinter needs a repetition of a short sprint that must be done many times each day. Therefore good cardiovascular endurance can certainly support these activities.

To determine young potential athletes including young talented sprinter or non sprinter, simply by including the results of anthropometric measurements, and physical fitness tests in the model of discriminant equations that have been designed. Therefore, trainers, teachers of Physical Education, and parents will be able to know whether their students are talented in sprinters or non sprinters. The discriminant analysis model that has been produced can be used as an instrument to classify the sprinter talent category or non sprinter for children aged 11-13 years.

Creating a discriminant analysis model requires special knowledge, therefore a program that facilitates trainers, teachers of physical education, parents, and coaches can be used easily and precisely. This model was created to identify young talented children aged 11-13 years. This is done to avoid measurement tests that are "too difficult" for children younger than 11 years old and "too easy" for children over 13 years old. If the child under the age of 11 years old is still in the stage of multilateral development. But if it is above 13 years old then it is in the stage of talent development. Furthermore, McComas says that the development of muscle and skeletal children aged 1 - 5 years will increase by 45%, aged 5-18 years increased by 60%, after the age of 18 years increased 20% [16]. Therefore, children aged 11-13 years in this study is the right age for growth and development of muscle and skeletal because of its relatively high increase.

The results of this research can be used to identify talent identification for children aged 11-13 years for sprinter using discriminant equation. Of course, these results can be helpful for coaches, coaches, parents, teachers of Physical Education in identifying young talented athlete for sprinters. Not every child has a talent in sprinters, therefore only certain children have potential in anthropometric and physical fitness to be nurtured and further developed. It is expected that the child can practice optimally so as to provide greater opportunities for the child to succeed as sprinters. Therefore, this talent identification model is a practical, effective and efficient to identify young talented sprinters.

4. Conclusion

Instrument tests that have been selected and a discriminant factor model that have been found can be applied to identify young talented sprinter. Model of Discriminant factor for identifying young talented sprinter was D = -24,497 + (0,155 sitting height) + (0,080 body mass) + (0,148 leg length) + (-1,225 Sprint 40m) + (0,563 MFT).

Acknowledgement

The authors would like to thanks to Director of Post Graduate Universitas Negeri Surabaya for supporting the research grant. In addition, many appreciate Dr. Sri Rahayu Vice Rector for Academic for giving the opportunity to present the article at the CAPEU 2017.

References

[1] JISS 2005 *Annual Report 2004* (Tokyo)
[2] Cooke C, Cobley S, Till K and Wattle N 2010 Searching for Sporting Excellence: Talent Identification and Development *Br. J. Sports Med.* **44**
[3] Yuan W 2004 Yuan Weimin's Speech on the Press Conference in Athens
[4] Abbott A and Collins D A Theoretical and Empirical Analysis of a “State of the Art” Talent Identification Model, High Ability Studies **13** 157–78
[5] Sport A 1993 *The Search Is Over* (Australian Sport Commission)
[6] Kusnanik N 2013 Anthropometrical Measurement, Physiological and Biometrical Testing in Identifying Young Talented Volleyball Athletes *Assiut J. Sport. Sci. Arts* 692–8
[7] Aouadi R, Jlid M, Khalifah R, Hermassi S, Chelly M, Van Den Tilaar R and Gabbett T 2012 Association of Anthropometric Qualities with Vertical Jump Performance in Elite Male Volleyball Players *J. Sports Med.* **52** 11–7
[8] Thakur V 2010 Talent Identification in Kabaddi *Br. J. Sports Med.* **44**
[9] Ballard R 2010 *Tennis Indonesia Youth Talent Identification Program* (Jakarta: PELTI)
[10] Reilly T, Bangsbo J and Franks A 2000 Anthropometric and Physiological Predispositions for Elite Soccer *J. Sports Sci.* **18** 669–83
[11] Reilly T, Williams A, Nevill A and Franks A 2000 2000b: A Multidisciplinary Approach to Talent Identification in Soccer *J. Sports Sci.* **18** 695
[12] William A and Reilly T 2000 Talent Identification and Development in Soccer *J. Sports Sci.* **18** 657–67
[13] Hoare D and Warr C 2000 , 2000: Talent Identification and Women's Soccer: An Australian

IOP Conf. Series: Materials Science and Engineering **296** (2018) 012055 doi:10.1088/1757-899X/296/1/012055

Experience, , Vol.18, No.9; *J. Sports Sci.* **18** 751–8
[14] Hoare D 2000 Predicting Success in Junior Elite Basketball Players – The Contribution of Anthropometric and Physiological Attributes *J. Sci. Med. Sport.* **3** 391–405
[15] Menpora 2003 *Indonesian Directorate General of Sport, Pemandıan Bakat* (Jakarta: Ditjen Olahraga)
[16] Belanger A and McComas A 1989 Contractile properties of human skeletal muscle in childhood and adolescence *Eur J Appl Physiol Occup Physiol* **58** 563

IOP Conf. Series: Materials Science and Engineering **296** (2018) 012056 doi:10.1088/1757-899X/296/1/012056

Abuse notation of improper integrals

E Arfi[1], L H Wiryanto[2]

[1]Study Program of Mathematic, Sumatera Institute of Technology, Lampung Indonesia
[2]Faculty of Mathematic and Natural Sciences, Bandung Institute of Technology, Ganesha Street 10 Bandung Indonesia

eristia.arfi@itera.ac.id

Abstract. In the definition of $\int_a^b f(x)\,dx$, it was assumed that the interval $[a, b]$ was finite. However, in many application in physics, engineering, economics, and probability we wish to allow a or b (or both) to be infinite or for the other case when f has an infinite discontinuity in$[a, b]$. In either case the integral is called improper integral. Definition of improper integral is the limit of a definite integral as an endpoint of the interval of integration approaches either a specified real number or ∞ or $-\infty$ or, in some cases, as both endpoints approach limits. Many of the first-year students in ITERA learning about improper integrals make an abuse notation like written improper integral just like standard definite integral. Based on this situation, we tried to change the habits of the students in the writing of improper integrals by providing sufficient understanding and considerable exercises. It turns out this way gives good result, based on the percentage of the value of the midterms to the problem of improper integrals give a fairly good percentage value.

1. Introduction

In addition to Physics and Chemistry, Calculus is one of the compulsory subjects to be taken by first year (TPB) Students of Institut Teknologi Sumatera (ITERA). Calculus can form a systematic, critical, logical, meticulous and consistent thinking pattern, and demands creative and innovative thinking.

Calculus has a big role in the advancement of science and technology because it has advantages such as language and rules in Calculus is well-designed and consistent, the structure of the network of mathematical information is very strong with a pattern of reasoning and systematic, Calculus acts as a way of approach to study Science and technology, Calculus is a tool in solving other field problems and through Calculus a problem can be seen in a compact, short and solid model [1]. Calculus is used in many fields, as in Newton's law expressed by the rate of change referring to the derivative and the second law of the Newton's law said force is the product of mass and acceleration while the acceleration itself is a derivative of the velocity. Another example, the function of vector has a big effect on the existence of a location in terms of the moving place. So, a vehicle can be known where and where is the location of the destination.

In the TPB, Calculus is divided into two parts that is Calculus I for the first semester and Calculus II for the second semester. Calculus I was discussed about Real Number System, Definitions and Various Functions, Definition of Limit and Continuity, The Derivatives and The Definite Integral and its application. Calculus II was discussed about techniques of integration, indeterminate form and improper integrals, infinite series, geometry in space and vector, derivative for function of two more variables and multiple integral. All these materials are the basic that should be given to TPB students as a base for them when they return to their study program.

In this paper we will discuss about the errors that often occur in improper integrals problem. Many of the students solve improper integral problems without writing notation of the improper integral correctly. Yet in Calculus the writing of the improper integral definition correctly is very important. For example, when a student does not write a correct definition of it will get an unusual result from the calculation.

$$\int_0^1 1 + \ln x \, dx,$$

In this paper also performed simple data analysis of some sample of student value in the problem of improper integral.

2. Improper Integrals

In Calculus I, students defined the integral $\int_a^b f(x)\,dx$ over a finite interval $[a, b]$. The function f was assumed to be continuous, at least bounded, otherwise the integral was not guaranteed to exist. However, in many application in physics, economics, and probability we wish to allow a or b (or both) to be ∞ or $-\infty$. We must therefore find a way to give meaning to symbols like

$$\int_1^{\infty} \frac{1}{1+x^3}\,dx, \quad \int_{-\infty}^{-1} \frac{1}{1+x^3}\,dx, \quad \int_{-\infty}^{\infty} \frac{1}{1+x^3}\,dx \tag{1}$$

The integrals are called improper integrals [2].

2.1 Definition (Improper Integral)

An integral is an improper integral if either the interval of integration is not finite (improper integral of type 1) or if the function to integrate is not continuous (not bounded) in the interval of integration (improper integral of type 2). We will give some example for each type of integral [3].

- $\int_1^{\infty} e^x dx$
  is an improper integral of type 1 sisnce the upper limit of integration is infinite.
- $\int_0^1 \frac{1}{x^3} dx$
  is an improper integral of type 2 because $\frac{1}{x^3}$ is not continuous at 0.
- $\int_{-\infty}^{-1} \frac{1}{x^2-1} dx$
  is an improper integral of type 2 because $\frac{1}{x^2-1}$ is not continuous at -1 and 1.
- $\int_{-\infty}^{2} \frac{1}{1-x} dx$
  is an improper integral of type 1 and also integral of type 2 because the lower limit of integration is infinite and also $\frac{1}{1-x}$ is not continuous at 1 and 1 is in the interval of integration.

Look at the interval of integration. If either the lower limit of integration, the upper limit of integration or both are not finite, it will be an improper integral of type 1.

2.1.1 Definition (improper integral of type 1)

1.

$$\int_{-\infty}^{b} f(x)\,dx = \lim_{a \to -\infty} \int_a^b f(x)\,dx \tag{2}$$

$$\int_a^{\infty} f(x)\,dx = \lim_{b\to\infty}\int_a^{b} f(x)\,dx \tag{3}$$

If the limits on the right exist and have finite values, then we say that corresponding improper integrals converge and have those values. Otherwise, the integrals are said to diverge.

2. If both $\int_c^{\infty} f(x)\,dx$ and $\int_{-\infty}^{c} f(x)\,dx$ converge, then we define

$$\int_{-\infty}^{\infty} f(x)\,dx = \int_{-\infty}^{c} f(x)\,dx + \int_{c}^{\infty} f(x)\,dx \tag{4}$$

The integrals on the right are evaluated as shown in 1.

2.1.2 Definition (improper integral of type 2)

Improper integral of type 2 are evaluated as follows;

1. If f is continuous on $[a,b)$ and not continuous at b then we define

$$\int_a^{b} f(x)\,dx = \lim_{t\to b^-}\int_a^{t} f(x)\,dx \tag{5}$$

Provided the limit exists as a finite number. In this case, $\int_a^b f(x)\,dx$ is said to be convergent (or to converge). Otherwise, $\int_a^b f(x)\,dx$ is said to be divergent (or to diverge).

2. If f is continuous on $[a,b)$ and not continuous at a then we define

$$\int_a^{b} f(x)\,dx = \lim_{t\to a^+}\int_t^{b} f(x)\,dx \tag{6}$$

provided the limit exists as a finite number. In this case, $\int_a^b f(x)\,dx$ is said to be convergent (or to converge). Otherwise, $\int_a^b f(x)\,dx$ is said to be divergent (or to diverge).

3. if f is not continuous at c where $a < c < b$ and both $\int_a^c f(x)\,dx$ and $\int_c^b f(x)\,dx$ converge then we define

$$\int_a^{b} f(x)\,dx = \int_a^{c} f(x)\,dx + \int_c^{b} f(x)\,dx \tag{7}$$

The integrals on the right are evaluated as shown in 1, and 2.

Evaluating an improper integral is exactly about limit problem and integral problem. Make sure that each integral is improper at only one place, that place should be either the lower limit of integration or the upper limit of integration.

- $\int_1^{\infty}\frac{dx}{x^2}$

This is an improper integral of type 1. We evaluate it by finding $\lim_{t\to\infty}\int_1^{\infty}\frac{dx}{x^2}$.

$$\int_1^{\infty}\frac{dx}{x^2} = \lim_{t\to\infty}\int_1^{t}\frac{dx}{x^2} = \lim_{t\to\infty}\left(1-\frac{1}{t}\right) = 1. \tag{8}$$

Many of student solve that's problem

$$\int_1^{\infty}\frac{dx}{x^2} = \left(1-\frac{1}{\infty}\right) = 1 \tag{9}$$

Although the answer was correct, didn't write improper integrals as a limit of definite integral is one of the abuse notation.

- $\int_0^\pi \sec^2 x\,dx$

This is an improper integral of type 2, because $\sec^2 x$ is not continuous at $\frac{\pi}{2}$. Thus,

$$\int_0^\pi \sec^2 x\,dx = \int_0^{\pi/2} \sec^2 x\,dx + \int_{\pi/2}^\pi \sec^2 x\,dx \tag{10}$$

$$= \lim_{t\to\frac{\pi}{2}^-} \int_1^t \sec^2 x\,dx + \lim_{t\to\frac{\pi}{2}^+} \int_t^\pi \sec^2 x\,dx = \tan t + \infty$$

Therefore $\int_0^\pi \sec^2 x\,dx$ diverges.

If we had failed to see that the above integral is improper, and evaluated it without using definition of improper integral, we could have obtained a completely different (and wrong) answer

$$\int_0^\pi \sec^2 x\,dx = \tan\pi - \tan 0 = 0 \text{ (this is not correct)} \tag{11}$$

- $\int_0^1 1 + \ln x\,dx$

If we didn't use right definition for the above integral we could obtained

$$\int_0^1 1 + \ln x\,dx = (x + x\ln x - x)|_{t=0}^{t=1} = \infty \tag{12}$$

the answer is not correct, since this is an improper integral of type 2, because $\ln x$ is not continuous at 0. Thus,

$$\int_0^1 1 + \ln x\,dx = \lim_{t\to 0^+}\left(\int_t^1 dx + \int_t^1 \ln x\,dx\right)$$
$$= \lim_{t\to 0^+}\left((x + x\ln x - x)|_{t=0}^{t=1}\right)$$
$$= \lim_{t\to 0^+}(1\ln 1 - t\ln t)$$
$$= 0 - \lim_{t\to 0^+} t\ln t = 0$$

3. Result

We take a sample of mid test score from several students and will be tested by using hypothesis test of one sample proportion.

We claims that 50 % of mid semester test results for the improper integral problem gets a good result. A random sample of 100 students has 61 students get a good result. We will test this claim at the 0.05 level of significance.

- State Hypotheses : $H_0: p = 0.5$
  $H_1: p > 0.5$
- Test Statistic : $z = \frac{\hat{p}-p_0}{\sqrt{\frac{p_0(1-p_0)}{n}}}$
  $z = \frac{61/100-0.5}{\sqrt{\frac{0.5(1-0.5)}{100}}} = 2.2$
- α : 0.05
- Critical Value : 1.645
- Decision : Because $z > 1.645$, so we decided to Reject H_0
- This test give conclusion that for 0.05 significance level more than 50 % students get a good result for improper integrals problem.

IOP Conf. Series: Materials Science and Engineering **296** (2018) 012056 doi:10.1088/1757-899X/296/1/012056

From this simple analysis data we can say that after providing sufficient understanding and considerable exercise, more than 50 % students can solve the improper integral problem correctly.

4. Conclusion

When given the problem of integrals we must know whether the integral is a definite integral or improper integrals. Incorrectly determining the type of integral will give very different results and possibly wrong answer. Be able to write an improper integral as the limit of definite integral is very important. Many of the students have been careless or unwise in this regard. For that reason, we provide a sufficient understanding of the integrals especially the improper integrals with adequate exercises on improper integrals problem. It turns out this way gives good result, based on the students mid test results.

5. Reference

[1] Martono, Koko 2007 *Hakekat Pembelajaran Matematika* ITB Bandung Indonesia
[2] Dale Varberg et.al 2007 *Calculus, 9th Edition* Prentice Hall Inc
[3] Laval, Philippe B 2005 *Improper Integral* Kennesaw State University

Acknowledgement
The author wants to acknowledge P3MI ITB Research Grant to supported this paper.